An Introduction to Mathematical Modelling

An Introduction to Mathematical Modelling

Neville D. Fowkes

and

John J. Mahony

Department of Mathematics,
University of Western Australia

JOHN WILEY & SONS
Chichester • New York • Brisbane • Toronto • Singapore

Baffins Lane, Chichester,
West Sussex PO19 1UD, England
National Chichester (0243) 779777
International +44 243 779777

Reprinted September 1994

Other Wiley Editorial Offices

John Wiley & Sons, Inc., 605 Third Avenue,
New York, NY 10158–0012, USA

Jacaranda Wiley Ltd, 33 Park Road, Milton,
Queensland 4064, Australia

John Wiley & Sons (Canada) Ltd, 22 Worcester Road,
Rexdale, Ontario M9W 1L1, Canada

John Wiley & Sons (SEA) Pte Ltd, 37 Jalan Pemimpin #05-04,
Block B, Union Industrial Building, Singapore 2057

British Library Cataloguing in Publication Data:

A catalogue record for this book is available from the British Library

ISBN 0 471 93422 4; 0 471 94309 6 (pbk.)

Produced from camera-ready copy supplied by the authors using LateX
Printed and bound in Great Britain by
Biddles Ltd, Guildford and King's Lynn

On 30 June 1992 John Mahony died. It was a gift to work with John on this, his final contribution to mathematics.

John asked me to dedicate this work to his father Patrick Mahony whose spirit of independence and sense of social justice inspired John throughout his life.

I'd like to dedicate this book to my parents who provided the opportunity and encouragement for me to pursue my interests. I'd also like to recognise those few individuals who by their example showed me how to think about mathematics. John was one of these.

Contents

Preface

Spurred on by past successes of mathematical modelling in the physical sciences, more recent successes in the biological sciences and the advent of cheap and powerful computing, the use of mathematical models has grown rapidly and spread to most disciplines. There is every reason to believe that this trend will continue to accelerate in the foreseeable future. The very recent growth of the use of models for examining problems arising in industry is especially marked. Industrialists have realized that mathematical models, when combined with computation, represent a *far* cheaper alternative to other options (experimentation etc.) for understanding many industrial processes. Basically, modellers and computing come cheap! It is hoped that this book will provide useful material for the training of students in the skills of mathematical modelling in general, and in industrial mathematics in particular. It's also hoped that the book will expose students to the challenges and fascination of mathematical modelling as a profession.

The attitudes and skills required for modelling are different from, but not incompatible with, those required for pure mathematics, and even differ from those normally emphasized in traditional applied mathematics courses. The authors believe that such attitudes and skills should be developed early in a student's background, and the only obvious way to do this is to expose students to real modelling problems. More specifically, students need to develop the ability to work with non-mathematical material presented in succinct form and to learn how to formulate models, determine solutions which are useful in context, and interpret the results. The need for such early modelling training has been generally recognized but, given the limited mathematical and contextual background of such students, and their limited exposure to the practical use of mathematics, it's not easy to design such courses. We've attempted to do this by carefully selecting the material (especially the models) and by using an algebraic package to provide technical assistance.

The book requires as mathematical background no more than second year calculus level (elementary differential equations, vector calculus etc.), together with the associated linear algebra, and assumes a small non-mathematical knowledge base. The non-mathematical background required is briefly discussed in the relevant chapters.

The models examined relate largely to industrial and (to a lesser extent) scientific questions that the authors have encountered recently, and which are seen to be of current and future practical interest. Particular situations are presented (often as presented to the authors by the customer) and what follows is a discussion of the alternative approaches which may be adopted, followed by an analysis of the relevant models and an appraisal of the practical usefulness of the results obtained. Techniques are introduced when needed, and the advantages and disadvantages of such techniques are indicated in context. Errors in approach are also part of the learning experience and are presented. In a real sense an apprenticeship approach is taken.

Additionally the material has been selected so as to:

- Illustrate the use of mathematical techniques which are efficient, practical, and frequently used.

- Reinforce significant ideas by meeting them in more than one context. The cohesiveness necessary to do this places restrictions on content. For this reason the large scale discrete optimization problems and the statistical problems which often arise in industrial contexts were excluded. Other computer intensive models were excluded for the same reason.

- Present as much of the material normally regarded as essential for continuum modelling as possible. It is in the manner of presentation of this (and other) material that this book differs from traditional presentations.

- Draw the student's attention to the features that make different systems behave in similar ways. An awareness of such features is essential for modelling; this is why dynamics, diffusion and wave propagation are fundamental topics. Of course it's this commonality feature that enables modellers to contribute significantly over such a broad front.

One major difficulty encountered by students is that of handling straightforward but intricate and tedious technical calculations. This is a difficulty experienced by all mathematicians; however, with experience, one learns to avoid getting bogged down in the technical details. For the apprentice and professional alike symbol manipulation packages are a Godsend. At the press of a button Fourier coefficients to any order are calculated, differential equations solved, plots drawn etc. This enables one to direct one's mind to major issues such as the appropriateness of the solution procedure or the relevance of the calculations in the modelling context. It should be pointed out that such packages are not just a convenient tool for performing calculations—the methodology of mathematics is rapidly changing as a result of their introduction. Apart from the fact that such packages must be regarded as essential for present day mathematics usage, the use of Maple (just one of the competitive packages) in this book has enabled a realistic working approach to modelling to be adopted; students are stepped through real calculations. Thus using Maple has enabled the authors to present techniques that are commonly used in practice (rather than those restricted by class-time constraints) and to examine realistic models. Also, students are expected to carry out (close to) real size investigations, rather than the artificially simplified (often superficially related) calculations necessitated by time constraints in the pre-algebraic package era. It would be helpful if students had some familiarity with Maple but it's not essential; students are introduced slowly to the commands in the first few chapters and more commands are introduced when needed. The authors have in practice found it useful to present one lecture illustrating the capabilities of the package and refer students to notes or any of the many good books on Maple (eg. Ellis, Johnson, Lodi and Schwalbe (1992), or Heck (1993)). Alternative algebraic packages can be used in association with the text.

An unusual exercise format is adopted to facilitate the handling of these real problems. The context of the problem is explained and then a series of directives is given to assist the student's own investigations. Hints are provided to enable the student to discover a sensible approach, and the student is expected to interpret results in context, and is encouraged to experiment and suggest extension work. Initially this approach is somewhat daunting for students (for starters many of the exercises are rather longer than usual and more open ended); however, it has been found that confidence quickly develops. The examples form an essential component of the text. They either elaborate the text or examine analogous situations arising in often entirely different contexts. The connecting features are thus emphasized, and also it's hoped that students will develop the confidence to handle situations outside their normal experience and build up a useful knowledge base.

In order to facilitate the apprenticeship approach an informal conversational style is used. As we all know, mathematicians don't think the way they write. They write like lawyers—and no one can think like a lawyer writes! In writing the way they think the authors hope that the intuitional models that enable them to see through situations might also prove to be of value to students. At the very least it's hoped that the approach will encourage students to develop their own intuitive way of looking at situations.

The book is an expanded version of a course given mainly by the authors and Malcolm Hood to second year university students at the University of Western Australia on mathematical modelling. Malcolm's imprint on a number of the exercises will be recognized by students. The course evolved from a standard course on applied mathematics into a course based on various scientific investigations we have personally been involved in (of local and global origin) and later, as our interests spread, on industrial problems. Our interest in industrial problems was stimulated especially by the work of the Oxford Industrial Mathematics group, and the annual Australian Mathematics-in-Industry meetings run until this year by CSIRO.

A great deal of this book was completed during the last year of John's life when he was struggling with cancer; a very difficult period for him and his family. In spite of very limited energy resources John insisted on continuing because of his belief in the importance of such a work. The example of his courage is at least as valuable as the work contained in this book. The work would not have been possible if it wasn't for Jocelyn's (John's wife) extraordinary support; for this we're all grateful.

Many friends and colleagues have offered encouragement, provided welcome advice, corrected errors etc. I'd especially like to acknowledge the encouragement and advice from Bill Pritchard and Glyn Davies, and to thank Jenny Hopwood for critically commenting on the book material. Peter Chapman also offered useful advice on various chapters of the book, and Grant Keady offered Maple advice. I'm also grateful to my good friends Roman Bogoyev and Con Savas who were most generous with their time and help.

Working with John during the last year of his life was distressing, and putting the material together since his death has been most difficult. Without the help of dear friends (especially Paula) I could not have managed. All I can say is thanks!

Nev Fowkes John Mahony

Bibliography

Ellis, W., Johnson, E., Lodi, E., Schwalbe, D. (1992). *Maple V Flight Manual,* Brooke's Cole Publishing Co., California.

Heck, André (1993). *Introduction to Maple,* Springer-Verlag, New York.

Chapter 1

Introduction and General Outline

1.1 Mathematics and Modelling

The use of mathematics to test ideas and make predictions about the real world has a long and distinguished record in the physical sciences; so much so that mathematics has become the basic language of physics and its applications in engineering. Given the success of Newton and his contemporaries, and the philosophical climate at the time, it's not surprising that he thought of his descriptions as referring to some objective reality. Today's scientists, however, would regard any description as just a *model*, of greater or lesser accuracy and range, which mimics certain aspects of observed behaviour, thus enabling useful predictions to be made. The work of Volterra (1926) during the first world war was particularly significant in this respect. He introduced models to *qualitatively* explain changes in the shark population in the Mediterranean under circumstances in which there was no possibility of obtaining reliable *quantitative* data. Nevertheless the resulting insights were most valuable and have changed attitudes towards modelling, as well as leading on to the very successful ecological models in current use. From this beginning the use of mathematical models, both qualitative and quantitative, has extended to an increasing range of disciplines utilizing a widening range of mathematical ideas and techniques. The potential usefulness of such models has been augmented by the developing power of computers and their greatly reduced cost. All the indications are that this trend will continue as a result of the recognition that the mathematical modelling approach represents an efficient and economic way of understanding, analysing and designing systems.

Mathematics in its own right has been becoming increasingly powerful because of the beautiful abstractions which allow one to concentrate on the essentials of what would otherwise be hopelessly complex arguments. Such abstract descriptions are extremely helpful to a mathematical modeller because they enable complex mathematical situations to be viewed in a way which simplifies the tasks in hand. The trend towards abstraction in mathematical education has, however, had the unfortunate side effect of tending to leave students quite inexperienced and inadequately prepared to undertake the mathematical tasks needed in the community. The fact is that there is a real difference between the attitude required for the pursuit of excellence in mathematics per se, where power, beauty and universality are the sole criteria for excellence, and that required for effective mathematical modelling. Some of the issues which require a difference in attitude will be briefly reviewed.

The Model

Modelling is as much an Art as a Science. The most important decision a modeller has to make concerns the choice of model. Of the very many model possibilities arising out of a particular application, very few can usefully illuminate the processes involved, and the useful models are not necessarily those of most intrinsic mathematical interest. Thus, although a thorough working knowledge of mathematics is absolutely essential for modelling, it is rare that the quality (i.e. usefulness) of a model is dependent only on the mathematical ingenuity displayed in its analysis. Obviously for a modeller the process is not complete with the presentation of the mathematical results. In the light of the inevitable modelling inadequacies, interpretation of the results is a major component of the study, and further studies are often indicated. It is often the case that the qualitative insights gained from modelling are more important than any quantitative results obtained. In a noteworthy summary paper describing pioneer work in fire research Howard Emmons stated that the challenge in fire modelling is "not to produce the most comprehensive descriptive model but to produce the simplest possible model that incorporates the major features of the phenomenon of interest"—very useful advice for the aspiring modeller! Knowing the limited range of available mathematical techniques one might surmise that it is technical limitations of a mathematical nature that necessitate such apparently limited aims. The problem is a much deeper and more interesting one, as was found by Richardson with his early pioneering attempts to understand weather patterns. He found that seemingly sensible attempts to include compressibility effects in meteorological models leads to ridiculous predictions—models predicted that weather fronts would move at the speed of sound. In fact, if one ignores compressibility effects entirely then sensible predictions can result. Thus an apparently less accurate model leads to better results! This case isn't special. In fact it is an essential part of modelling to direct the analysis to aspects which are relevant to the context and to omit other aspects of the real world situation which often lead to spurious results. The precision normally required for purely mathematical investigations is thus not necessarily appropriate for modelling.

The Mathematics

To the undergraduate, mathematics often appears to be a tool of limitless power. This mistaken impression arises largely because traditional mathematics courses confine their attention to simple systems (usually linear) for which fairly complete results are available. The real world is both complex and nonlinear. There are no generally useful techniques available for such systems, and purely numerical techniques are of very little use for exposing the underlying processes. It's a challenging situation in which recipes just don't work! An inventive ad hoc approach (based on an intuitive understanding of the processes involved) is often required. Although such approaches are not as universal in application as one might hope it's often the case that approaches found useful in one application can be modified to handle others. Thus there is a large body of knowledge which can be drawn on; however, this information is usually not discussed at an undergraduate level. Even in situations addressed by both pure and applied mathematicians the differing aims lead to reliance on different techniques. Thus the convergent series that can be usefully used to establish the existence and uniqueness of solutions to a class of well defined mathematical problems rarely converges rapidly enough to be of practical use. Often, in fact, pure mathematicians will avoid entirely the process of "constructing" a solution to display existence by using non-constructive Fixed Point Theorem techniques—very neat theoretically, but useless for practical evaluation! Perhaps even more surprising to the uninitiated is that the solution technique one brings to bear on answering a particular question about a well defined mathematical model depends on the actual question being asked. Thus the mathematics cannot be treated in isolation from the model context.

Economic Reality

Further complications of a real world nature influence the modeller's work. An unnecessarily accurate solution which is costly in time/money is useless, as is a model requiring expensive or unavailable data input. In this regard (although one might hope for a broadly useful result) one might settle for a result that is of little use outside the parameter range of specific interest. Often, in fact, simple parameters (to be experimentally determined) can be introduced to avoid complications not central to the mechanism of interest (the friction coefficient of mechanics is a case in point).

The ultimate aim of many investigations is to control a process under operating conditions. Under such circumstances it's clearly necessary for any computation to be simple enough to be performed in real time; often a major constraint.

Not all of the above factors apply in any specific situation but it is rare for a number of them not to apply. Any mathematician who wishes to become a modeller must learn how to cope with such factors.

1.1.1 Modelling Skills

It was once said (in jest) of a friend and colleague George Carrier: "George can only solve boundary layer problems; but there's no problem he can't convert into a boundary layer problem." At the time George had made a number of major contributions to science by skilfully using techniques referred to as boundary layer techniques. Now, there is no area of applied mathematics that I know of that George doesn't have a complete command of and he is renowned for his modelling skills, but the statement has some truth. A good modeller can recognize the application of ideas across disciplines. The skills to do this can be developed; in fact, both the authors (and many others) owe a great deal to George for the lessons he taught by example.

Rarely is it the case that a modeller will be faced with a problem in which he can claim expertise and for which the standard approach works. In problems arising out of industrial applications, for example, engineers would have inevitably creatively explored ways of handling the situation of interest and these would have failed before the modeller was brought in. It's thus almost an advantage not to be burdened with the conventional wisdom. The modeller doesn't start from scratch in such circumstances, however. Apart from mathematical knowledge (knowing which procedures are likely to produce useful results etc.) the modeller can bring to bear on such problems knowledge gained from other contexts. There are general principles (both mathematical and physical) that can be observed in operation in many areas. Thus, for example, all diffusion problems share features in common with the archetype problem in the area; the heat equation. Very often it's useful in fact to examine first an analogous heat conduction problem before attempting the real diffusion problem of interest. The training of a modeller thus involves a training to recognize such patterns and principles. In this text we've tried to focus on these common features both by drawing the reader's attention to them in the course of model investigations and by pointing out their application in different areas.

1.1.2 Computation

The role of computational techniques in modelling calls for further comment. The ability of the computer to handle large calculations makes it an ideal tool for handling data that requires little interpretation, but as a modelling and scientific tool its value is highly overrated. The modeller's and the scientist's aim is to *understand* processes so that predictions can be confidently made. In the industrial context, for example, often the aim is to increase the operating range of a particular process, or to understand why unexpected difficulties have arisen. The numerical output of a particular calculation (even if the calculation is sensibly based) is not much help in this respect; one needs to know how the (usually many) various features of the problem interact to produce the outcome. This is why analysis is necessary, and why less than 1% of scientific *understanding* can be ascribed to computational work. Once the process is understood the computer takes over. There are other more subtle reasons for not immediately having recourse to numerical techniques which relate back to comments made above about the necessity to filter out inessential details of a model. More often than not seemingly innocuous errors in the mathematical model can swamp the answer. One therefore needs to process the model equations to arrive at a sensible mathematical model for computational purposes. To do this the understanding that comes from analysis is essential. In this regard algebraic packages represent a major advance for the modeller, enabling him to produce analytic results in situations that could not be dealt with by hand. Of course the reality of practical mathematical modelling is that virtually all mathematical models are completed using number crunching techniques, but the work of a modeller is almost finished when the preprocessing is complete.

1.2 Book Outline

In the absence of any real understanding of the physics of a process (beyond the describing equations) it's still possible for the modeller to proceed using scaling analysis. We illustrate this procedure in Part I Chapter 2 by examining a problem which students will have no experience of—the problem of determining conditions under which a flagpole will survive earthquake vibrations. Using a variety of approximations suggested by scaling arguments, solutions are produced and results plotted and analysed using Maple. The mathematical, experimental, and modelling implications of the results obtained are explored. The ideas developed in this chapter are central to the text.

In Part I Chapter 3 we examine an imaginary situation in which a manufacturer seeks to design tables which won't rock. The mathematics associated with this problem is elementary, enabling us to concentrate on modelling issues that are both subtle and of common occurrence.

In Chapter 4 an optimization problem associated with the mooring of the giant mother ships used in the oil industry is examined. The variational approaches so useful for examining complex modelling situations are thus introduced. The techniques introduced are returned to in Chapter 5 and in later sections of the text.

Part II is devoted to diffusion problems. The physical and mathematical fundamentals are discussed in Chapter 6, and then the classical fundamental solution and Fourier series techniques are introduced in Chapters 7 and 8 in the context of surface heating processes. The boundary integral methods so useful for computational work are also introduced in these chapters. There are then three chapters devoted to specific applications. Inverse problems are discussed in the cooking context in Chapter 9. In Chapter 10 aspects of the greenhouse problem are studied. This work provides some insight into how one can attempt to understand very complex processes. In the remaining chapter a fairly complete preliminary case study, undertaken to assess the viability of a potential continuous steel casting process, is presented. This should provide students with an insight into how such a study proceeds.

Part III examines vibrational and wave propagation problems. Effects associated with resonant systems are examined, and the various amplitude limiting mechanisms are discussed in Chapter 12. A specific example (a car transmission problem) brought to an industrial study group is examined in Chapter 13. Multiscaling and averaging ideas are discussed in this context. In the final chapter signal speed and shock wave ideas are discussed in the traffic dynamics context.

The three parts of the book can be dealt with in any order and in isolation, and chapters from each of the parts can be combined to provide a usefully broad introduction to the various modelling areas. At the University of Western Australia, for example, Chapters 2 and 3 of Part I; Chapters 6, 7 and 10 of Part II; and Chapter 14 of Part III are used for an introductory modelling unit. The table fable of Chapter 3 is usually discussed first because it provides a good illustration of the attitudes required for effective modelling in a gentle mathematical context.

The exercise set is extensive and hints are provided both in the main body of the text and in the the Hints and Answers Chapter at the end of the book. Many of the problems require the students to use Maple. Generally speaking, the first few examples of a set are straightforward (often students are asked to reproduce results developed in the text) and are designed to enable students to understand and become familiar with important aspects of the physics or mathematics. Following these examples are more substantial examples which serve to illustrate the use of the physical and mathematical ideas developed in the main text in other contexts. Starred examples indicate more difficult problems or problems requiring more time. Starred sections of the book indicate optional material that may be omitted without loss of continuity.

For further (more advanced) work on mathematical modelling in an industrial context the authors recommend Tayler (1986), which describes a fascinating range of industrial problems that have arisen out of the Oxford Study Groups with Industry over the past 20 years.

Bibliography

Tayler, A. B. (1986) *Mathematical Models in Applied Mechanics,* Clarendon Press, Oxford.

Volterra, V. (1926). *Variazione e fluttauzini del numero d'individui in specie animali conviventi* Mem. Accad. Nazionale Lincei (ser. 6) 2, p 31-113.

Part I

Mechanical Systems

Chapter 2

Scales, approximations and solutions

Dynamic similarity
Dimensionless groups
Perturbation methods

The ability to simplify the mathematical description of a physical situation without losing too much accuracy is a fundamental requirement for mathematical modelling. In order to simplify this description it's necessary to be able to obtain informed estimates of the numerical significance of either including or excluding various physical effects in the model. In the absence of a detailed knowledge of the area of interest the most effective way for a mathematician to obtain the necessary estimates is to assess the relative size of various terms as they appear in the mathematical description. The scaling arguments described here lead to the required estimates. The process is often referred to as *inspectional analysis.* The mathematics used in this process is simple, but the arguments are subtle.

After obtaining the required estimates it then makes sense (at least in the first instance) to simply remove the relatively small terms from the mathematical system. Subsequently one should run checks on the validity of the resulting approximations.

The scaling and approximation procedures described here are fundamental to modelling and will be used throughout the book. Although an understanding of the background of the particular problem can be helpful, the scaling technique requires no such understanding. As the text develops the description of the technique, which will be detailed here, will gradually become more terse. With experience the assessment can be performed without writing down the mathematical equations describing the physical situation.

2.1 Scaling and Inspectional Analysis

In dealing with any equation arising out of a model of real world processes it is important to recognize two basic facts:

Fact 1. Precisely because it describes a real world process, the equation must involve the addition and equating of things of the same nature (therefore possessing the same dimensions) expressed in the same units. Additionally the results obtained in real physical terms must be independent of the units used or the particular mathematical formalism used to present the physics.

Fact 2. The terms arising in the equation cannot be regarded as being either large or small in isolation; it's the ratio of comparable terms that matters.

These facts will be elaborated on in the work to follow. The conventional square brackets notation will be used to designate the dimension associated with a quantity. For example if l is a length, the notation $[l]$ denotes the dimension of l which is length $[L]$. Thus the dimension of a velocity is given by $[L]/[T]$, i.e. a length divided by a time. A passing familiarity with the ideas of dimensions and units will suffice for what follows and the required material will be introduced when needed. For the moment all that is needed is the knowledge that only combinations of the fundamental dimensions $[L],[T]$ and mass $[M]$ are required to specify the dimensions of *all* the quantities of interest in mechanical problems. More information about the ideas of dimensions, and an account of the dimensional analysis ideas often used in engineering and physics will be provided at the end of the chapter.

It makes no sense to talk about the size of a particular physical quantity possessing dimensions (e.g. the height of a particular person with dimensions $[L]$) without referring to the particular units used in the description, and the actual number associated with the description can be deceptively large or small. Thus the actual number expressing the person's height is large if measured in units of millimetres and small if measured in kilometres. In order to make comparisons between the heights of different people it is necessary to adopt a common unit of measurement, and in most parts of the world the metre is the accepted standard. Although such standards for measurement are invaluable because of their almost universal acceptance, within the context of a particular model we will see that it's useful to work with the *ratios* of comparable terms based on units that are *specific to the context*. Because such ratios are the ratios of terms of the same dimension they are themselves without dimension and are said to be *dimensionless* and *unitless* i.e. they are just real numbers. Apart from the advantage that this enables one to avoid the need to keep in mind the units attached to the various terms, the approach enables one to see more clearly the comparative importance of terms in the model equations so that rational decisions concerning approximations can be made. The advantages are, however, *much* more far reaching, as will be seen.

2.1.1 The flagpole

As an illustration of this approach, we consider ***the flagpole problem.*** This problem arose out of the question of survival of a tall flagpole under earthquake conditions. Basically the question was: how thick does a tall flagpole have to be to resist earthquake damage? Consider the following partial differential equation

$$\rho\frac{\partial^2 u}{\partial t^2} + Ek^2\frac{\partial^4 u}{\partial x^4} = 0, \text{ in } 0 < x < l. \tag{2.1}$$

The aim is to find a solution, u, of this partial differential equation, which also satisfies the conditions:

- that it be periodic in the variable t with prescribed period $T = 2\pi/\omega$,
- and satisfies the subsidiary boundary conditions,

$$u(0,t) = a\sin\omega t, \quad \tfrac{\partial u}{\partial x}(0,t) = 0, \tag{2.2}$$

$$\frac{\partial^2 u}{\partial x^2}(l,t) = 0, \quad \tfrac{\partial^3 u}{\partial x^3}(l,t) = 0. \tag{2.3}$$

The equations describe the lateral vibrations, of amplitude $u(x,t)$, of a beam of length l when one end, $x = 0$ (the ground), is forced to move with a prescribed periodic motion of frequency ω (period T) and amplitude a, and the other end is free to move, see Fig. 2.1. For a derivation of these equations see Segel (1977). The flagpole connection is evident. However, it should be realized that with modifications the same equations model the vibrations of very large structures such as skyscrapers, moderate structures such as fishing poles, and the very small beam structures that are often encountered in biological contexts.

The Young's modulus E (with dimension $[M][L]^{-1}[T]^{-2}$) and the density ρ ($[M][L]^{-3}$) are properties of the material of the beam and k ($[L]$) is a property of the cross-sectional shape known as the radius of gyration. It's not necessary to understand the meaning of these words; in fact the investigations to follow require no real understanding of the underlying physics as expressed in the above equations. All that is needed, apart from a general idea of what the problem is about, is the specification of the dimensions associated with those quantities given above, and some typical data values.

Input data. We'll suppose the pole is a cylindrical steel pipe structure of diameter 15 cm and 10 metres high. It's a tall pole! (We'll in fact examine a range of heights ranging from 10 m to about 20 m) For a cylindrical

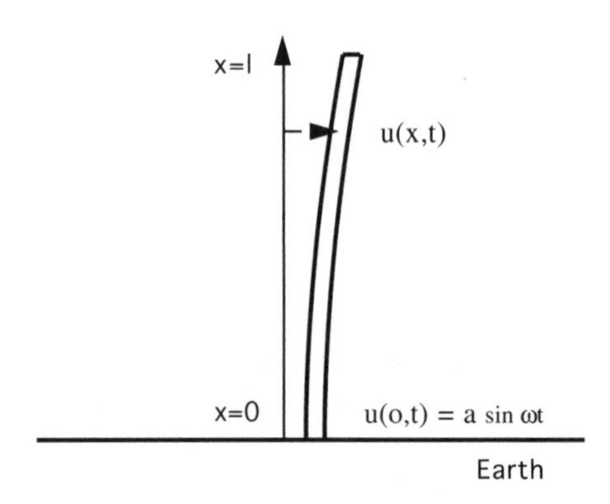

Figure 2.1: The flagpole.

pipe structure $k = \alpha d$, where d is the diameter of the pole and $\alpha \approx 1/(2\sqrt{2})$ for a relatively small thickness pipe. This gives $k \approx 5.3$ cm for the steel pole of interest. Tables give for the Young's modulus of steel the value $E_s = 2.0 \times 10^7$ kg/m/s^2; and for its density the value $\rho_s = 7.8 \times 10^3$ kg/m^3. Earthquakes produce ground oscillations with amplitudes of size micrometres to centimetres, with periods ranging from fractions of seconds to 20 or 30 seconds. We'll use 2 seconds as a representative value, with corresponding angular frequency $\omega = \pi$ radians/s.

Unfortunately, although there is now almost universal acceptance of the S.I. system in professional scientific and engineering journals, this system is not the one in everyday use everywhere, for example in the United States. There is also substantial usage of other units in certain technical areas as the result of the carryover of past practices. Thus it is frequently necessary to convert from one set of units to another in order to obtain data in the appropriate form. The required conversion is straightforward but tedious. Thus to obtain E and ρ in terms of feet and pounds (denoted by lbs) one simply applies the equivalences,

- 1 m = 3.281 ft and
- 1 kg = 2.205 lb

in the way indicated by the specification of the units. Thus the Young's Modulus E in the alternative system is given by

$$\frac{(2.0 \times 10^7 \text{ kg/m/s}^2) \times (2.205 \text{ lb/kg})}{(3.281 \text{ ft/m})} = 1.34 \times 10^7 \text{ lb/ft/s}^2.$$

For those with a UNIX system the conversion can be performed automatically by issuing the system command, *units*, without the need to supply any of the conversion factors; hopefully such features will become standard on computer systems.

Which parameter values should be inserted into the above mathematical system? The answer is that either is OK, provided that one *consistently* uses one system or the other when substituting all the parameter values into the model equations. The solution obtained must also, of course, be interpreted in terms of the particular system used. Different systems of units will give rise to equations that look different (with different coefficients multiplying the various terms), and these equations will have solutions that superficially look different. When comparisons are made, however, (with the appropriate conversion factors applied to the parameters and variables) the results obtained will be seen to be equivalent. This is of course no accident and is a direct consequence of **Fact 1**. It's clear on physical grounds, for example, that the actual physical deflection predicted for the flagpole *must* be independent of the system of units used—if this wasn't the case then the partial differential equation with the prescribed boundary conditions could not be accepted as a possible model of the real world. Thus when one converts completely (both parameters and variables) from any one system to another, the same combined conversion factor must multiply all individual terms of the equation and so cancel out. In the flagpole case the terms in the partial differential equation (2.1) share the units $\mathrm{kg/m^2/s^2}$ in the SI system, and the combined conversion factor in common that arises when transforming the equation to the ft, lb, s units system is $2.205/3.281^2$, which cancels from all terms.

The mathematical invariance associated with changing from one set of units to another in such real world systems suggests that there might be an underlying mathematical description of such systems which is free from this dependence on units. This turns out to be so. One way to obtain this description is to opt for (non-standard) units called **scales** that are *specific for the particular problem* in hand.

2.1.2 Scaling the flagpole problem

In the flagpole problem there is an obvious natural length scale for the problem; the length of the pole. Thus it makes sense to specify position along the pole in terms of ξ defined by

$$x = l\xi.$$

Now $0 \leq \xi \leq 1$ becomes the domain on which the deflection of the pole is defined. Note that ξ is the ratio of two quantities with dimension $[L]$, so that its value is independent of the units chosen to describe lengths i.e. it's *dimensionless*. There is also a natural time scale set by the frequency ω of motion, so we introduce a new *dimensionless* time-like variable τ such that

$$t = \tau/\omega.$$

The natural length scale for the deflection is a, the amplitude of the forcing motion, see (2.2), so we use

$$u = av.$$

We now examine what happens if the problem is restated in terms of these new *dimensionless* variables. The chain rule gives

$$\frac{\partial u(x,t)}{\partial t} = \frac{\partial av(\xi,\tau)}{\partial t} = a\frac{\partial v(\xi,\tau)}{\partial \tau}\frac{d\tau}{dt} = a\omega\frac{\partial v(\xi,\tau)}{\partial \tau},$$

and

$$\frac{\partial u(x,t)}{\partial x} = \frac{\partial av(\xi,\tau)}{\partial \xi}\frac{d\xi}{dx} = \frac{a}{l}\frac{\partial v(\xi,\tau)}{\partial \xi}$$

etc., so the partial differential equation (2.1) becomes

$$(\rho a\omega^2)\{\frac{\partial^2 v(\xi,\tau)}{\partial \tau^2}\} + (\frac{Ek^2 a}{l^4})\{\frac{\partial^4 v(\xi,\tau)}{\partial \xi^4}\} = 0.$$

Now the terms in curly brackets $\{\cdots\}$ are combinations of dimensionless variables, and so must also be dimensionless. The coefficients of these terms, in round brackets $(\cdots)$, must therefore have the same dimensions; a check reveals that both terms have the dimensions $[M][L]^{-2}[T]^{-2}$. Thus, if one divides through by the factor $\rho a\omega^2$, the equation reduces to the completely dimensionless or unit invariant form

$$\frac{\partial^2 v}{\partial \tau^2} + J\frac{\partial^4 v}{\partial \xi^4} = 0, \qquad (2.4)$$

where

$$J = \frac{Ek^2}{\rho\omega^2 l^4} \qquad (2.5)$$

is again a dimensionless combination of the parameters of the problem, often referred to as a *dimensionless group* or *dimensionless parameter.* The subsidiary conditions become:

- v should be periodic with period 2π,
- and the boundary conditions become:

$$v(0,\tau) = \sin\tau, \quad \frac{\partial v}{\partial \xi}(0,\tau) = 0, \qquad (2.6)$$

$$\frac{\partial^2 v}{\partial \xi^2}(1,\tau) = 0, \quad \frac{\partial^3 v}{\partial \xi^3}(1,\tau) = 0. \qquad (2.7)$$

This transformed mathematical system involves just the single dimensionless parameter J. The value of the parameter J serves to specify the invariant form of the mathematical model and therefore serves to characterize the motion of ***all such*** beams under similar forcing. This mathematical problem, if it has a unique solution, defines the scaled displacement v solely in terms of J, ξ and τ, and thus generates a solution V of the scaled problem of the form

$$v = V(\xi, \tau, J).$$

Thus, by transforming back to unscaled variables, we can obtain an expression for the (unscaled) deflection $u(x,t)$ in the form

$$u(x,t) = aV(x/l, \omega t, J). \tag{2.8}$$

There are a number of advantages to the above unit invariant form (2.4,2.6, 2.7) of the mathematical statement of the problem:

- **For analytical work.** The analysis involves fewer symbols. The simpler appearance often makes it easier to see any structure, and hence organize calculations and understand results. Also the reduced form makes the estimation of errors involved in omitting any particular term from the equation possible, because the coefficients are real numbers (not dimensional quantities) for which smallness and largeness can be given some meaning.

- **For numerical work.** If a numerical approach to the solution is to be used the number of calculations to be performed is reduced enormously, since one only needs to consider a range of values of the *single* parameter J and not the larger initial parameter set $(\rho, E, k, l, a, \omega)$. The number of calculations increases roughly like n^r, where n is the number of evaluation sites for r parameters, so the above simplification is *essential* for either numerical or tabulation purposes.

  Aside: Mathematical tables normally only attempt to present data on functions of a single variable for a few values of one additional parameter (for example $f(x,a)$). The sheer size becomes too great for tabulation if further variables or parameters are introduced.

- **For experimental work.** The unit invariant form provides a sensible basis for experimental design. To elaborate: The behaviour of an experimentally manageable 2 metre wooden rod will be dynamically similar to that of the (large) 15 cm, 10 metre steel pole of interest, providing the two have the same J value. Tables give $E_{wood} \approx E_s/20$, $\rho_{wood} \approx \rho_s/13$, so this can be realized for the same vibration frequency ω, see (2.5), if $k_{exp}^2 = (20/13)5^{-4}k_{pole}^2$, i.e. if $k_{exp} = 0.05k_{pole}$. After adjustments are made for the difference in cross-sectional shape (the experimental rod is solid whereas the pole is hollow) one finds that the 2 metre pole would have to be 1 cm in diameter to *simulate* the behaviour of our flagpole. The word simulate is used here because we chose to use the same value of ω in the experimental set-up as for the real situation; thus the wooden pole will behave the same way as the flagpole when subjected to the same oscillation (or when subjected to a combination of oscillations). There is, however, no necessity to duplicate ω in this way. Thus experimentally one could use, within reason, ***any*** available experimental pole, and simply determine the frequency required to duplicate J; the results obtained could then be used to determine the behaviour of our flagpole, again using equation (2.5).

Aside: In the same way a reasonably sized experimental model could be used to investigate the behaviour of minute "beams", for example the hair cells in the cochlea of the human ear (25μm long, 6μm diameter).

In the above work we chose to use different length scales, l and a, for the position along the beam x and for the displacement u. What would have happened if the same length scale l had been used for both x and u; so that the variable w defined by $u = lw$ had been used? Since both a and l have the dimensions of length no dimensional inconsistency would be introduced by this alternative scaling, and in fact from a *purely mathematical* point of view the two formulations are equivalent in the sense that both would produce identical *exact* results. Explicitly the solution for w would be given by $w = \frac{a}{l}V(\xi, \tau, J)$. From a practical point of view, however, the two formulations are *very* different. A second parameter (a/l) appears in the boundary conditions in the second formulation, so the complexity in this case is much greater. (In this particular problem linearity arguments can be used to simplify this second formulation; for nonlinear systems such simplifications are not usually available.) Also from an approximation point of view the two formulations are very different. While both choices of scales are mathematically correct, the one we have chosen is the useful form when it comes to approximation; basically because a is the true representation of the size of the deflection, so the estimates of the terms of the equations based on this scale are correct. The message is that it is a good general policy to *look carefully when choosing scales.*

As indicated above, even at this early stage a great deal has been achieved and much can now be said about the flagpole problem. Thus, even if the problem was analytically intractable, one could either use a standard numerical package to determine the solution for the J value of interest, or perform a convenient experiment. Of course much more understanding about the general situation will result if one can proceed further with the analysis, as we will see. Before continuing in this way we'll make a few preliminary observations/speculations. Such observations can help

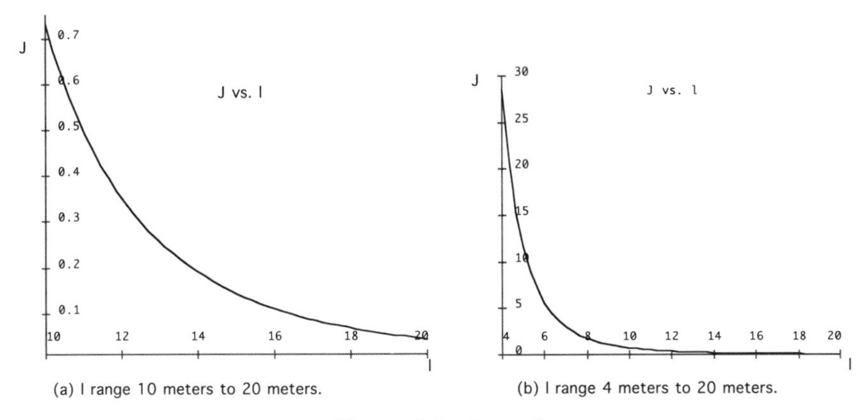

(a) l range 10 meters to 20 meters.

(b) l range 4 meters to 20 meters.

Figure 2.2: J vs. l.

direct further work. First note that the relatively short, fat and stiff poles which one would expect to be safe under earthquake conditions correspond to relatively large J values, see (2.5). Of course it's the J value that determines what one means by small in this context. We would expect the long thin poles corresponding to small J values to be in danger under earthquake conditions. Also note that the value of J decreases *very* rapidly with increasing pole length l (like $1/l^4$); so a plot of J vs. l with the other parameters fixed (by the given data) would be informative. The Maple commands:

```
T:=2/(2*Pi);                                    %Pi ≡ π
J:=E*k^2*T^2/(rho*l^4);                         %rho ≡ ρ
e:=2*10^7;
rho:=7.8*10^3;
k:=15/(2*sqrt(2));
plot(J,l= 0..30, title='J vs.  l');
evalf(subs(l=10,J));
evalf(subs(l=15,J));
evalf(subs(l=20,J));
```

produce Fig. 2.2, and give J = 0.73, 0.14, and 0.04 when l=10 m, 15 m, and 20 m.

Note that the value of J varies over a 10^4 range for l in the range $2 \text{ m} \leq l \leq 20$ m! One would thus speculate that the pole's behaviour is likely to vary rapidly over the range of pole lengths of interest. It's also worth observing:

- for fixed values of (ρ, E, ω), J remains fixed if k/l^2 is kept constant. Thus the same dynamic response will be realized from poles of different length providing the pole diameter is increased in proportion to the square of the length. Thus the cost of producing a safe pole is likely to increase rapidly with the flagpole height.

- J varies with $1/\omega^2$, and since it's the small values of J that are a worry, it's the high frequency earthquake vibrations that are likely to be most damaging.

The above observations are a little premature since we don't really know anything about the solution behaviour at this stage, however, it's as well to always keep the broad picture in mind.

2.2 Approximations

Assuming that the scales chosen above are truly representative, then one might reasonably expect that both the scaled function $v(\xi, \tau)$ and its derivatives with respect to the scaled variables ξ and τ (eg. $\partial v/\partial\xi, \partial v/\partial\tau$) etc. to be of unit order. If this is so then the relative sizes of the terms in the beam equation (2.4) are represented by the numerical coefficients attached to these terms (1 and J). Thus if the parameter range of interest is such that the dimensionless group J is either small or large, significant simplifications are possible.

2.2.1 The large J solution

In the large J (stiff pole) range it's sensible to write the partial differential equation (2.4) in the form

$$\frac{\partial^4 v}{\partial \xi^4} = -\frac{1}{J}\frac{\partial^2 v}{\partial \tau^2} \tag{2.9}$$

and, providing one is willing to ignore terms of order $1/J$, the simpler equation

$$\frac{\partial^4 v_0}{\partial \xi^4} = 0,$$

should accurately represent the physics and thus should enable us to determine a good approximation v_0 for v. To obtain this approximation integrate this equation once with respect to ξ to give

$$\frac{\partial^3 v_0}{\partial \xi^3} = C(\tau),$$

where $C(\tau)$ is an arbitrary function of τ. The boundary condition (2.7) at $\xi = 1$ then gives

$$C(\tau) = 0.$$

A second integration then gives

$$\frac{\partial^2 v_0}{\partial \xi^2}(\xi, \tau) = 0,$$

after applying the remaining boundary condition at $\xi = 1$. The general solution of this partial differential equation is

$$v_0 = A(\tau) + B(\tau)\xi,$$

where $A(\tau)$ and $B(\tau)$ are as yet undetermined functions. From the boundary conditions (2.6) at $\xi = 0$ it is easy to show that

$$v_0 = \sin\tau.$$

This provides us with an approximation with a relative accuracy initially estimated as being of order $1/J$. If this level of error is regarded as acceptable then the usual practice is to proceed no further. Even under such circumstances, however, it's wise to obtain a better estimate, even if only to provide some check on the accuracy of the first approximation. A more accurate estimate can be obtained by iteration using the arrangement (2.9). The idea is simple: if v_0 is a reasonably good estimate for v, then $(-1/J)\partial^2 v_0/\partial\tau^2$ should provide a reasonable estimate for the right hand side of (2.9) so that

$$\frac{\partial^4 v_1}{\partial \xi^4} = -\frac{1}{J}\frac{\partial^2 v_0}{\partial \tau^2} = \frac{1}{J}\sin\tau, \tag{2.10}$$

with the same boundary conditions as for v i.e.,

$$v_1(0,\tau) = \sin\tau, \quad \frac{\partial v_1(0,\tau)}{\partial \xi} = \frac{\partial v_1^2(1,\tau)}{\partial \xi^2} = \frac{\partial v_1^3(1,\tau)}{\partial \xi^3} = 0, \tag{2.11}$$

should provide a better estimate v_1 for v. The process can be repeated by successively substituting improved approximations for v into the right hand side of (2.9) to obtain

$$\frac{\partial^4 v_n}{\partial \xi^4} = -\frac{1}{J}\frac{\partial^2 v_{n-1}}{\partial \tau^2}, \text{ for } n = 1, \ldots, \tag{2.12}$$

again with the boundary conditions

$$v_n(0,\tau) = \sin\tau, \quad \frac{\partial v_n(0,\tau)}{\partial \xi} = \frac{\partial v_n^2(1,\tau)}{\partial \xi^2} = \frac{\partial v_n^3(1,\tau)}{\partial \xi^3} = 0. \tag{2.13}$$

Equation (2.10) for v_1 can be integrated immediately and, after applying the above boundary conditions, one gets

$$v_1 = \{1 + (1/J)[\xi^4/24 - \xi^3/6 + \xi^2/4]\}\sin\tau. \tag{2.14}$$

If one looks ahead to the next approximation one can see that the solution will contain an additional term of order $1/J^2$. Thus we'd expect v_1 to be accurate to order $1/J$, with an error estimated to be of order $1/J^2$. We also expect the above approximation process to be reasonable providing $1/J$ is small, with an error of order $(1/J)^{n+1}$ after n iterations of (2.12). Of course the algebra quickly becomes tedious if performed by hand, however, it's precisely this type of algebraic process that Maple is designed to handle.

Comment: The belief underlying the above work is that relatively small terms in an equation will have little effect on the solution to that equation; an assertion that is often true, but is not true in general. By calculating higher order terms we can at least ensure that there is some consistency in the approach, but the fact that relatively large terms don't arise late in the procedure doesn't ensure that the results are correct. There are many cases in which the process described above fails to produce accurate solution estimates; with experience one can recognize the cases where care may be needed. The fundamental justification for the use of the arguments presented above is the success obtained throughout the history of science and technology which has followed from their adoption. The needs of modelling often call for approximation procedures that cannot be justified rigorously.

It is instructive to interpret the results we've obtained for the pole. We've obtained the result that $v(\xi, \tau) = v(0, \tau) + O(1/J)$, so that for large values of J the pole undergoes a horizontal displacement almost equal to that of the ground. Thus to order $1/J$ there is no bending of the pole, and therefore there is no tendency for it to break *above* the ground. Under such circumstances any failure must occur below the ground in its foundations. Thus it is only for moderate to small values of J that failure of the pole above ground needs investigation; a result that we anticipated earlier.

2.2.2 The small J solution?

The other extreme case where J is small might appear to lead to something of a paradox in that the approximate partial differential equation obtained by putting $J = 0$ is

$$\frac{\partial^2 v}{\partial \tau^2} = 0,$$

which suggests that the solution is a linear function of τ. Clearly this is inconsistent with the requirement that the solution vary periodically in time, and also with the sinusoidal boundary condition (2.6) at $\xi = 0$. Thus, assuming there is a solution of the required periodic form, there must be something wrong with the arguments that lead us to the above result. We'll return later to examine in detail the above approximation arguments after having determined the exact solution in the next section.

Aside: One might well ask why bother with approximate solutions when exact solutions are available? If the exact solutions are easily *computable* and *understandable* then there is no justification within the context of the particular problem. Often, however, in order to build up understanding of a mathematical procedure or a physical process, it's useful first to look at simple problems that can be solved exactly. Our present aim is to understand the strengths and weaknesses of commonly used procedures, so the exactly solvable pole problem suites our purposes. The approximation procedures being described here are *very* useful for circumstances which cannot be dealt with using the exact methods described below (eg. for nonlinear problems).

2.3 The Exact Solution

In the large J case we found that the solution was of the form of a function of ξ times $\sin\tau$. This suggests that we should look for a solution of the form

$$v(\xi,\tau) = \chi(\xi)\sin\tau, \tag{2.15}$$

for the general (i.e. arbitrary J) solution to our problem—at least it's worth a try. The sinusoidal boundary condition (2.6) also suggests that such a form is appropriate.

Comment: This may seem to be a somewhat ad hoc approach; however, it's very usual in dealing with real world problems (where non-standard equations often arise) to guess a solution form, and then check to see if the form is correct by direct substitution into the equation. By the end of this book the reader will be very familiar with this approach.

A direct substitution of this sinusoidal form into the equation (2.4) for v gives, after cancelling out the common $\sin\tau$ factor,

$$J\chi''''(\xi) - \chi(\xi) = 0, \tag{2.16}$$

and the boundary conditions become

$$\chi(0) = 1, \quad \chi'(0) = \chi''(1) = \chi'''(1) = 0. \tag{2.17}$$

The Maple commands:

```
deq:=J*(D@@4)(X)(xi)=X(xi);                          %X(xi) ≡ χ(ξ)
bcs:=X(0)=1,D(X)(0)=0,(D@@2)(X)(1)=0,(D@@3)(X)(1)=0;
Xsoln:=dsolve({deq,bcs},X(xi));
amp:=rhs(");                                          %amp≡ χ(ξ),
```

determine the solution for $\chi(\xi)$. Notice that, because $\sin\tau$ varies between $+1$ and -1, $v(\xi,\tau)$ varies between $+\chi(\xi)$ and $-\chi(\xi)$ as τ varies, so that $|\chi(\xi)|$ may be interpreted as being the amplitude of vibration of the pole as a function of ξ. It is usual to ignore the modulus operation and simply refer to $\chi(\xi)$ as the amplitude of motion along the pole. Although Maple very conveniently performs the tedious algebra for us it's as well not to work

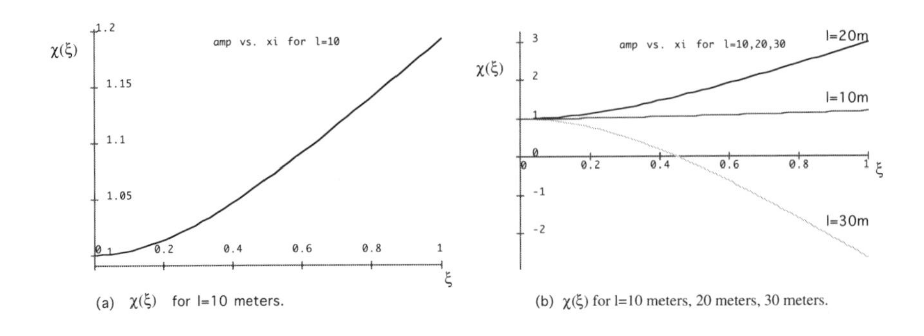

(a) $\chi(\xi)$ for l=10 meters.

(b) $\chi(\xi)$ for l=10 meters, 20 meters, 30 meters.

Figure 2.3: Beam amplitude.

blindly with such packages; partly because they are not completely reliable at this stage, but also because it's necessary to *understand* the results. In the present context checks should be made that `amp` is of the right form and satisfies the equation `deq` and the boundary conditions `bcs`. Thus differential equation theory tells us to expect solutions of the form $e^{\lambda\tau}$ where $J\lambda^4 = 1$ i.e. $\lambda = \pm 1/J^{1/4}$, $\pm i/J^{1/4}$; so we should check to see that the solution involves linear combinations of $e^{\xi/J^{1/4}}$, $e^{-\xi/J^{1/4}}$, $\sin(\xi/J^{1/4})$ and $\cos(\xi/J^{1/4})$.

Comment: Be warned that working with algebraic systems can be particularly frustrating at this stage of their development, so that unexpected difficulties can arise, for example when trying to check that the solution obtained above in fact satisfies the conditions of the problem. Usually the best plan is to experiment first and then, if necessary, seek help from fellow students.

Having made the basic checks, we now proceed to extract useful information. Firstly to get some feel for the results it's useful to plot the $\mathcal{X}(\xi)$ for poles of various lengths in the range of interest. The Maple command:

```
plot({subs(l=10,amp),subs(l=15,amp),subs(l=20,amp)},xi=0..1);
```

does this for pole lengths l = 10 m, 15 m, 20 m, after inputting $J(l)$ with the other parameters prescribed as above. The results are displayed in Fig. 2.3.

Although l is a sensible parameter to use to get a feel for the results, it's really not the appropriate parameter to use for presenting information in the most succinct form. The relevant parameter to employ is of course the dimensionless parameter J. Thus l = 10 m corresponds to $J = 0.73$ for the data values used, so it would be more useful to label the curve corresponding to l=10 m with its J value. The relative amplitude displayed on this curve corresponds to any of the possible $(\omega, l, k, \cdots)$ combinations that give rise to $J = 0.73$—thus, *much* information is contained in this single curve. The curves *suggest* that for large to medium values of J the maximum pole displacement occurs at the free end, with a 19% relative displacement (compared with the amplitude of the earth's movement) occurring for the chosen data conditions (use `xamp1 := evalf(subs(xi = 1,l = 10,amp));`). Notice that the $l = 20$ m ($J = 0.045$) curve $\chi(\xi, J)$ crosses the $\chi = 0$ axis at $\xi \approx 0.47$. The implication of this is that this point must be a node for the motion, i.e. it must remain fixed in space while the rest of the pole moves. You'll recall that the solution form contains oscillatory functions of $\xi/J^{1/4}$, so amplitude oscillations *might* be expected for (small) values of J such that $1/J^{1/4} > \pi$, with more and more nodes appearing as J gets smaller. Maple can be used to settle many of the questions raised by the above observations. The interested reader might like to use Maple to determine the location and size of the maximum amplitude as a function of J. Some of these questions are raised in Exercise 2.1. To determine the probability of the pole breaking one would need to enquire about the fracturing properties of the material of the pole—under many circumstances the pole will snap if the bending stresses exceed a critical value; the curvature of the pole provides a measure for bending stresses.

2.3.1 The small J solution

Now that we have the exact solution we can investigate the difficulty that arose earlier with the attempt to determine a solution approximation for small J values. You'll recall the basis for the attempt: for small enough values of J it was assumed that

$$|J(\frac{\partial^4 v}{\partial \xi^4})| \ll |\frac{\partial^2 v}{\partial \tau^2}|,$$

so that equation (2.4) could be replaced by

$$\frac{\partial^2 v}{\partial \tau^2} \approx 0.$$

The exact solution for $v(\xi, \tau)$ consists of a linear combination of exponential and oscillatory functions of $\xi/J^{1/4}$. Now for small values of J these functions vary rapidly in ξ and thus have large derivatives with respect to ξ, for example

$$\frac{\partial^4 \sin(\xi/J^{1/4})}{\partial \xi^4} = \frac{\sin(\xi/J^{1/4})}{J}.$$

The implication is that the apparently small term $J\partial^4 v/\partial\xi^4$ is not small, but of unit order; and therefore cannot be neglected by comparison with the retained term $\partial^2 v/\partial\tau^2$. The suggested approximation thus doesn't work. We would of course like to understand the difficulty at a deeper level so we can avoid similar difficulties arising in other situations. A correct but relatively uninformative observation that one could make is that, although there exist sensible solutions to the partial differential equation for very small values of J, such solutions are not close to the solution corresponding to $J = 0$ i.e. the limit is *discontinuous* or *singular*. (Recall that for $J = 0$ we obtained a solution that was linear in time.) More usefully one can note that an incorrect assessment of the size of the terms of the partial differential equation arose because of the assumption we made that the length scale for variations of $u(x, t)$ was the length l of the beam; whereas for small J values the actual length scale for variations is the smaller value $(J^{1/4}l)$. In physical terms, oscillations of wavelength of order $(J^{1/4}l) < l$ are generated by the ground motion.

Comment: Perturbation Techniques The technique described in the above work for the large J solution approximation is often referred to as a *regular perturbation* technique—in our case we perturbed about the $J = \infty$ solution, and the limit was continuous. The small J case is referred to as a *singular perturbation* problem and a reasonable understanding of such problems has been built up and techniques have been developed to handle such problems. We'll examine such techniques later in the Vibrations Chapter. An excellent (and elementary) account of these methods and regular perturbation methods can be found in Carrier and Pearson (1968). Collectively these techniques are referred to as *asymptotic* techniques and it's true to say that it's these techniques, rather than the convergent series techniques commonly studied in analysis courses, that have been found to be more useful for understanding real world phenomena. Thus, for example, virtually all the theories of physical science represent asymptotic approximations to to the real world. It is unfortunate that pure mathematicians have generally turned their backs on this most important area.

2.4 Dimensional Analysis

Any statement about the magnitude of a physical quantity implies an accepted method of comparison of like magnitudes, leading to the statement that one of them is a certain multitude of another one, which may serve as a *unit*. Derived physical units are defined in such a way to give a simple form to the physical laws of basic importance. Since all equations of physics are derived either from these physical laws or compatible definitions; consistency of units in any equation describing physics is required by the way the structure is set up. A dimensional formula for a derived quantity shows the relationship of the derived unit to the fundamental units in terms of which it is defined. Thus dimensional consistency is required of any equation expressing physics. For historical reasons the units of mass, length and time, with dimensions denoted by $[M]$, $[L]$ and $[T]$ are regarded as fundamental. Speed is defined to be the change in distance divided by the time taken for this change, and so has units such as metres/s and is said to have the dimensions $[L][T]^{-1}$. Similarly acceleration, from its definition, has dimensions $[L][T]^{-2}$. From Newton's Second Law of Motion, force is obtained as the product of mass and acceleration, and so has dimensions $[M][L][T]^{-2}$. To the list of basic dimensions has been added the dimensions of charge $[Q]$ and temperature $[\Theta]$. For mathematical purposes angles are measured in radians and are defined by the basic arc length vs. radius relationship $a = r\theta$. Dimensional consistency is thus preserved only if $[\theta] = [1]$, i.e. if angle is regarded as a *dimensionless* quantity. All quantities currently known to physics have been defined in terms of, or derived from (via physical laws), the above basic list. Thus the dimension of any physical quantity can be expressed in the form

$$[M]^{\alpha}[L]^{\beta}[T]^{\gamma}[\Theta]^{\delta}[Q]^{\epsilon},$$

where the indices are integers. The fact that the quantities can be built from such a small list is powerful evidence of the detailed structure whereby new physical quantities have been introduced, and the fact that only integers arise is due to the care with which physicists have defined such new quantities. Thus if absolute temperature had been defined as the square of the expression presently adopted, then fractional powers would appear in the dimensions of some physical quantities.

The dimensional consistency requirements enable one to make certain statements about the mathematical form the outcome of a physical model can take, and *dimensional analysis* uses this. Thus if

$$y = aF(x_1, x_2, \ldots, \alpha_1, \alpha_2, \ldots)$$

represents an expression for the quantity y in terms of the variables x_i and parameters α_i obtained from some model, and $[y] = [a]$, then it must be possible to write F in a form that involves only dimensionless combinations β_j (so $[\beta_j] = [1]$) of the (x_i, α_i). Thus, for example, the deflection of the beam according to the model we used above depends only on the parameters $(E, l, k, a, \rho, \omega, x, t)$ defined by the model and so can be written thus

$$u(x,t) = aF(E, l, k, a, \rho, \omega, x, t),$$

where F only contains dimensionless combinations of its arguments. Dimensional analysis seeks out the possible dimensionless combinations. To do this one can examine all combinations

$$E^m l^q k^n a^p \rho^s \omega^r x^y t^z,$$

and by writing down the dimensions of such combinations

$$([M][L]^{-1}[T]^{-2})^m [L]^q [L]^n [L]^p ([M][L]^{-3})^s [T]^{-r} [L]^y [T]^z,$$

and seeking out those that are dimensionless, one can determine the allowed

$$(m, q, n, p, s, r, y, z)$$

combinations. This is a reliable but *very* tedious process; rarely used. A few simple observations usually enable one to determine more quickly the solution form. Thus, since ρ and E are the only quantities in the above list containing $[M]$, they can only occur in the combination E/ρ. This combination has the dimensions $[T]^{-2}[L]^2$, so that by throwing in any combination with dimensions $[T]^2[L]^{-2}$ (for example $1/(l^2\omega^2)$) one ends up with a dimensionless combination ($E/(\rho\omega^2 l^2)$ in the above case). One might then anticipate that the variables (x, t) should be scaled in the obvious way $(x/l, \omega t)$. Also, because of linearity, a must only occur as a multiplicative factor in the expression for u. In this way one ends up with the following description

$$u = aF(E/(\rho\omega^2 l^2), k/l, x/l, \omega t).$$

Understandably one might be a little concerned that some required combination might be left out; care is required to ensure this doesn't happen. It should be pointed out that the prescription for F is not unique, and that some descriptions are more useful than others. Thus for example if one had *lucked in* on the combination J that we discovered to be the defining parameter for the problem, follow up work (for example experiments) would be *greatly* simplified.

The advantages of *inspectional analysis* over *dimensional analysis* are evidently:

- the inspectional analysis approach enables one to determine the most appropriate dimensionless groups
- the inspectional analysis approach, because it works with the defining equations, enables one to assess the relative importance of the various processes involved so that one can proceed to further simplifications.

Clearly, *if at all possible*, one should use inspectional analysis. Sometimes, however, it's so difficult to understand the processes involved in a particular situation that inspectional analysis is not possible.

2.5 Summary

We've seen that *scaling* procedures enable one to identify the particular combinations of the parameters of the problem that characterize its behaviour. These dimensionless combinations, referred to as *dimensionless groups*, provide a basis for all follow up work; whether it be mathematical or experimental. If some of the dimensionless groups are either small or large (and this is often the case) *iteration* or *perturbation* techniques can often be used to obtain useful approximate results.

2.6 Exercises

Exercise 2.1 *Loose Ends*

Using Maple reconstruct the amplitude vs. location curves displayed in Fig. 2.3 for the flagpole. Maple is a very convenient tool for examining various aspects of this problem. Investigate aspects of the problem that you find interesting. A few suggestions are:

(a) Choose a particular value of J and plot out the scaled displacement $v(\xi, \tau)$ as a function of ξ for various values of τ, covering a full cycle of the pole's motion (eg. $0 \leq \tau \leq 2\pi$). In this way obtain "snapshots" of the pole's motion. Observe the waves travelling along the pole. Experiment with different J values.

(b) Plot out the amplitude of motion of the free end of the pole as a function of J over the range of interest and comment on the results. Is the maximum displacement realized at the free end of the pole?

(c) The accuracy of the large J solution can be investigated by comparing the approximate solution v_1 with the exact solution v. The amplitude of motion of the free end of the pole provides a convenient comparison tool. Plot $\{\chi(1),\ v_1(1,\tau)/\sin\tau\}$ vs. J and comment on the range of usefulness of the large J approximation.

(d) ⋆⋆ The longitudinal curvature of the pole when in motion is likely to be a critical feature as far as snapping is concerned. As a function of J, at what location along the pole is the maximum curvature reached?

Exercise 2.2 *Perturbations*

The iteration procedure described for determining better solution estimates in the large J case is inefficient in that at each iteration level all the earlier approximation terms are reevaluated. Thus, for example, when we solve equation (2.10) for v_1 we obtain (see (2.14) the $\sin\tau$ term, which we recognize as being simply v_0. This is a minor issue unless one is calculating many iterates; in which case the additional algebraic complexity introduced as a result of this inefficiency can be a major problem (even if using an algebraic package). A simple reorganization of the iteration procedure, referred to as a *perturbation* procedure, avoids this duplication of effort. We simply recognize that the iteration procedure is producing a solution expansion of the form

$$v(\xi, \tau) \simeq V_0 + \frac{1}{J}V_1 + (\frac{1}{J})^2 V_2 + \cdots = \sum_{i=0}^{m} \frac{1}{J^i} V_i,$$

i.e. a series expansion in the small parameter $(1/J)$.

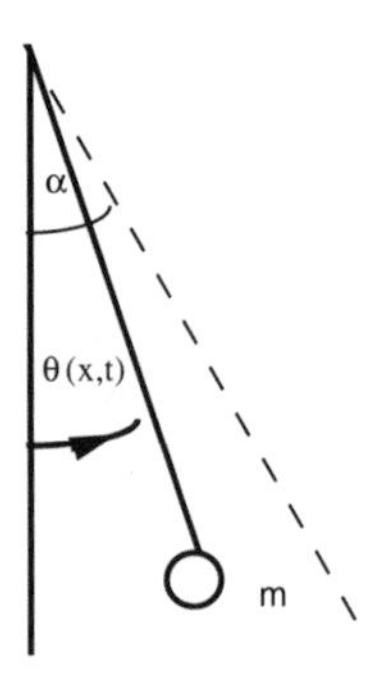

Figure 2.4: The pendulum.

(a) Substitute the above expansion form into the equation (2.4) for v and its associated boundary conditions and by equating coefficients of $1/J^n$ to zero, show that $V_0 = \sin\tau$ and that the equations to be satisfied by V_n are given by

$$\frac{\partial^4 V_n}{\partial \xi^4} = -\frac{\partial^2 V_{n-1}}{\partial \tau^2},$$

with the boundary conditions

$$V_n(0,\tau) = 0, \quad \frac{\partial V_n(0,\tau)}{\partial \xi} = \frac{\partial^2 V_n(1,\tau)}{\partial \xi^2} = \frac{\partial^3 V_n(1,\tau)}{\partial \xi^3} = 0, \;\; \text{for } n = 1\ldots.$$

Compare these equations with those obtained using the iteration procedure.

(b) Determine V_1 by hand.
Hint: Look for a solution of the form $V_1 = \mathcal{V}_1(\xi)\sin\tau$.

(c) Use Maple to determine V_1 (check with the result you obtained by hand). Determine V_2 using Maple.

(d) Using Maple, plot out $\sum_{i=0}^{m}(\frac{1}{J})^i V_i$ for $m = 0, 1, 2$ and $J = 10$, and make the appropriate comparisons.

Exercise 2.3 *The Pendulum*

We seek to determine some information concerning the period τ of pendulums made up of different length cords and different mass bobs, using scaling arguments, see Fig. 2.4. The equations describing the motion of a pendulum of mass m released from rest at an angular displacement α (with $\alpha < \pi/2$ so the cord remains taut) are given by

$$ml\ddot{\theta}(t) + mg\sin\theta(t) = 0,$$

with initial conditions,

$$\theta(0) = \alpha, \ \dot{\theta}(0) = 0,$$

where $\theta(t)$ is the angular displacement of the pendulum from its equilibrium position. The time taken for the pendulum to first reach the zero angular displacement position is $\tau/4$. Thus the smallest value of τ such that

$$\theta(\tau/4) = 0,$$

determines τ.

(a) Scale the above equations appropriately (you will need to determine an appropriate time scale) and thus reduce them to the simplest non-dimensional form

$$\alpha\ddot{\Theta}(t') + \sin\alpha\Theta(t') = 0,$$

with initial conditions,

$$\Theta(0) = 1, \ \dot{\Theta}(0) = 0,$$

where Θ and t' are the new scaled variables. Check to see that the time scale thus determined has the required dimensions, and by simply putting in typical values for string length etc. check to see if this time scale fits experience. What does the above scaling work tell us about the dependence of the period of the motion on the length of the string and the mass of the bob? Note that to determine the dependence of τ on the amplitude of motion it's necessary to investigate the above scaled equation further.

Exercise 2.4 *The Simple Pendulum*

The standard simple pendulum approximation is obtained by replacing $\sin\alpha\Theta$ by $\alpha\Theta$, (the first term of its Taylor expansion) and this approximation is likely to be valid if α is small enough. The equation then reduces to the familiar harmonic equation

$$\ddot{\Theta} + \Theta = 0.$$

(a) Obtain the solution to this equation satisfying the prescribed initial conditions, and thus determine an approximation for τ.

Exercise 2.5 *The Nonlinear Pendulum*

The simple pendulum approximation used above suggests that, for small enough α, $\tau(\alpha) \approx \tau(0)$ i.e. the period is approximately amplitude independent. This result is correct, but small departures of $\tau(\alpha)$ from $\tau(0)$ can be of critical importance under resonant forcing conditions, as we will see in Chapter 12. For use in such circumstances it's necessary to obtain a better approximation for $\tau(\alpha)$.

(a) If Θ_0 denotes the simple pendulum solution obtained in Exercise 2.4, show that an improved description of the pendulum equation is given by

$$\ddot{\Theta} + \Theta = \alpha^2 \Theta_0^3 / 6.$$

(b) Using Maple (use **dsolve**) obtain the solution of this equation with appropriate initial conditions and, by plotting out this solution for $\alpha = 0.1, 0.2$, exhibit the dependence of the period on the amplitude of the motion for small values of α.

(c) ⋆ Those very familiar with Maple may wish to process the results further and thus plot out the $\tau(\alpha)$ relationship for small α.

(d) ⋆⋆ Adventurous students may like to examine the case when α is not necessarily small.

Exercise 2.6 *Sky Diving*

Sky divers manage to link up into formations in the sky even though they jump out of the plane at different times. To do this individual divers must (to a limited extent) be able to modify the time taken to fall a given distance. To determine the strategy necessary to do this we examine the model equations

$$m\dot{v} = mg - kCSv^2, \quad \text{with} \quad v(0) = 0,$$

describing the velocity of vertical fall of an individual diver mass m, accelerating under the influence of gravity, and experiencing an air drag force proportional to the surface area S the diver presents to the vertical, and the square of his velocity $v(t)$; $C \approx 0.5$ is a dimensionless form drag coefficient which is relatively independent of the diver's shape, and $k = \rho/2.0$ where ρ is the density of air, see Landeau and Lifshitz (1959).

(a) In C.G.S. units (i.e. cms, gms and secs) $k \approx 0.6 \times 10^{-3}$. What are the dimensions of k that ensure the above equation is dimensionally sensible?

(b) Divers presenting a fixed S are observed to increase speed initially over a time scale τ (say), and eventually move with a constant speed (known as the terminal velocity) of scale U (say). This observation suggests that we should choose U and τ so that all three terms of the above defining equation balance. Show that this choice leads to the result

$$U = \sqrt{\frac{mg}{kCS}}, \qquad \tau = \frac{U}{g},$$

with the dimensionless description of the problem being given by

$$\dot{V} = 1 - V^2, \qquad V(0) = 0.$$

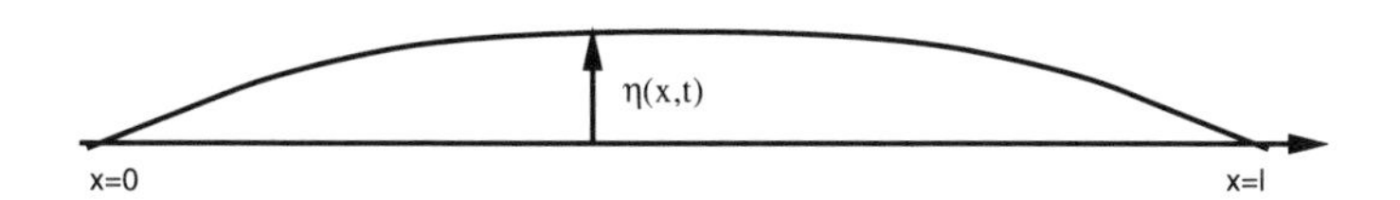

Figure 2.5: The damped vibrating string.

(c) Check to see that the dimensions of U and τ as given above are correct, and determine estimates for these scales for typical divers based on the above data and estimates you make for m and S. Based on these scaling results how should divers effect a velocity change required to link up with other divers?

(d) By solving the scaled equation of motion and plotting solutions, check to see that the equations do produce solutions that match observations. Estimate how long it takes for a diver to reach terminal velocity and the corresponding distance of fall.

Exercise 2.7 *The Damped Vibrating String*

A model set of equations for the lateral displacement $\eta(x,t)$ of a vibrating string of length l, mass per unit length m, tension T, and with an initial displacement $Ax(l-x)$ (see Fig. 2.5) is the partial differential equation

$$m\eta_{tt} + \mu\eta_t = T\eta_{xx}, \qquad 0 < x < l, \;\; t > 0,$$

subject to the subsidiary conditions

$$\eta(0,t) = 0 = \eta(l,t), \quad \eta_t(x,0) = 0, \quad \eta(x,0) = Ax(l-x).$$

Here $\mu\eta_t$ is a damping term. Assume that the lengths η, l are measured in metres, m is measured in kg/m, and t in s.

(a) For consistency what units should the quantities A, T and μ be measured in?

(b) If the system is such that the partial differential equation can be approximated by putting $\mu = 0$ what is the time scale of the motion?

(c) By scaling the full equations determine the dimensionless parameter that measures the size of the damping term relative to the other terms. Under what conditions might you expect the $\mu = 0$ approximation to produce accurate results (so that damping effects are negligible)?

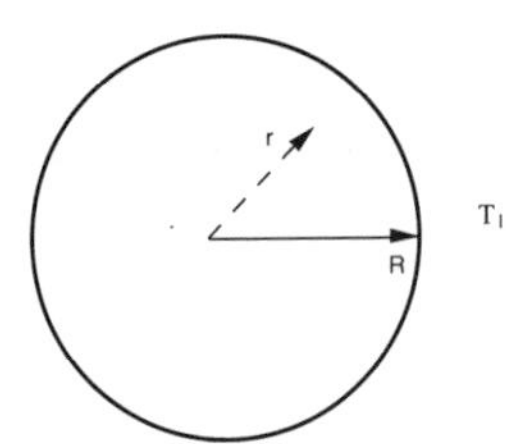

Figure 2.6: Meteorite cooling.

Exercise 2.8 *Meteorite Cooling*

The loss of heat from a sphere of radius R initially at temperature T_0 into an environment at temperature $T_1 < T_0$ is often modelled by the heat equation

$$T_t(r,t) = \kappa(T_{rr}(r,t) + \frac{2}{r}T_r(r,t)), \quad 0 < r < R, \quad t > 0,$$

where r is the distance from the sphere's centre, see Fig. 2.6, with initial condition

$$T(r,0) = T_0, \qquad 0 < r < R,$$

and the boundary condition

$$kT_r(R,t) = -\beta(T(R,t) - T_1), \quad t > 0.$$

This boundary condition simply states that the heat loss rate per unit area from the boundary is proportional to the temperature difference between the surface of the body and the environment. The parameters κ, k (the diffusivity and the conductivity) characterize the heat transfer process within the sphere, and β (the heat transfer coefficient) characterizes the heat loss from the sphere's surface. We wish to determine the dependence of the cooling time $t_c(R, \kappa, k, \beta)$ on these parameters.

The radius R seems a sensible length scale to employ for the problem. It also seems sensible to choose T_1 as a datum for temperature measurements and to scale the temperature in such a way that the scaled temperature varies from 1 to 0 as t increases to ∞. Thus the scaling

$$T - T_1 = (T_0 - T_1)T'$$

is appropriate.

(a) Using these scales determine the time scale t_0 that reduces the heat equation to its simplest non-dimensional form

$$T'_{t'} = T'_{r'r'} + 2/r' T'_{r'}$$

and show that the heat loss condition becomes

$$T'_{r'}(1,t') = -\gamma T'(1,t'),$$

in non-dimensional form, where $\gamma = R\beta/k$. Check for dimensional consistency.

(b) Using the above scaling results write down a functional expression relating the cooling time t_c to the appropriate dimensional and nondimensional parameters that arise.

(c) Comment on what these results imply about the cooling time for different size spheres, with different thermal properties, under different cooling conditions.

(d) Assuming that the time scale determined above does provide a crude estimate for the cooling time, estimate roughly (seconds, days, years?) how long it takes for a meteorite made of iron (diffusivity $\kappa = 1\text{cm}^2/\text{s}$) radius 100 metres to cool to earth's temperature after falling to earth.

Comment: One should be cautious about the time estimates obtained using the above arguments. The time for cooling will in fact be mainly determined by the *slowest* heat transfer process; so the value of γ could play an important role. Thus the extent that the meteorite is buried in the earth could strongly affect its cooling time..

Exercise 2.9 *The Forced String*

The small amplitude vibrations of an elastic string of length l with both ends clamped, with mass per unit length ρ and tension T, under the influence of an applied force per unit string length of $p \sin \omega t$, is governed by the equation

$$\rho\eta_{tt} - T\eta_{xx} = p \sin \omega t$$

with

$$\eta(0) = 0, \;\; \eta(l) = 0,$$

where $\eta(x,t)$ is the lateral displacement, see Fig. 2.7.

(a) Using the time scale associated with the imposed vibration, scale the equations and by choosing the scale η_0(say) for η appropriately, reduce the string equation to the non-dimensional form

$$\mu\eta'_{t't'} - \eta'_{x'x'} = \sin t',$$

where the primes refer to scaled quantities. Identify $\mu(\rho, l, T, \omega)$.

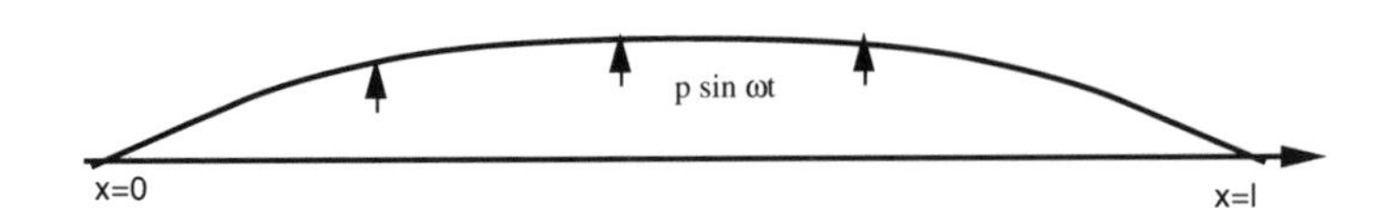

Figure 2.7: The forced string.

(b) What do the scaling results tell us about the size of the vibration induced by the applied force? Using the scaling results write down a dimensional statement relating the maximum amplitude of vibration of the string to the parameters of the problem.

(c) If the frequency of forcing is relatively small (compared with what?) introduce an approximation based on the smallness of μ, and show that the resulting equations integrate to give

$$\eta' = -\frac{x'(x'-1)}{2}\sin t',$$

and thus determine an accurate estimate for the maximum amplitude of vibration in the small μ case.

(d) $\star$ One can feel confident about this result only if the next iterate yields a small change. Use Maple to obtain a solution that's accurate to order μ and thus explicitly determine the functional dependence of the amplitude of vibration on the parameters of the problem for situations in which the forcing is in the small frequency range.

Exercise 2.10 *Plate Vibrations*

$\star\star$ The lateral vibrations of a square plate of side l clamped around its edges when subject to an oscillating applied pressure (i.e. force per unit area) of $p(x,y)\sin\omega t$ (see Fig. 2.8), are described by the partial differential equation

$$mw_{tt} + D[w_{xxxx} + 2w_{xxyy} + w_{yyyy}] = p(x,y)\sin\omega t,$$

with boundary conditions

$$w = 0 = \mathbf{n}\cdot\nabla w,$$

around the edges, where $\mathbf{n}$ is the unit vector normal to the edges. Here m is the mass per unit area of the plate, $w(x,y,t)$ is the lateral displacement

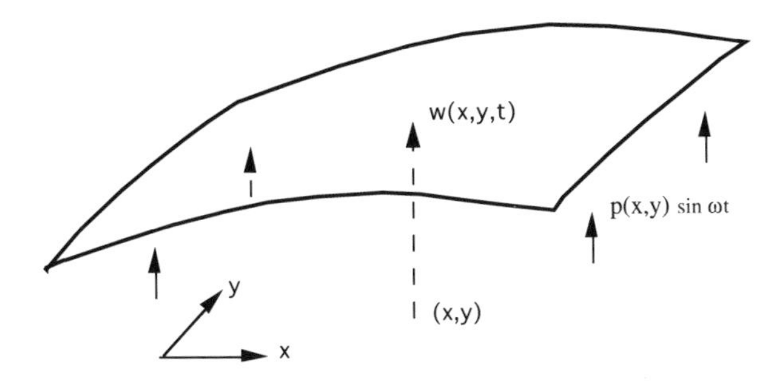

Figure 2.8: Plate vibrations.

of the plate, and D is a physical quantity specifying the resistance of the plate to deflection (related to its thickness and elastic properties). The position co-ordinates of a point initially (before deflection) at $(x, y, 0)$ on the surface of the plate are given approximately by $(x, y, w(x, y, t))$ at a later time t.

(a) What dimensions must D have?

(b) If the typical pressure is of size P, and the response scale is determined by a balance of the pressure term and the plate resistance term, what will be the scale of the deflection?

(c) What dimensionless parameter measures the significance of the transverse acceleration term mw_{tt} ? Under what circumstances would one reject the hypothesis that the transverse scale of deflection was basically determined by the plate resistance term ?

Exercise 2.11 *For Guitar Enthusiasts*

Comment: The situation to be examined here is sufficiently complicated that it's hard to see how to set up defining equations and thus use inspectional analysis. Thus one needs to resort to dimensional analysis procedures (supplemented with experiments) to examine the situation.

⋆⋆ The objective is to apply dimensional analysis, the results of certain experiments, and some background general knowledge to understand how the frequency of vibration of a guitar string is affected by various physical factors. The materials used for guitar strings of course vary. Two commonly used materials are nylon and steel-wrapped nylon, and the strings vary in thickness. Various notes are made by changing the effective length l of a vibrating string by fingering on the frets. Musical harmony requires careful control of the frequency Ω of the vibration of the strings. An instrument is tuned by adjusting the tension T of the strings.

An elementary physical model of a vibrating string, see Exercise 2.7, gives the result that the frequency of vibration is given by

$$\Omega = \frac{c}{l}\sqrt{\frac{T}{m}}, \tag{2.18}$$

where c is a dimensionless constant. Although this model is correct to first order, strings with the same tension, length, and mass per unit length produce slightly different sounds; so the simple theory is not exact. A short list of physical string characteristics that could play a role in determining Ω includes:

- Elastic properties usually defined by the string's Young's modulus E ($[M][T]^{-2}[L]^{-1}$) and the dimensionless Poisson's ratio σ (the ratio of relative lateral contraction to the relative extension when the string is uniformly stretched).

- The mass per unit length of the string m ($[M][L]^{-1}$), and its diameter d ($[L]$).

(a) What other physical variables do you think could be important? List them.

(b) For the purposes of the following analysis we'll assume only those physical characteristics itemized above matter, so that

$$\Omega = fn(T, m, l, d, E, \sigma).$$

Use dimensional arguments to show that the modified form of (2.18) given by

$$\Omega = \frac{1}{l}\sqrt{\frac{T}{m}}gn(\sigma, \frac{El^2}{T}, \frac{d}{l})$$

is appropriate.

(c) Limited observations on guitar strings show:

- Ωl is constant for a given string with fixed T,
- $\Omega\sqrt{T}$ is constant for a given string with fixed l,
- $\Omega\sqrt{m}$ is not constant (for different strings) with fixed l and T.

These observations suggest that the value of function gn referred to above is sensitive to changes in the values of one of the non-dimensional parameters. Which? What further information would be necessary to enable you to draw firm conclusions?

Exercise 2.12 *A Fishing Pole*

⋆⋆⋆⋆ Real flagpoles are normally tapered, as are fishing poles. The partial differential equation

$$\rho A(x)\frac{\partial^2 u}{\partial t^2} + E\frac{\partial^2}{\partial x^2}[A(x)k^2(x)\frac{\partial^2 u}{\partial x^2}] = 0$$

provides a model for the dynamic behaviour of a tapered pole, see Segel (1977), where $A(x)$ denotes the sectional area at location x along the pole and where $k(x) = \alpha d(x)$ is the local radius of gyration of the pole; here $d(x)$ is the diameter of the pole. Using this equation, together with the same boundary conditions as used for the uniform pole, one can obtain information about the effect of tapering on the pole's vibration. The ideas used to extract the solution for the uniform pole case readily extend. Thus a solution of the form $\chi(\xi)\sin(\tau)$ is to be expected. However, χ will now satisfy a non-constant coefficient linear differential equation, so it's unlikely that exact solutions are available except in very special cases. This does not change things greatly; it just means that inexact (probably numerical) methods need to be used to extract relevant information. Experience suggests that the amplitude of the motion of the small diameter tip of the pole will be exaggerated by tapering (witness the whip of the fishing rod), so this is what one should look out for when computing results. Assuming uniform tapering and using parameter values that fit poles of interest, compute the displacement of the free end of the pole and make appropriate comparisons. *Comment:* Although Maple can be used to numerically determine the displacement of the end of the pole for particular values of the pole's parameters, it would be better to use a numerical ordinary differential equation package for such work.

Bibliography

Carrier, George F. and Pearson, Carl E. (1968). *Ordinary Differential Equations*, Blaisdell.

Landau, L.D. and Lifshitz, E. M. (1959).*Fluid Mechanics*, p. 171 Pergamon Press.

Segel, Lee A. and Handelman, G.H. (1977).*Mathematics Applied to Continuum Mechanics*, p. 207, p. 518, Macmillan, New York.

Chapter 3

A table fable

Lessons in modelling

3.1 The Situation

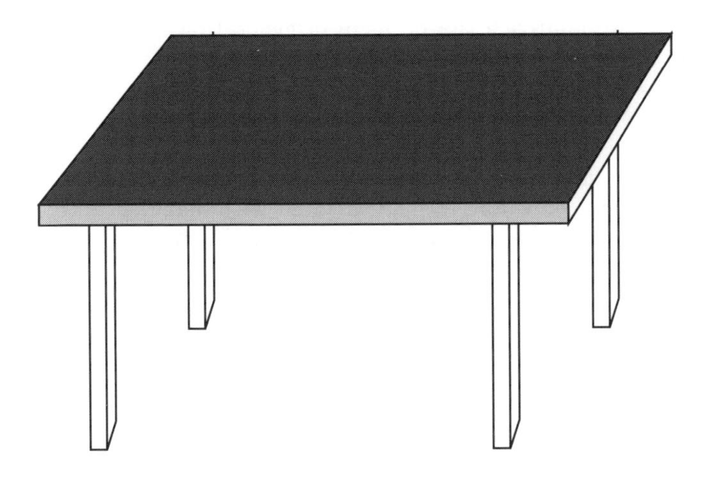

Figure 3.1: The table.

It is not uncommon for a table to rest on only three of its four legs and to rock to and fro in response to very small forces applied to the table top. The phenomenon is most likely to be observed with the robust tables used in public places, particularly when the tables are placed on concrete floors.

A manufacturer wishes to base his sales campaign on the fact that his tables will not have this disconcerting property, and asks a mathematical modeller for design criteria that will ensure that his claims can be substantiated.

This is a fable; no manufacturer has approached the authors with such a request; however, the problem raises a number of important mathematical modelling issues in a context that only requires the use of simple and elementary mathematics, and so serves our purpose of introducing these issues. Some of the detailed considerations that would need to be addressed if the situation were real will be avoided; however, it is hoped that the reader will emerge with a clear idea of what needs to be done to complete the story.

3.1.1 Preliminary discussion

As a means of deciding on a line of attack our first step is to pose appropriate questions. Consider the following questions:

A Why do tables rock?

B What aspects of the table's structure contribute towards this rocking tendency?

C How can we approach the problem of generating design criteria which will eliminate rocking?

We'll consider these questions in turn.

A If the table is perfectly manufactured and standing on a perfectly flat floor then all four legs will maintain floor contact, and rocking will not occur. Rocking is therefore to be associated with manufacturing imperfections and/or unevenness of the floor. One has only to recognize that three points define a plane to conclude that on a level floor any difference in the lengths of the four legs is likely to lead to one leg not making contact with the floor. Under such circumstances a small unbalanced force will cause the table to rock. Other imperfections will lead to a similar outcome and since manufacturing imperfections are inevitable and floors are uneven it is perhaps a surprising observation that *most* tables *don't* rock; even on uneven floors! There seems no explanation of this observation other than that tables avoid rocking by deforming under their own weight, thereby compensating for manufacturing imperfections. Of course the floor may also compensate; for simplicity we'll assume this doesn't happen. The above explanation is further confirmed by the observation that it is robust tables placed on firm floors that have the greater tendency to rock. These remarks based on everyday experience make it clear that any attempt to answer the manufacturer's problem will have to address the question of the deformation of the table under its own weight. This may come as a surprise because, on the basis of visual evidence, we would normally assume that tables are rigid (i.e. they have a fixed shape); in fact in elementary Physics courses rigidity is taken as axiomatic. In many contexts the rigidity assumption is in fact acceptable; however, it's clear that in our case this is not so. It's also clear that *very small* displacements are crucial here.

B Apart from unequal leg lengths, what other manufacturing imperfections could contribute to the tendency of a table to rock? The legs may not be squarely cut or set squarely to the top. The top may not be flat. The floor inevitably will not be smooth. A long list could be drawn up. Clearly at this stage it would be sensible for the modeller to obtain some information (hopefully statistical) from the manufacturer about the variability of the manufactured product and the floors of concern. At the same time he or she should enquire about the improvements in manufacture and quality control that are realistically possible. Given this information the modeller should then be able to decide on the appropriate accuracy requirements for any model he or she may set up. In the present case, given that the geometric imperfections of interest are relatively small, the data would be difficult to obtain and the expense of collecting it difficult to justify. The manufacturer would probably supply limited crude data and would assuredly reject any model outcome that could not be easily and relatively cheaply implemented. For simplicity of presentation of this fable we will proceed on the assumption that, on the basis of available information, it is the unevenness in lengths of the legs which is the significant manufacturing imperfection. We'll return later to discuss this further.

C To obtain appropriate design criteria will require model analysis; however, before attempting this we should identify the physical variables that are relevant as far as the assessment of performance of the system (the table) is concerned, and also decide on what we should be trying to extract from an appropriate model. We've identified manufacturing imperfections as being the basic cause of rocking and clearly the differences in leg lengths provide us with an appropriate measure for the quality of manufacture of tables. It's not clear, however, how we should quantify the rocking tendency of the table. Here guidance can be obtained from the following considerations. When one of the legs is off the floor it cannot be carrying any share of the table's weight to the floor. Also if the share being carried by a leg is small, then it will only require a small force (suitably applied) to rock the table. Thus the loads transmitted to the floor by the legs serve to provide a measure of the likelihood of the table rocking. If any of these loads is zero when the table is on a horizontal floor, then the table will rock easily, and if any of the loads is close to zero then it would take only a small inopportune unevenness in the floor to make the table prone to rocking. It's now clear that the initial aim of our models should be to predict the loads carried to the floor as a function of the length of the legs.

Our earlier discussion has suggested that a table will rock if the manufacturing imperfections are too large to be compensated for by the deformation caused by the table's weight. The weight will cause the legs to deform and the top to distort, so the next step in a real modelling investigation would be to assess the relative significance of the deformation of the top of the table and the legs under forces of the size of the weight of the table. The inspectional analysis introduced in Chapter 2 is appropriate for this purpose, and an experienced modeller would proceed along these lines. However, unlike the flagpole situation investigated in Chapter 2, in this problem there are *many* relevant small parameters and corresponding dimensionless groups, and it's not easy to see how one can write down the appropriate equations. To obtain a better understanding of how to proceed it's therefore sensible to examine the simplest possible case, even if the situation envisaged is unlikely to be of significance in context. With this in mind we'll first examine the situation in which the deformation of the top is insignificant in comparison with that in the legs; an assumption which in fact is rarely likely to be true. To further simplify matters we'll also assume that the weight of the table is located solely in the top, i.e. the legs are weightless. The hope is that this choice will lead to simple calculations which in turn will lead to the insights that are necessary for our purposes. We should of course keep in mind that eventually we will want to return to examine a more realistic model, so we should choose a mathematical framework that is capable of the required extension.

The preliminary discussion is now complete. It can be seen that before putting pen to paper there is much to be done! Different modellers would attack the problem in different ways (perhaps choosing a different initial model), but the above approach is fairly typical. Modelling is as much an art as a science. There are many possible model choices, very few of which are useful, so it's as well to spend time on such preliminaries.

3.2 Background Information

The following background knowledge is needed for the material that follows. The level of mastery of this material required for both the models and the exercises presented in this chapter is not high.

Statics

1. The motion of a rigid body can be described and analysed in terms of its translational and rotational velocity components.

2. If the vector sum of the forces acting on the body vanishes, its translational velocity will remain unchanged.

3. The vector moment or **torque M** of a force **F** about a point P is defined to be

$$\mathbf{M} = \vec{QP} \times \mathbf{F}, \tag{3.1}$$

where Q is any point on the line of action of the force.

4. The rotational (angular) velocity of a body about a point P will remain unchanged if the sum of the vector moments of the forces acting upon it about P vanishes. If the net force on a body vanishes, the net moment about *any* point will vanish if it vanishes about any one *particular* point.

5. The implication of the above is that a body will remain at rest only if the net vector force and the net vector moment (calculated about *any* point) vanish, i.e.

$$\sum_i \mathbf{F}_i = \mathbf{0}, \quad \sum_i \mathbf{M}_i = \mathbf{0}. \tag{3.2}$$

Thus 6 independent scalar component equations are required to specify a static equilibrium situation.

6. Any body experiences a vertically downward force called its weight. For symmetric bodies it acts through the centre of symmetry of the body. In other cases a straightforward centre of mass calculation is required to determine the "effective" location of the body force.

7. The force that any body exerts upon another is equal and oppositely directed to that exerted by the second body on the first.

8. The forces acting on one body due to contact with a second body may be described in terms of two components:

   - The **normal reaction** $\mathbf{N} = N\hat{\mathbf{n}}$. This component acts normal to the tangent plane of contact between the two bodies and acts on each body with the magnitude, and in the direction, required to prevent mutual penetration. If the bodies are tending to move apart then (in the absence of an adhesive sticking the surfaces together) there is no normal component of force exerted. Thus

   $$N \geq 0. \tag{3.3}$$

   - The **frictional force** $\mathbf{F} = F\hat{\tau}$. This tangential component of size F opposes relative sliding motion between the bodies and acts in the direction $\hat{\tau}$ opposite to that in which relative motion would occur if there were no frictional force acting. In the absence of slipping its magnitude is *just* sufficient to prevent slipping.

If, however, the magnitude of the frictional force F required to prevent slipping exceeds μN, where μ is a property of the contacting surfaces called the coefficient of limiting friction, then slipping will occur and thus the bodies will move relative to each other. Static equilibrium is thus possible only if

$$F \leq \mu N. \tag{3.4}$$

Hooke's Law

The change in length, δl, of a body of length l and cross-sectional area A when subject to a force F acting along its length is given by

$$\frac{F}{A} = E\frac{\delta l}{l}, \tag{3.5}$$

where E is a property of the material of the body, known as the Young's modulus. There is a sign convention to be attached to this result; namely that forces tending to extend the body are positive. It is frequently convenient to work with the stiffness k of the body defined by

$$F = k\delta l, \text{ so that } k = EA/l. \tag{3.6}$$

For more details see Fowles (1970).

3.3 The Simplest Model

We are now in a position to investigate a simple table model. Let's consider a table, weight W, with a rigid rectangular top of size (a, b), with weightless line legs placed at the corners. This table is perfect except that the legs, of length l_i', $i = 1 \ldots 4$, are not necessarily equal in length, and it is resting on a horizontal rigid floor, see Fig 3.2. Note that l_i' is the *compressed* length of the ith leg *with* the table resting on the floor. Recall that our aim is to determine the load borne by the legs. The four pieces of information about the table which have to be converted to mathematical statements in order to generate the mathematical model are:

A The table is at rest.

B The legs deform under the forces applied to them.

C The tops of the legs are in contact with the plane rectangular top.

D The table rests on a horizontal floor.

One has further to deal with the possibility that one of the legs may not be in contact with the floor. Since this is a situation one is hoping to ensure won't happen, we assume initially that all four legs make floor contact. The other case will be dealt with when the form of the analysis is clearer.

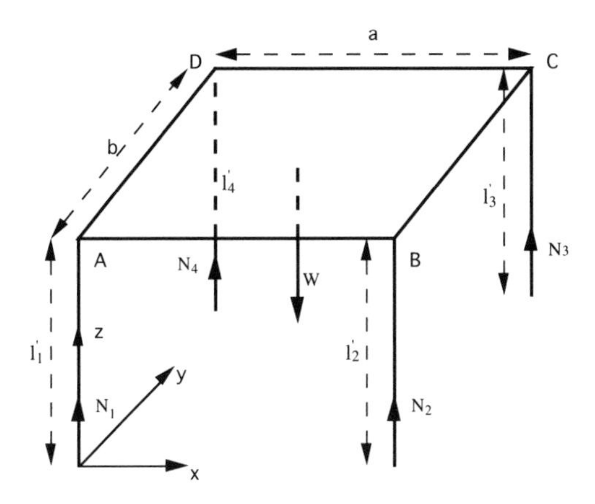

Figure 3.2: The model table.

A Table at rest

In spite of the fact that we've reduced the complexity of the problem as much as possible the situation remains complicated. Static equilibrium will be ensured only if the six scalar force and moment equations expressing static equilibrium, see (3.2), are satisfied. However, it is the weight of the table that drives the system and this acts vertically, and also, the legs will be *almost* pointing in a vertical direction for cases of interest (the difference in leg lengths will be minute compared with the actual average leg length). Thus, intuitively one might expect that an approximation based on the assumption that all forces act vertically should work. Under such circumstances translation could only occur in a vertical direction, so the only non-trivial force balance equation is in the vertical direction. Also, since vertical forces are only capable of producing rotation about horizontal axes, it would be sufficient to ensure that there is no net moment about two independent horizontal directions.

Assuming the above, if N_i is the load carried by the ith leg, see Fig. 3.2, then no translation is ensured if and only if, see (3.2)

$$\sum_1^4 N_i - W = 0. \tag{3.7}$$

To ensure there is no rotation we need to equate the moments about *any* two independent horizontal directions to zero. Although alternative choices simply give rise to equivalent equation sets, a careful selection can lead to a simpler equation set. In particular if one takes moments about a diagonal of the rectangular top, the forces acting through this diagonal will not appear in the moment equation and a simple equation will result. Thus if the diagonals AC, BD of the table top are chosen (see Fig. 3.2) we obtain, see (3.2, 3.1),

$$N_2 = N_4, \quad \text{and} \quad N_1 = N_3, \tag{3.8}$$

since the pairs of forces act at equal perpendicular distances from the relevant diagonal.

The above three static equilibrium conditions (3.7, 3.8) guarantee that there will be no movement of the table. We can use the moment equations to put the force balance equation in the more revealing form

$$N_1 + N_2 = W/2. \tag{3.9}$$

Simply interpreted these equations tell us that diagonally opposite legs bear the same load, but the proportion of the table weight born by adjacent legs is as yet undetermined. The diagonal symmetry exhibited in this result is an interesting feature of the problem.

Comment: static indeterminacy It is interesting to note that static equilibrium requirements led to just **3** equations (3.7, 3.8) for the **4** unknown loads N_i. *The implication is that elementary static equilibrium considerations do not uniquely determine these loads*; additional information is necessary to obtain the unique solution we expect on physical grounds! Although such indeterminacy problems are usually avoided in elementary courses on statics, they are in fact a very common and fascinating feature of statics, of current research interest. In the present case we will see that deformation issues resolve this apparent paradox. This is not always the case. Often in elementary textbooks symmetry arguments are invoked to resolve the indeterminacy. Thus the solution $N_i = W/4$ is presented. Given that real tables rock, it's clear that this solution is nonsense.

Comment: approximations By only taking into account vertical forces we've been able to reduce the order of the system (from 6 to 3 equations) . Can we justify this? In technical terms we might expect the above simpler equations to provide a first approximation to the complete set of static equilibrium equations. Although one could formally justify this assumption using the perturbation procedures introduced in Chapter 2, this would not normally be done unless unexpected difficulties arose. *A certain cavalier but aware attitude is necessary if one is going to get to grips with real world problems.*

B Deformation of the legs

The loads carried by the legs with the table resting on the ground cause the legs to compress by an amount proportional to these loads, and this is quantified by Hooke's Law (3.6). Thus if l_i is the length of the ith leg before compression then

$$l_i' = l_i - N_i/k, \tag{3.10}$$

where k is the stiffness of the legs which, purely for simplicity of display, is assumed to be the same for all legs. Recall that l_i' is the length of the ith leg with the table in position, i.e. sitting on the floor.

C,D The table top is flat

Let $(Oxyz)$ denote a set of rectangular cartesian axes chosen so that the origin is at the foot of the leg indexed by unity, the x-axis in the direction of the second leg and the z-axis vertical, see Fig. 3.2. Then the four corners of the table A, B, C, D, (see Fig. 3.2), specified by $(0, 0, l_1')$, $(a, 0, l_2')$, (a, b, l_3') and $(0, b, l_4')$ are required to be coplanar. This will be so providing there exists four constants α, β, δ and γ, not all zero, such that the plane

$$\alpha x + \beta y + \delta z + \gamma = 0$$

passes through each of the corners. A substitution of the above co-ordinates into this equation leads to four homogeneous equations for the constants α, β, δ and γ. The associated coefficient matrix is

$$\begin{pmatrix} 0 & 0 & l_1' & 1 \\ a & 0 & l_2' & 1 \\ a & b & l_3' & 1 \\ 0 & b & l_4' & 1 \end{pmatrix} \tag{3.11}$$

and a straightforward Gaussian reduction leads to the consistency condition

$$l_1' + l_3' = l_2' + l_4'. \tag{3.12}$$

Again the diagonal symmetry of the problem is exhibited. Simply interpreted this equation states that, with the table resting on the floor, the total length of diagonally opposite pairs of legs must match; in retrospect an obvious result. You might like to sketch a few tables to see that this result makes sense.

There are simpler ways to derive this condition using vector or geometric methods but the way illustrated here readily generalizes to bending top and skew leg situations that will need to be looked at later.

3.3.1 The solution

The use of equations (3.8) and (3.10) to eliminate the l''s in favour of the l's, and N_3 and N_4 in favour of N_1 and N_2 in the consistency condition (3.12), leads to the equation

$$N_1 - N_2 = k\Delta/2, \tag{3.13}$$

where the parameter, Δ, is defined by

$$\Delta = (l_1 + l_3) - (l_2 + l_4). \tag{3.14}$$

From the two equations (3.13, 3.9) the loads N_1 and N_2 may be determined and the complete results obtained are:

$$N_1 = N_3 = W/4 + k\Delta/4, \tag{3.15}$$

$$N_2 = N_4 = W/4 - k\Delta/4, \tag{3.16}$$

where use has been made of the symmetry results of equation (3.8). It's worth noting that the only geometric feature that matters in this problem is Δ, the difference in the *combined uncompressed* lengths of *diagonally opposite* pairs of legs! Note that by choosing to label one of the legs corresponding to the larger pair length with the index 1 we can assume that the parameter Δ is non-negative.

Floor contact ?

The above solution was predicated on the assumption that all four legs are in contact with the floor; and so can only be believed if all the normal reactions N_i are non-negative, i.e. if

the stability condition

$$W \geq k\Delta \tag{3.17}$$

is satisfied, see (3.15, 3.16). For the parameter range defined by this condition all the physical constraints of the problem are satisfied, so the above is the required solution. What happens if

$$W < k\Delta?$$

Often it happens that the solution obtained using a particular argument extends beyond the parameter range for which the argument is valid; however, this is not the case in our problem since negative normal reactions are physically unacceptable.

A little thought about the physical situation gives us the required answer for the $W < k\Delta$ range. Note that in the borderline case in which $W = k\Delta$, the above solution gives $N_2 = N_4 = 0$ and $N_1 = N_3 = W/2$; thus *all* the load is borne by the diagonal pair of legs with the greater combined length. In this case the leg compression of this longer length pair caused by the table's weight is such that the shorter diagonal pair of legs is *just* able to touch the floor. For larger values of Δ it's clear that the leg compression will be insufficient to enable the shorter pair to reach the floor, so the load borne by this shorter pair will remain zero. Of course the full load will be born equally by the other legs. Thus explicitly

$$N_2 = N_4 = 0, \quad N_1 = N_3 = W/2, \quad \text{for } W \leq k\Delta,$$

and the table will rock.

Comment: a complete mathematical formulation The reader will note that to determine this solution we found it useful to step outside a strictly mathematical framework and use physical arguments to guess the required solution. This is a very common feature of real world investigations; often physical insight can greatly assist the solution process. *Any available tool should be brought to bear on the problem at hand.* We could, in fact, have chosen to set up a complete mathematical formulation. The equations (3.7, 3.8, 3.10), and a slightly modified (3.12) (to take into account that all legs don't necessarily make floor contact) need to be supplemented with the surface contact conditions $N_i \geq 0$, see (3.3). Thus we arrive at the following complete mathematical description of the problem:

Find N_i, l_i' with $N_i \geq 0, l_i' \geq 0$ such that

$$\sum_1^4 N_i - W = 0,$$

$$N_1 = N_3, \quad \text{and} \quad N_2 = N_4,$$

$$l_i' = l_i - N_i/k,$$

$$l_1' + l_3' \geq l_2' + l_4'.$$

Note that both the equality and inequality constraints are linear so that *in theory* obtaining a solution set is straightforward. In practice (without insight) it's not so easy to extract the solution we obtained earlier. The reader may wish to check Maple's abilities in this regard. Standard packages can be used to obtain *numerical* solutions for *particular* values of the parameters; however, such specific results are of little use to us. In order to proceed to further models we need to *understand* the results; the stability criterion obtained above, namely $W \geq k\Delta$, is understandable, whereas the numerical solution for a range of values of W, k and the various lengths l_i is not useful in this sense. One *might* be able to extract the desired simple algebraic result from the numerical information by using powerful modern software for plotting multidimensional data and examining various projections, but the message is clear; if understanding is the aim then *if at all possible extract analytic results.*

3.3.2 Solution interpretation

The required solution has been obtained, so it's time to understand the results, draw conclusions, and think about how the results can be extended to deal with real tables. The essence of the above results is contained in the stability condition $W \geq k\Delta$, which provides us with an explicit quantification of the results anticipated in the preliminary discussion (i.e. the table won't rock providing it deforms sufficiently to compensate for any imperfections). Explicitly this result tells us that the particular combination of the relevant table parameters (a, b, l_i, k, W) that matters as far as rocking is concerned is the *dimensionless stability parameter* combination

$$r = k\Delta/W. \tag{3.18}$$

In particular if **the stability criterion**

$$r < 1 \tag{3.19}$$

is satisfied, then the table is sufficiently flexible to prevent rocking. If $r > 1$, then in order to avoid rocking one can either decrease k (for example by decreasing the sectional area of the legs) or increase W. It is interesting to note that the stability parameter depends only on the *difference* in combined length of diagonal pairs of legs; one might have expected the actual leg size typified by l_1 to play a role. This explains why the rocking table appears paradoxical to a human observer. As far as the human senses are concerned what is observed is the deformation of a table relative to its size and shape, and little change occurs on this scale. In the notation above this is described by small numbers typified by $\epsilon = \Delta/l_1$, the fractional deformation parameter. What matters, however, as far as the uneven distribution of the loads carried by the legs is concerned is the stability parameter r, which is *independent* of ϵ.

Comment: model approximations The fact that ϵ is small means that the approximations made in setting up the equilibrium equations (3.7, 3.8) are valid. In the derivation of the geometrical condition (3.12) stating that the table top is flat, however, it is *essential not* to assume that the normal to the table top points in the vertical direction; if one did this then the asymmetry that is the essence of the problem (and causes rocking) would be modelled out of the problem! The actual normal direction has not been explicitly determined but can be obtained from the equation of the plane after completing the evaluation of $(\alpha, \beta, \gamma, \delta)$. This illustrates a very significant feature of approximation work. One might have thought that to be consistent when approximating a mathematical system it would be sensible (and perhaps even essential) to make the same type of approximation in all equations; in our case to replace all almost vertical directions with vertical directions. This suggested consistency is misleading, not appropriate, and can lead to incorrect results.

A little thought about the above model results may lead one to the following explanation of these results, and advice to the manufacturer. The legs are likely to change lengths by an amount of the order of W/k when loaded. Of course with one leg off the floor, just two legs will carry the load so $W/2k$ provides a better estimate. Suppose the manufacturer estimates that the manufacturing process is likely to lead to legs that differ in length by an average amount δ. Now if the deformation $W/2k$ were large in comparison with δ one could be reasonably sure that the table would not rock, if not the advice would be to use less rigid legs. Often a useful simple model leads to an explanation of the above type that is in retrospect obvious; so much the better! ***One should always look for such simple explanations*** because such explanations provide further insight, and often enable one to avoid many calculations. In the present case this explanation enables us to see what's required to deal with other manufacturing defects. Thus, if one knew how to estimate how much the table top would deflect under forces typified by the weight, then it would be possible to obtain an estimate of the resulting deflection and in particular decide on whether one or both types of the deformations so far discussed (leg shortening or top bending) ought to be considered for tables of interest. Having done this one could then proceed to extract numbers from the correct mathematical model. We'll return to this work later.

Comment: scaling ideas In a slightly disguised form the above arguments are the same *scaling* arguments we used in the flagpole problem of Chapter 2. The setting was more formal and mathematical in the flagpole problem; whereas here the discussion was more physical. In both cases the aim was to compare the relative sizes of competing effects. Without further analysis it's clear that such arguments can only give crude size estimates; however, in many situations more precise information may not be warranted. To obtain the exact numbers it's necessary to perform analysis as we did in the flagpole problem and as we did when we extracted results for our simple table model.

In the light of the above work our hope might be that scaling arguments, together with the analysis suggested by these arguments, might lead to a generalized stability parameter which would provide the basic information required by the manufacturer. Before undertaking this work it's worth thinking about how such calculations might be used by the manufacturer.

The manufacturer's guarantee

Obviously it's the risk of failure of his guarantee (that his tables won't rock) that concerns the manufacturer; so we need to determine the statistical probability of rocking from information concerning the variability of the manufacturing process and the floors that the tables are likely to be used on. Let's examine the probability of rocking of our simple model table on perfect floors.

We'll assume the leg length is normally distributed about an average value l_0 with standard deviation σ, so that the probability density function is given by

$$p(l,\sigma) = \frac{1}{\sigma\sqrt{2\pi}} \exp -\frac{1}{2}(\frac{l*}{\sigma})^2, \text{ where } l* = l - l_0. \tag{3.20}$$

The following result of normal distribution theory provides the information we need:

Theorem 3.1 *Summed Normal Variates* *Let $X_i (i = 1 \ldots, n)$ be a sequence of mutually independent normally distributed random variables with parameters (m_i, σ_i); then their weighted sum*

$$L = \sum_{i=1}^{n} a_i X_i$$

follows a normal distribution with mean and variance given by

$$E[L] = \sum_{i=1}^{n} a_i m_i,$$

$$\sigma^2[L] = \sum_{i=1}^{n} a_i^2 \sigma_i^2.$$

Thus the expected value for Δ is zero and its standard deviation is 2σ. You'll recall that the table will rock if $k\Delta < W$. Thus, examining tables of the error function, we can see that the manufacturer can expect less than a 5% failure rate if $2k\sigma < 1.38W$. From such information the manufacturer would be able to assess the likely cost of guaranteeing the performance of his tables on floors satisfying certain evenness criteria.

3.4 Summary

A seemingly straightforward problem gave rise to formulation difficulties and subtle modelling issues. It was not even clear initially what would constitute an acceptable model, given the manufacturer's objectives and the impracticability of collecting or using accurate data. Clearly in this regard an accurate computer simulation of the table's behaviour would be *worse* than useless! We've come a long way towards understanding the issues involved by examining a simple model table. In particular we've been able to identify the relevant dimensionless stability parameter r (see (3.18)) for this (admittedly highly idealized) situation. Thus, at the very least, simple experiments could be set up to identify the *effective* k for tables, and the stability criterion (3.19) could be used to determine the allowable tolerance in manufacture given the table's weight. With further modelling one might hope for an even more useful outcome.

Although conceptually we can now see what's required to extend the above work to real tables, formidable technical difficulties arise in setting up the required defining equations. Thus, for example, an examination of the displacement size resulting because the legs are not squarely set into the top shows that such effects could be of major importance. To allow for the legs no longer being set perpendicular to the top one would be forced to deal with the full set (6) of force and moment equations. Also frictional forces acting on the legs couldn't be ignored; so there would be in total 12 force components to determine (an additional 2 frictional force components are required at each leg). If additionally the legs were not rigidly set into the top then the unit vectors defining the directions of the legs would need to be determined and bending moments at each joint would also need to be determined, requiring the introduction of a **further 20** variables. *Clearly this is getting out of hand!* This is even before we take into account the most important aspect of the problem; the bending of the top! But the problem of a table standing upon a floor is really a very simple situation in comparison with practical problems facing engineers working with certain structures. This has led to the development of rather more efficient methods for handling such situations. These techniques have been found to be easier to implement and more *robust* in application than the Newtonian approach used above in such cases. (By robust we mean here that the results obtained are rather less sensitive to minor aspects of the situation.) There are a number of different formulations which have proved useful but they share the common feature that they are based upon minimizing the energy of the system while satisfying the conditions of the geometry and loading. We'll use such a technique when we return later to examine more realistic table models.

3.5 Exercises

The first few exercises represent attempts to model various defects and common preventive measures associated with unstable table behaviour. The challenge is always to produce a revealing model that's simple enough to analyse and understand. It's important to interpret the results you obtain.

Exercise 3.1 *Table Loading*

One measure often used to prevent rocking of an unstable table is to place a weight on top of one of the legs of the table that's prone to rocking eg. the leg indexed #4 in the above work. Consider an idealized table of the type studied in the text (weightless legs set properly into the perfect rectangular top etc.) which is liable to rock. A force, P, acting vertically downwards is applied over leg #4.

(a) Assuming the legs remain in contact with the floor, show that the loads borne by the legs are given by

$$N_1 = (W + P + k\Delta)/4 = N_3,$$

$$N_2 = (W - P - k\Delta)/4,$$

$$N_4 = (W + 3P - k\Delta)/4.$$

(b) Show that the effect of such a force is to increase the likelihood of one leg being off the floor, but to reduce the likelihood of rocking.

Hint: Which leg is most likely not to touch the floor? Under what circumstances will an external force cause the table to "shift" from leg #2 to leg #4 ?

(c) Is applying the force over one of the legs corresponding to a longer combined diagonal pair length effective in preventing rocking ? Quantify this result.

Exercise 3.2 *Leg Packing*

Packing the legs is often effective in preventing rocking of poorly made tables. We'll first examine the situation in which rigid i.e. inelastic packers of known thickness $t_i' \ll l$ are used. The legs no longer make floor contact so the results obtained in the text need *minor* modification. It's sensible to work with a floor based rather than the legs based co-ordinate system used in the main text.

(a) Modify the results in the text to determine the loads carried by the legs of the table in this case.

Hint: With little change (amounting to a minor change in the definition of Δ) the calculations of the main text carry across.

(b) What does this result tell us about the value of placing some packing under one leg? Does it have to be the shortest leg?

Exercise 3.3 *Leg Sleeves*

⋆⋆ One common way of dealing with solid tables (in particular tables made out of steel tubing) is to slip relatively elastic rubber cushions over the ends of the legs. These cushions serve the dual purpose of protecting the floor and preventing rocking. The obvious question to ask is: what sleeve thickness is required for this purpose? Set up an appropriate model to examine this situation. Make calculations based on your experience of this situation and tabulated values of the Young's modulus for appropriate materials.

Exercise 3.4 *Real Legs*

In any actual table the legs do not contribute a negligible fraction of the total weight of the table. Suppose each leg weighs w and the top weighs W. What effect will the weight of the legs have on the model outcome?

To assess the situation we need to determine the loads N'_i carried by the legs to the floor, and we need to account for any modification of the geometry that may result. Let N_i denote the forces the legs exert on the table top. Follow this sequence to carry out the required calculations:

(a) Show that, with the N_i's defined as above, the equilibrium equations are the same as obtained in the weightless legs case examined in the main text.

(b) The forces acting on leg i are N'_i, $-N_i$, and $-w$. Use the requirements of static equilibrium to relate N'_i to $-N_i$.

(c) ⋆ We now determine the length change brought about in an individual leg by the floor and table forces and the weight of the leg. Let the force acting on the leg below location z (measured from the bottom end of the unstressed leg) due to that portion of the leg above z be denoted by $-F(z)$ (so that $F(z)$ is positive). By considering the forces acting on the leg below z show that

$$F(z) = [N' - w\frac{z}{l}],$$

where N' is the force due to the floor and l is the unstressed leg length.

(d) Now if du is the change in length of an element of the leg of unstressed length dz located at z, then Hooke's Law, see (3.6), implies that

$$\frac{du}{dz} = -\frac{F(z)}{EA} \equiv -\frac{F(z)}{kl}.$$

Using this result determine the total change in the leg's length.

(e) Thus show that the table's *stability condition* is unaltered if the real legs are replaced by the weightless legs we used in our simple model but with W replaced by $W + w$. An interesting result! Can you offer a simple explanation for this result?

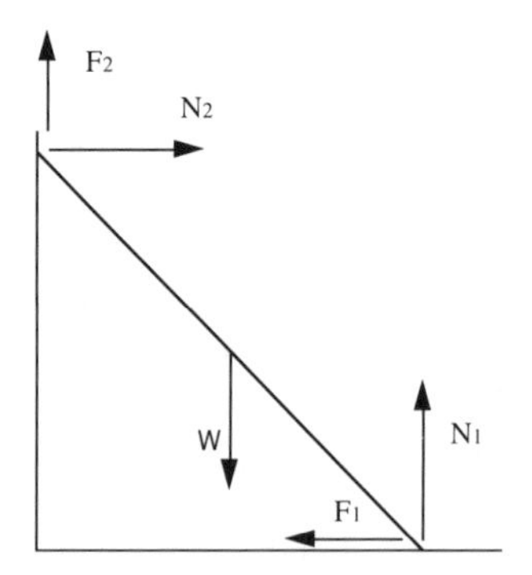

Figure 3.3: The ladder.

Exercise 3.5 *Symmetry Arguments*

In elementary books on physics, symmetry arguments are used to overcome the problem that the static equilibrium equations alone are not sufficient to determine the solution of the table problem.

(a) In terms of the relevant dimensionless group define the circumstances under which the symmetric solution $N_i = W/4$ provides an accurate approximation for the correct solution to our simple model table problem. Why do the symmetry arguments fail for the table problem under most circumstances?

(b) Discuss the various scenarios that can result for very small values of Δ and very large values of k, and discuss the implications.

Hint: Note that

$$\lim_{k\to\infty}(\lim_{\Delta\to 0} r) \neq \lim_{\Delta\to 0}(\lim_{k\to\infty} r).$$

Care is required when changing the order of limits!

Exercise 3.6 *The Ladder Problem.*

⋆⋆ Experience tells us that if the floor is slippery it is not possible to lean a ladder against the wall, so clearly friction must play some role in determining the conditions under which a ladder will stay in place. We'd like to determine these conditions, and also determine the loading on the ladder under equilibrium conditions. This is an exercise of major *general* importance. The lecturer may wish to provide guidance.

It's sensible to reduce the geometric complexity, so suppose the ladder is at $\pi/4$ to the wall (and floor) and that the centre of mass is at the midpoint of the ladder.

(a) Show the static equilibrium equations, see Fig. 3.3, are given by:

$$N_2 = F_1, \quad F_2 + N_1 = W, \text{and} \quad N_2 + F_1 + F_2 = N_1.$$

(b) Note that there are 3 equations in 4 unknowns; so that a one parameter family of solutions is expected. Thus if $F_1 = t$ (say) specifies the free parameter, a complete solution can be written down in terms of t. Obtain this solution.

(c) The physics of the problem, see (3.4) , requires also that

$$-\mu N_i \leq F_i \leq \mu N_i, \quad i = 1, 2,$$

where $\mu < 1$ is the limiting friction coefficient, assumed for simplicity to be the same for both wall and floor. This completes the *classical* (see later) mathematical specification of the problem. We need simply to find solutions obeying the above relations.

(d) Although the equations completely specify the physics of the problem it pays to keep the physics in mind when solving the equations—significant simplifications can result, and also one can often anticipate the best way to find the solution. In this problem experience, physics, and the above equations, suggest:

- The required solution lies in the $t \geq 0$ range. Why?
- The ladder will slip for small μ values and stay in place for large μ values. Thus we'd expect to find solutions only if μ is large enough. Are there solutions if $\mu = 0$?

(e) Note that W provides us with a sensible measurement scale for the various wall forces. Scale the equations and show that W disappears from the problem. Notice also that the ladder length doesn't occur in the static equilibrium equations. What are the theoretical and experimental implications of these results?

(f) Given the above observations a sensible question set is:

(i) Is there a region in the (t, μ) plane with $t \geq 0$, such that all the inequality constraints are satisfied ? Recall that the t value identifies the complete (N_1, N_2, F_1, F_2) solution.

Hint: Using Maple plot out the various inequality relations in the (μ, t) plane, and thus determine the permissible region. Clearly indicate the floor slip and wall slip critical curves on your plot.

(ii) Using Maple determine the minimum value of μ (μ_{min} say), required for static equilibrium. Comment on the physical significance of your results.

(iii) Is there a unique solution for $\mu > \mu_{min}$? Choose a particular value of $\mu > \mu_{min}$ and check to see that the solutions you've obtained satisfy all the requirements of the problem.

(iv) ⋆⋆ The fact that there are many solutions for $\mu > \mu_{min}$ is interesting and deserves further examination. One might speculate that the ladder falls either because the wall/ladder contact is not good enough or that the floor/ladder contact is not good enough. Examine your results and comment.

(g) ⋆⋆ Note how the $\mu = \mu_{min}$ solution relates to the critical slip curves in the (μ, t) plane and use this observation to speculate on what determines the stability criterion in the general case (with different μ_{wall} and μ_{floor} values and ladder angles $0 \leq \theta \leq \pi/2$ to the wall). Obtain the critical slip condition.

Returning to the non-uniqueness question: We'd expect the ladder forces to be determined ***uniquely*** in the static equilibrium case, so what's gone wrong with our solution procedure? One might speculate (correctly) that, as in the table case, elastic effects need to be taken into account. However, the uniqueness question is much more difficult to resolve in the ladder problem and turns out to be of a type that's of general interest throughout science. In order to understand some of the relevant issues an energy formulation of the problem is required. We'll examine such formulations in Chapter 4, and return to re-examine this question in Chapter 5 when we take another look at the table problem.

Exercise 3.7 *Climbing a Ladder?*

A man climbs up a ladder resting against a wall at an angle of $\pi/4$. We'll assume in this problem that the weight of the ladder is small in comparison with that of the person who is climbing it; so that it's reasonable to neglect the weight of the ladder. The coefficients of friction μ at both ground and wall are assumed to be equal and less than 1. We will show that it is not possible for the person to climb to the top of the ladder.

The calculations are made easier if the weight of the person is used to scale all forces, and the length of the ladder is used for the length scale.

(a) Draw a diagram of the various force components acting upon the ladder. Be careful to show the friction forces acting so as to oppose the direction experience would say the ladder would slide.

(b) Write down suitable force and moment equations for the equilibrium of the ladder.

(c) Show that all forces are not uniquely determined and obtain expressions for all the forces in terms of the vertical friction force at the wall.

(d) Show that it is not possible for the person to climb to the top of the ladder without the ladder slipping.

There are other interesting issues concerning the ladder that are beyond the scope of this book. For example experience suggests that the movement of the climber can generate waves travelling up the ladder that can affect its stability.

Bibliography

Fowles, Grant R. (1970). *Analytical Mechanics,* Holt Rinehart and Winston Inc., New York.

Chapter 4

Moorings

The calculus of variations
Energy principles and methods
Optimization

In the table fable problem we found that real difficulties arose in setting up the static equilibrium equations for imperfections of importance. In that chapter we indicated that an alternative approach based on energy ideas has proved to be successful in such circumstances. In this chapter we'll develop these ideas in another context. Never fear! We will return to the table problem later.

4.1 The Situation

The problems considered in this chapter derive from offshore oil field developments. Until recently offshore oil fields have been located in shallow water so that the rigs could sit on the sea bed. The value of oil and the technical understanding is now such that drilling takes place in water too deep for the rig to be grounded. The oil is transported from the oilhead to an enormous ship referred to as a "mother ship" via a flexible hose. This ship is used as a storage vessel from which other tankers can be filled. The mother ship needs to be moored in such a way that it doesn't move too far from the well-head on the sea floor (thus threatening the hose). Tides, currents and wind can cause such a drift. To achieve this it's secured by long chains connecting it to the ocean floor at some distance from the well-head. In order to provide a useful restoring force to movements of the ship it is necessary to have the chains provide a significant force even when the ship is at rest over the top of the well. The question we shall first consider is how to get the required level of force at minimum cost.

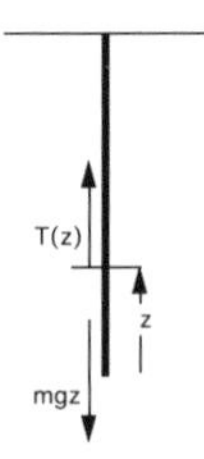

Figure 4.1: The vertical hanging chain.

4.1.1 Preliminary discussion

The chains have to be strong enough to withstand the tension forces experienced without breaking. The strength of a chain at any location is proportional to its cross-sectional area at that location. Thicker chains are of course stronger and heavier, and thus more costly. For chains of interest the cost of the chain is directly related to its weight.

A simple picture illustrates the nature of the problem. Consider the vertical, uniform diameter hanging chain shown in Fig. 4.1. Static equilibrium requires that the tension in the chain at any location be sufficient to support the weight of the chain below that point. Thus $T(z) = mgz$ where m is the mass per unit length of chain and z is the length of chain below the point of measurement of the tension $T(z)$. The tension in this uniform hanging chain will thus be greater at the top of the chain than at the bottom. Clearly a weight saving could be made if the chain were made thinner at the bottom. Notice also that, since any surplus weight at the lower end of the chain will have to be borne by all the chain above that point, inefficiencies arising from using too much material will be multiplicative; thus one might expect major savings to be possible. The first question we ask is:

How should one distribute weight along the chain so as (with an appropriate safety margin) to minimize the amount of material used, while still generating the required level of restoring force?

It has in fact been found possible to save hundreds of thousands of dollars in the cost of the cables used in such moorings by using optimally proportioned cables, rather than settling for uniform chains.

For the vertical hanging chain it is a simple matter to show that the cross-sectional area of the chain should exponentially decrease with distance from the hanging point to achieve minimum cost, see Exercise 4.1. The moorings problem is more difficult than the vertically hanging chain problem in that the way the tension varies along the chain depends upon the shape of the curve taken up by the chain in the water, which in turn depends on the weight distribution along the chain and where the chain is anchored, as well as other factors. Thus the problem of determining the shape of the curve the chain takes up in the water and the optimal cross-sectional area problems are intimately connected. Problems of this type are addressed by the classical theory of the Calculus of Variations, which will be described briefly in Section 4.3.

Of course the simple optimization objective stated above hides the major issues. The moorings must obey the laws of physics. When a ship moored by a number of such chains moves, each chain will be capable of

- being stretched so that its length is changed,
- changing the form in which it hangs in the water and thus changing the force it will exert on the ship,
- gaining kinetic energy through its movement and thus modifying the force it can exert on the ship.

If all these effects matter, the way in which the ship will move under the action of a disturbing force acting on it, such as a wave field, will be extremely complex. Although Newton's Laws can be brought to bear on such problems to produce the governing equations, it is often true that an alternative approach based on variational principles is much more efficient for such complex problems, both for producing the governing equations and for producing approximations. We will give a brief outline of the basic results in this area in Section 4.4, and then consider their application to the dynamics of moored ships. Firstly, however, some background information concerning Lagrange multipliers and rigid body dynamics will be listed.

4.2 Background Information

Lagrange multipliers Sometimes one wishes to find the stationary values of a function f of n variables x_i which are not independent. Suppose that there are $m < n$ constraint conditions $G_j = 0$ where the G's are functions of the variables x_i. One obvious way to attack this problem is to attempt to eliminate m of the variables x_i by solving for them in the constraint conditions and then substituting for them into f; an unconstrained stationary value problem would then result. Often, however, it's just not possible to perform the required eliminations so this direct approach is ineffective. There is a very neat way of avoiding this difficulty due to Lagrange. The device is to introduce a set of m parameters λ_j, and then seek the stationary values of

$$K(x_i, \lambda_j) = f(x_i) - \sum_{j=1}^{m} \lambda_j G_j(x_i),$$

with respect to the x_i, treating the variables x_i *as if they are independent.* The constraint conditions are then appended. A little thought convinces one that the required conditions for stationarity of f are satisfied subject to the constraints. The method leads to n equations obtained by equating each of the partial derivatives of K with respect to x_i to zero, and with the m constraints appended, this gives $m+n$ equations for the $m+n$ variables x_i, λ_j.

To see why the procedure works note that the required conditions for stationarity on f are met if

$$0 = df = \sum_{i=1}^{i=n} \frac{\partial f}{\partial x_i} dx_i$$

for all allowable changes dx_i in the x_i variables. The allowable changes dx_i are restricted by the conditions

$$0 = dG = \sum_{i=1}^{i=n} \frac{\partial G_j}{\partial x_i} dx_i, \text{ for } j = 0 \cdots m.$$

If one compares these stationarity conditions with those associated with K, it can be seen they are equivalent. For an excellent account see Strang (1986). The device also works in situations involving inequality constraints.

Rigid Body Dynamics An understanding of the derivation of the following rigid body dynamics results is not necessary for the work that follows. Any of the standard texts on mechanics derives these results, eg. Goldstein (1950), or Fowles (1970).

Newton's Laws for a particle extend readily to systems of particles, and major simplifications result if relative motion between particles can't occur; in particular for rigid bodies. The results obtained are:

The dynamics of rigid bodies can be described in terms of a translation of the centre of mass and a rotation about this centre of mass. Specifically:

1. Translation of the centre of mass. If the position vector of the centre of mass is denoted by $\bar{\mathbf{x}}$ and the forces acting on that body are denoted by $\mathbf{F}_i$ and m is its mass then

$$m\ddot{\bar{\mathbf{x}}} = \sum_i \mathbf{F}_i. \tag{4.1}$$

2. Rotation of the body. Let the angular velocity of the body be $\boldsymbol{\Omega}$ and I be the inertia matrix (with entries $I_{ij} = \int x_i x_j dm$), then

$$I\dot{\boldsymbol{\Omega}} + \boldsymbol{\Omega} \times [I\boldsymbol{\Omega}] = \mathbf{M}, \tag{4.2}$$

where the **torque** $\mathbf{M}$ is given by

$$\mathbf{M} = \sum_i \mathbf{x}_i \times \mathbf{F}_i.$$

Here $\mathbf{x}_i$ is the position vector of the point of application of the force $\mathbf{F}_i$, relative to the centre of mass. The complexity of these equations is greatly reduced if special "symmetric" principal axes fixed in the body are adopted, in which case I reduces to a diagonal form, with elements $I_{ii} \equiv \mathrm{I}_i$ (say) referred to as the principal moments of inertia of the body. For planar motion these equations reduce to the single equation

$$\mathrm{I}_i \dot{\Omega}_i = M_i.$$

For the simple bodies of interest to us I_i can be looked up in standard texts.

The kinetic energy of a rigid body may be calculated as the sum of:

1. the kinetic energy of a particle of the same total mass m undergoing translational motion with the velocity of the centre of mass $\bar{\mathbf{v}}$ of the body, so that

$$\mathcal{E}_t = \frac{m}{2} \bar{\mathbf{v}}^2. \tag{4.3}$$

2. the rotational energy of a similar rigid body fixed at the centre of mass. This is equal to

$$\mathcal{E}_r = \sum_{i=1}^{3} \mathrm{I}_i \Omega_i{}^2/2, \tag{4.4}$$

where Ω_i are the components of the angular velocity relative to principal axes with associated inertia moments I_i.

4.3 The Calculus of Variations

4.3.1 The archetype problem

We begin by considering the archetypal problem in this field; namely that of finding the function $y(x)$ defined on the fixed interval $a \leq x \leq b$ such that

1. $y(a) = A$, $y(b) = B$, where A and B are given constants,

2. the integral J defined by

$$J(y) = \int_a^b F(x, y(x), y'(x))\, dx$$

achieves the least possible value amongst all *admissible* functions $y(x)$.

Here F is a specified function of three arguments, which for any given function $y(x)$ is to be converted to a function of x by substituting $y(x)$ for the second argument and $y'(x)$ for the third. The value of the above integral J of course depends on the specific function $y(x)$ and so is a function of this function, and for this reason is referred to as a *functional.* To a pure mathematician ***admissible*** means a specified class of functions (eg. functions with continuous second order derivatives may be considered) but to a mathematical modeller ***admissible*** is defined by the application in a way that normally can't be specified in such precise mathematical terms. Usually an ***admissible*** function is simply one for which it makes sense to consider the integral as defined. *This difference in perspective is often critical.* By restricting the class of functions considered to "well behaved" functions one often eliminates the physically relevant solution.

How may we find such a function? Suppose that we have a candidate function $Y(x)$ which we think is the minimizer. Then, if it is in fact the minimizer, we will not be able to reduce the value of the functional by changing from the function $Y(x)$ to another (neighbouring) function $Y(x)+\epsilon\eta(x)$ where ϵ is a small positive parameter and $\eta(x)$ is any function keeping the combination $Y + \epsilon\eta$ within the admissible class. From the boundary conditions it follows that

$$\eta(a) \;=\; \eta(b) \;=\; 0.$$

If the function F admits of a second order Taylor expansion in its variables y, y' then

$$J(Y + \epsilon\eta) = J(Y) + \delta J + O(\epsilon^2),$$

where

$$\delta J = \epsilon \int_a^b [F_y(x, Y, Y')\eta + F_{y'}(x, Y, Y')\eta']\, dx. \tag{4.5}$$

Comment: Some care is required when evaluating the above integral. One needs first to treat the arguments x, y, y' of F as *independent* variables, and determine the indicated partial derivatives. One then substitutes the values $y = Y(x)$, $y' = Y'(x)$, corresponding to the candidate function, into the resulting expressions and evaluates the integral with respect to x. Thus, for example, if

$$F = y^2 + y'^{\,2} + yx^2,$$

and $Y(x) = \sin x$ is the candidate function, then

$$F_y(x, Y, Y') = 2\sin x + x^2, \text{ and } F_{y'}(x, Y, Y') = 2\cos x,$$

and these expressions can be substituted into δJ.

Returning to expression (4.5) for the change in J due to the $\epsilon\eta$ variation, an integration by parts on the $F_{y'}$ term yields the result

$$\int_a^b F_{y'}\eta'\, dx = [F_{y'}\eta]_a^b - \int_a^b \eta(x)\frac{d}{dx}F_{y'}\, dx. \tag{4.6}$$

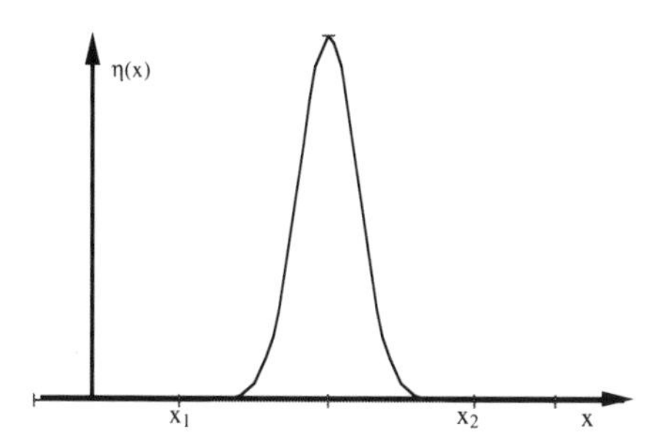

Figure 4.2: The Euler-Lagrange condition.

Comment: Note that the *total* derivative d/dx of $F_{y'}(x, Y(x), Y'(x))$ is required here. In other words the substitution $y(x) = Y(x)$ and $y'(x) = Y'(x)$ is first made, and then the derivative with respect to x is found. The contribution from the end points on the right hand side vanishes since η vanishes there. Thus the expression (4.5) for δJ reduces to

$$\delta J = \epsilon \int_a^b \eta(x) \left[F_y - \frac{d}{dx} F_{y'} \right] dx.$$

If the candidate function $Y(x)$ is in fact the minimizer then the first order change δJ in J must vanish for ***any*** admissible $\eta(x)$. Now for any given function $Y(x)$ the expression

$$L(x) = F_y - \frac{d}{dx} F_{y'} \tag{4.7}$$

that occurs in δJ is a known function of x. It's clear that δJ vanishes if $L(x) \equiv 0$. It's not so obvious that this is the ***only*** way to achieve a zero δJ. However, given that δJ must vanish for ***all*** $\eta(x)$, one might suspect that the only possibility is $L(x) \equiv 0$. This result is easy to display and prove:

Suppose that L were non-zero and one signed on some (***any***) sub-interval (x_1, x_2) *however small.* Then if $\eta(x)$ were chosen, as shown in Fig. 4.2, to be

- sufficiently smooth,
- one signed and of opposite sign to L on that interval,
- and zero outside the interval,

then δJ for this η would be negative, and thus $J(Y(x) + \epsilon\eta(x)) < J(Y(x))$ to first order for this η; so Y does not minimize J. One such η is given by

$$\eta(x) = \begin{cases} k(x - x_1)^2 (x - x_2)^2 & \text{for } x_1 < x < x_2 \\ 0 & \text{for } x < x_1,\ x > x_2. \end{cases}$$

Thus Y can only be a candidate for the function sought if $L \equiv 0$ on the interval (a, b). Thus a *necessary* condition, usually termed **the Euler-Lagrange condition**, for a function Y to be a minimizing function is that

$$F_y(x, Y(x), Y'(x)) - \frac{d}{dx}\left[F_{y'}(x, Y(x), Y'(x))\right] = 0. \tag{4.8}$$

If the differentiation with respect to x is executed then one will obtain a second order ordinary differential equation for $Y(x)$. This, in conjunction with the two boundary conditions imposed by the required values of Y at a and b, provides some basis for attempting to determine Y. It should be noted that, unlike initial value problems, boundary value problems for ordinary differential equations often do not have solutions, and when they do they may have more than one—thus one can expect special difficulties and interesting questions to arise out of variational problems. Also, solutions of physical interest may not have the smoothness properties required for the derivation of the Euler-Lagrange equation. In spite of these potential difficulties the Euler-Lagrange equation offers a very powerful starting position to commence the search for solutions.

Obviously arguments of the above type can be used to discuss the maximum J case, so the word extremum should replace minimum in the above statements.

Some special situations occur sufficiently often to deserve mention:

- Trivially, if the function F does not involve explicit dependence on y, the Euler-Lagrange condition (4.8) can be integrated to yield the simpler form

$$F_{y'} = \text{const.} \tag{4.9}$$

- What is not so obvious and is *very* useful is that it's possible to obtain an explicit integrated form in the case in which F has no *explicit* dependence on x. In such cases the Euler-Lagrange equation may be rearranged as one involving the function $X(y)$ on each sub-interval in which the function $Y(x)$ is invertible. On each such sub-interval the integral J can be written in the form

$$\int F(y, y')dx = \int F(y, 1/x'(y))x'(y)\, dy$$

where x' denotes $dx(y)/dy$. Clearly the integrand in this expression is independent of x so the Euler-Lagrange equation for the minimizer $X(y)$ can, as above, be integrated to give

$$\{F(y, 1/X')X'\}_{x'} = \text{const},$$

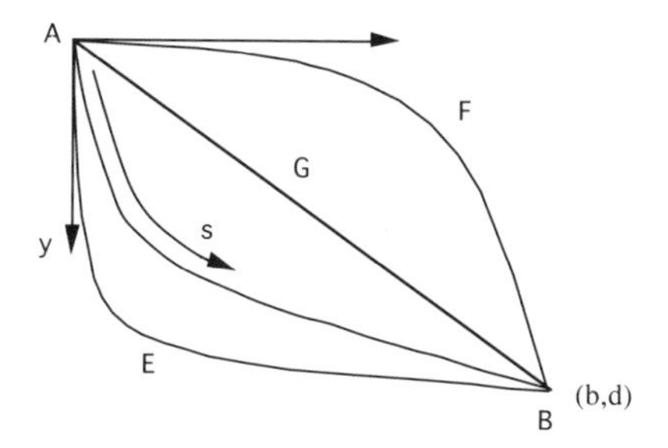

Figure 4.3: The minimum time path problem: possible particle paths.

and thus if we change back to our original variables (using the chain rule) we get

$$\text{constant} = -F_{y'}(y, 1/X')/X' + F = -Y'F_{y'}(Y, Y') + F(Y, Y'). \tag{4.10}$$

Example 4.3.1 *The Particle Path Problem*

As a simple illustration of the use of the above ideas consider the first problem to be solved using variational methods. The problem addressed was: "Find the path joining two given fixed points A and B so that a particle moving along this path from A reaches B in the shortest time, friction and air resistance being neglected." This path is referred to as the *brachistochrone*.

First we should check to see if the problem is likely to have a "sensible" solution. (It is often the case that seemingly sensible variational questions either don't have an answer or have an answer with unexpected properties.) We might as well choose the point A as the origin and measure y vertically downwards. Let (b, d) be the location of the path terminus B, then alternative paths are shown in Fig. 4.3. Now the speed of travel of the particle is $\sqrt{2gy}$ where g is the acceleration due to gravity, so we'd expect relatively higher average speeds on the paths that are initially steep, such as AEB. Such paths are long and so won't be minimum time contenders. Paths such as AFB are obviously long distance and small speed paths and so are also not possible minimum paths. The shortest distance path AGB is an obvious candidate, but the average speed on this path is relatively small. Thus one might expect the required path to lie somewhere between AEB and AGB, and be represented by a well behaved function of x. Of course one could make explicit calculations for particular paths between A and B and this would increase one's understanding of the problem (eg. quadratics could be examined). This would be simple to do using Maple; however, for this particularly simple problem this is hardly worthwhile. Let's proceed with the exact analysis.

Let s be the arc length along the path measured from A, then the time taken to reach B will be given by

$$\int_A^B \frac{ds}{\sqrt{2gy}}.$$

Now $ds^2 = dx^2 + dy^2$, and the multiplicative factor is irrelevant for the determination of the path, so the objective is to find the function $y(x)$ that minimizes the integral

$$\int_0^b \sqrt{\frac{1+y'^2}{y}}\, dx.$$

The integrand does not explicitly involve x so the Euler-Lagrange equation integrates to give the form $F - y'F_{y'} = C$, see (4.10). Thus after simplification we obtain

$$\frac{1}{\sqrt{y(1+y'^2)}} = C,$$

or $y(1+y'^2) = c$ (where c is an arbitrary constant of integration). Solving for y' gives

$$y' = \sqrt{\frac{c-y}{y}},$$

so we require the solution to this first order nonlinear differential equation satisfying the boundary conditions $y(0) = 0, y(b) = d$. The features to note are:

- y' is real and finite for $0 < y < c$ and $y' \to 0$ as $y \to c$,

- Near $y = 0$, $y' \approx \sqrt{c/y}$, , which integrates to give a solution of the form $c_2 x^{\frac{2}{3}}$ near $(x, y) = (0, 0)$; so the equation singularity at $y = 0$ is physically acceptable and also *interesting*; the optimum path drops vertically from A. (This will come as no great surprise to the skier.)

- There are enough arbitrary constants to satisfy the required boundary conditions.

It's clear from this analysis that there exists a sensible unique solution to the required equations. All that remains is to tidy up using Maple or if necessary (it's not) using a numerical package, *and* interpret the results, see Exercise 4.2.

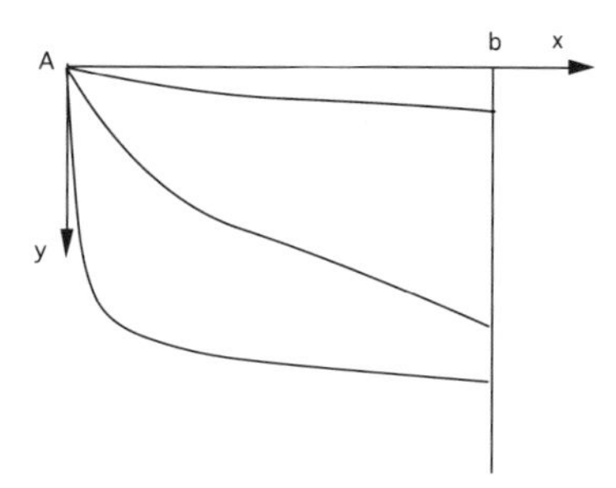

Figure 4.4: The minimum time to reach a barrier: possible paths.

4.3.2 Generalizations

The approach used on the archetypal problem considered above may be extended to deal with other situations. The variety of situations that arise from natural contexts is so great that it is necessary to determine the appropriate Euler-Lagrange conditions in context and from first principles, rather than refer to general results. It's sensible, however, to have some idea of what can be done, so a short summary of extensions will be given. The exercises at the end of this chapter and later in the book will expand on the ideas presented here. For a detailed discussion of the extensions see Courant and Hilbert (1957), or Akhiezer (1962).

- ⋆ **Free boundary conditions.** In the brachistochrone problem the two ends of the path were specified, and this provided the right number of boundary conditions for the second order Euler-Lagrange equation. We'd expect the above theory to work for any sensible conditions on y providing there were the required number (two) of boundary conditions. Problems arise naturally, however, in which there are too few prescribed conditions or, worse still, too many prescribed conditions. For example, clearly a sensible question to ask is: which path results in a minimum transit time of a particle from A to a vertical barrier located at a horizontal distance b from A, see Fig. 4.4. In this case $y(b)$ is not prescribed and needs to be determined as part of the solution process. There's also a very important class of problems (eg. phase change problems) in which the x interval is unknown and needs to be determined as part of the solution process.

  As our basis of illustration of how to attack such problems we assume that the value $y(a) = A$ is specified where a is fixed, but the physics of the situation demands only that $f(x, y(x)) = 0$ at the other end point $x = b$ of the domain, where the actual value of b needs to identified as part of solution process—very often it's the most interesting feature of the problem. Suppose our candidate function Y satisfies the condition with $b = B$, so that $f(B, Y(B)) = 0$. When we add our trial function $\epsilon\eta$; we will need, as before, to require $\eta(a) = 0$. The other end of the domain won't, in general, be at B but will be close to B for the small variations of interest. Suppose the domain for our variation is (a, x_1). Now, using Taylor's expansion, we can determine the effect of this small change in boundary location on the boundary condition. Thus

$$f(x_1, y(x_1)) = f(B) + f_x(B) \cdot (x_1 - B) + f_y(B) \cdot \epsilon\eta(B) + O(\epsilon^2),$$

where $f(B)$ denotes $f(B, Y(B))$, $f_x(B)$ denotes $f_x(B, Y(B))$ etc. Using the boundary condition $f(B) = 0$ and neglecting squares of ϵ, we obtain

$$f_x(B) \cdot (x_1 - B) + f_y(B) \cdot \epsilon\eta(B) = 0, \tag{4.11}$$

which provides us with a condition on $(x_1 - B, \eta(B))$ that must be satisfied if the boundary constraint is to remain satisfied. The change in the functional may be written in the form

$$\delta J = \epsilon \int_a^B [F_y \eta + F_{y'} \eta'] \, dx + \int_B^{x_1} F \, dx,$$

where the additional term (see (4.5)) again arises because of possible variations of the boundary location. To first order in ϵ the additional term is approximated by $F(B)(x_1 - B)$, using the same notation as above. As in the archetype problem an integration by parts on the first term enables one to break the variation up into a domain contribution and a boundary condition contribution. Thus

$$\delta J = \epsilon \int_a^B L(x)\eta(x) dx \; + \{[F_{y'}\epsilon\eta]_a^B + F(B)(x_1 - B)\},$$

where L is the Euler-Lagrange expression obtained earlier, see (4.7), and the arguments used earlier establish that both components must vanish independently if the first order variation is to vanish. Thus we again require the Euler-Lagrange condition (4.8) to be satisfied, but to ensure that the contribution from the endpoint B also vanishes (the contribution at $x = a$ vanishes as before) we will need to require

$$F_{y'}(B) \cdot \epsilon\eta(B) + F(B)(x_1 - B) = 0. \tag{4.12}$$

We therefore have two linear equations (4.11,4.12) in $\eta(B)$ and $(x_1 - B)$ to be satisfied if the variation is to vanish. In general two such homogeneous conditions will be satisfied only by the trivial zero solution. If however

$$f_x(B)F_{y'}(B) - f_y(B)F(B) = 0,$$

then the end conditions do not contribute to the variation. Such boundary conditions are termed ***natural*** for the particular variational problem, and provide the required additional conditions for the problem. They're called natural because, in the absence of prescribed boundary conditions, these are the conditions that need to be met by the minimizer. We'll examine some applications of these ideas in the exercises, see Exercise 4.3.

- **Additional Constraints.** Sometimes additional constraints are imposed on the required function; for example only positive functions with a given area $\int_a^b y(x)\,dx = \Gamma$ may be permissible. The approach used for the above free boundary condition problem can also be used in this case but the **Lagrange multiplier method** (discussed in the background information Section 4.2) provides a very neat and simple way of dealing with such subsidiary conditions. The technique is of quite general applicability. In the present case one simply introduces a parameter λ (the Lagrange multiplier), and looks for stationary values for the modified functional

$$\int_a^b [F + \lambda y]dx - \lambda\Gamma.$$

  Evidently the $(y(x, \lambda), \lambda)$ combination that minimizes this functional and satisfies the area constraint is the desired minimizer. The resulting simplification is major, as we'll see later.

  Equality or inequality constraints on the solution may also arise. Thus, for example, a slippery slide manufacturer would not be happy with the fast slide brachistochrone solution—the initial vertical drop may put sliders off. The manufacturer thus may be interested in determining the fastest slide that's not too steep; so $y'(x) < \alpha$ for some α may present an appropriate constraint. In such cases the Lagrange multiplier method is again appropriate and the solutions can be generated by patching together Euler-Lagrange solution curves and curves expressing the constraints. Such problems need to be attacked individually, and intuition often leads to the solution. A few simple examples given in the exercises at the end of the chapter will provide some insight.

- **Many functions.** If more than one function is involved, provided these functions can be varied independently, the extension is straight forward. Thus, for example, if the function in the integrand depends on $y(x)$ and $z(x)$ which can be varied independently, then the analysis is more or less the same but there are now two terms which will have to vanish to achieve stationarity. In such a case, corresponding to the integrand $F[x, y(x), y'(x), z(x), z'(x)]$, there will be two Euler-Lagrange equations

$$\frac{d}{dx}[F_{y'}] - F_y = 0, \tag{4.13}$$

$$\frac{d}{dx}[F_{z'}] - F_z = 0. \tag{4.14}$$

- **Higher Order problems.** If the function F also involves the second derivative of y (so $F = F(x, y, y', y''))$ then a second integration by parts is necessary to separate the domain and boundary contributions to the variation, and it is easy to show that the corresponding Euler-Lagrange equation becomes

$$F_y - \frac{d}{dx}[F_{y'}] + \frac{d^2}{dx^2}[F_{y''}] = 0.$$

  Of course additional boundary conditions would normally be expected.

- **Multiple integrals.** Here we indicate the results for three-dimensional space but the extension to higher dimensions is evident. Consider the functional

$$J(\phi) = \int_D F(x_i, \phi(x_i), \frac{\partial \phi(x_i)}{\partial x_i})\, dV_{\mathbf{x}},$$

  where $dV_{\mathbf{x}}$ is a volume element in three-dimensional space and D is the domain of integration. In applying the variational argument we need only recognize that the generalized form of integration by parts is Gauss's Divergence theorem to show that the corresponding Euler-Lagrange equation becomes

$$F_\phi - \sum_{i=1}^{3} \frac{\partial}{\partial x_i}[F_{\phi_i}] = 0, \tag{4.15}$$

  where F_{ϕ_i} means the partial derivative of F with respect to $\partial\phi/\partial x_i$. This must hold in the interior of the domain and again appropriate boundary conditions need to be imposed.

4.4 Variational Principles in Dynamics

The motion of a single particle of mass m under the influence of a force $\mathbf{F}(\mathbf{x})$ is described by Newton's equation

$$\mathbf{F} = m\ddot{\mathbf{x}}.$$

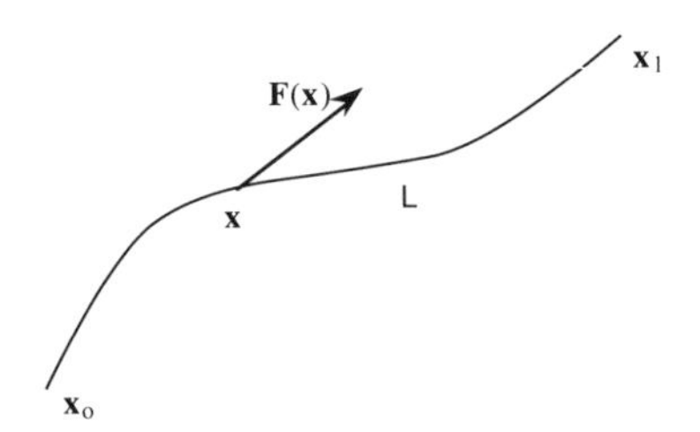

Figure 4.5: The energy equation. Work.

Multiplying the equation through by $\dot{\mathbf{x}}$, and integrating along the particle path L, see Fig. 4.5, from its initial location $\mathbf{x}_0$ to its final location $\mathbf{x}_1$, we obtain the energy equation

$$\frac{m}{2}(\dot{\mathbf{x}}_1^2 - \dot{\mathbf{x}}_0^2) = \int_L \mathbf{F}(\mathbf{x}) \cdot d\mathbf{x}.$$

The left hand side defines the change in **kinetic energy** T, the right hand side the **work done** by **F**, and simply interpreted this states that the work done by the force on the particle results in an increase in kinetic energy. In general the work done will depend on the path L; however, in a very important class of situations the force field is such that the value of the integral is independent of the path L, and depends only on the position of the end points. Such forces are called *conservative* and for these cases there exists a scalar *potential* V such that $\mathbf{F} = -\nabla V$. The prototype example is the force due to gravity. In this case $V = mgz$, where z is the height above any convenient reference level. It's evident from the energy equation that forces acting at right angles to the path will not cause a kinetic energy change. Thus, if the forces acting on a body are either conservative or act at right angles to the motion, the total (potential plus kinetic) energy remains fixed and this result often enables one to avoid the process of solving Newton's equations for all points along the particle path. The range of ideas that stem from this humble origin is remarkable. Potentials play a major role throughout the physical, chemical, and biological sciences.

Now all the above work extends to systems of particles and continuous systems (for example elastic bodies and fluids). Thus:

- The kinetic energy T of the system $\mathcal{S}$ is simply obtained by forming the sum (or integral) of the individual kinetic energy components of the system. Thus
$$T = \int_{\mathcal{S}} \frac{1}{2}\mathbf{v}^2 \, dm.$$
- Forces acting at right angles to the motion of a body at the point of contact perform no work.

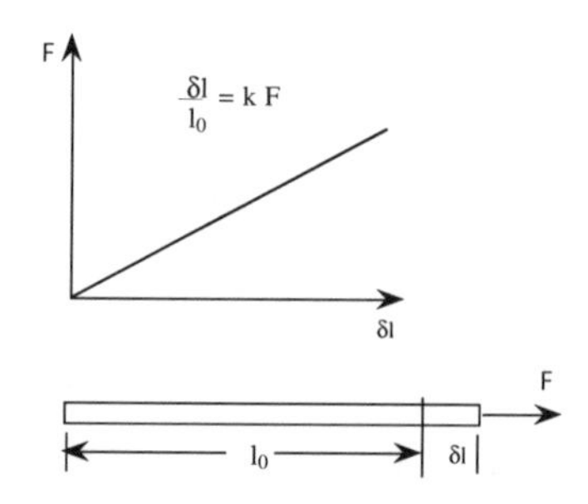

Figure 4.6: A rod being stretched: Hooke's Law.

- Forces acting within systems (for example elastic forces) can be identified as being conservative if any external work performed on such systems is (in theory) recoverable in the form of kinetic energy.

In calculating the potential energy of a system we need to include **Internal Potential Energy** terms. To illustrate these ideas we'll determine the internal elastic energy stored in a rod that's either stretched or compressed by the application of an applied force; for example the leg of the table in the Table Fable of Chapter 3.

Elastic Energy

Hooke's Law

$$F = k\delta l, \quad \text{with} \quad k = EA/l, \tag{4.16}$$

see equation (3.6) in Section 3.2, provides us with the required experimental information about the change in length δl caused by the application of an applied load F, where l is the unstretched length of the rod, A is the sectional area of the rod, and E is Young's modulus of the material the rod is made of. For ease in presentation we'll describe the $F > 0$ rod stretching case, see Figure 4.6.

The first thing to notice about the Hooke's Law formula is that it's single valued, so that it indicates a *unique* relationship connecting displacement and force. The implication is that the process being described is conservative. If the internal forces acting were non-conservative then some of the work performed on the rod by the application of the stretching force would be converted to other non-mechanical forms of energy (typically displaying itself in the form of heat), so that after the cyclic application of the force from an initial non-zero displacement the rod would not return to its initial displacement. Thus, see Fig. 4.7, if $(F_i, \delta l_i)$ represents the initial status (A in the figure), and the force is first increased so the rod stretches further (B in the figure), and then the force is reduced until it again reaches

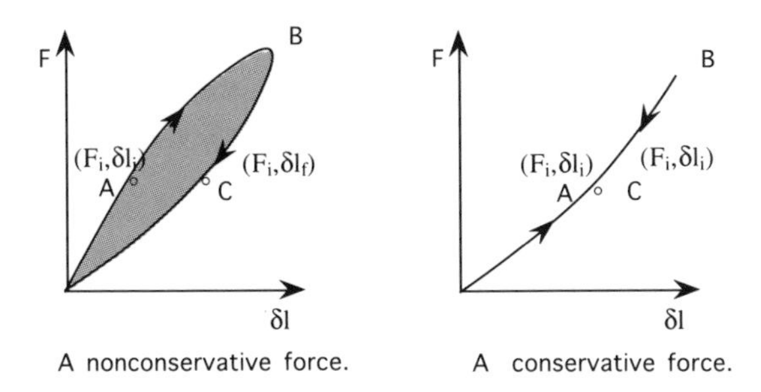

Figure 4.7: Cyclic force application to a rod: conservative and non-conservative forces.

the value F_i (the point C); a new status $(F_i, \delta l_f)$ will result if the force is non-conservative whereas the initial state $(F_i, \delta l_i)$ will be returned to if it's conservative. Non-conservative forces are said to exhibit ***hysteresis***, and the energy lost over a cycle is simply the shaded area in the figure, as one can see from the definition of work. Usually hard materials (for example metals) exhibit Hooke's Law type behaviour for small displacements, whereas softer materials (for example rubber) exhibit marked hysteresis.

The elastic energy stored in a rod carrying a load F (corresponding to an extension F/k) is calculated as follows: When the load on an initially unstretched leg has increased to αF, where $0 \leq \alpha \leq 1$, the resulting extension will be $\alpha F/k$. Now let the load be fractionally increased to $(\alpha + d\alpha)F$. The work done by the load in bringing about this change is given by the product of the force αF and the distance $d\alpha\, F/k$ it moves its point of application. Assuming the process described takes place without causing a net increase in kinetic energy the work done will be stored within the body in the form of elastic energy and thus the elastic energy associated with the extension is given by

$$\mathcal{E}_e = \int_0^1 [F\alpha][F/k]\, d\alpha = F^2/(2k). \tag{4.17}$$

In terms of the displacement δl we can write this in the form, (using Hooke's Law (4.16))

$$\mathcal{E}_e = k(\delta l)^2/2. \tag{4.18}$$

Note that here is no need to know anything about the actual mechanism of storage of elastic energy in order to make the required calculations—all one needs is the experimentally determined force vs. displacement relationship. The implication is that one can understand macroscopic behaviour (for example the vibration of the flagpole in Chapter 2) without knowing about the microscopic details of how the energy is stored within the material. The above result depends only on the Hooke's Law relation and so can be used in any circumstances in which a linear relation between applied force and displacement holds. Thus it can be used for compression as well as stretching situations, and for springs and even for complex structures. With obvious modifications the results extend to nonlinear force/displacement situations not exhibiting hysteresis.

4.4.1 Hamilton's principle

An alternative procedure to the Newtonian procedure for describing the behaviour of dynamical systems is based on **Hamilton's Principle**. Hamilton's principle for a conservative system states:

Theorem 4.1 *Of all possible paths by which a conservative system can change from an initial configuration at time t_i to a final configuration at time t_f, the actual path taken is such that the line integral*

$$J = \int_{t_i}^{t_f} L\,dt \tag{4.19}$$

where

$$L = T - V, \tag{4.20}$$

is an extremum.

L is referred to as the **Lagrangian** of the system.

Of course the variational work of the previous section is immediately applicable to this situation. Thus if there are n position functions q_i, together with their time derivatives q_i', specifying the kinetic and potential energy components, and these are capable of being varied independently, then the Euler-Lagrange equations, see (4.8), give

$$\frac{d}{dt}\left[\frac{\partial L}{\partial q_i{}'}\right] - \frac{\partial L}{\partial q_i} = 0, \quad \text{i=1,2}\cdots n. \tag{4.21}$$

These equations are known as **Lagrange's equations**. Lagrange first developed their use for the solution of complicated dynamical systems.

For a single particle the equivalence of the Hamilton approach and the Newton approach is evident—the Lagrange equations simply reduce to Newton's equations. For systems of particles, if appropriate assumptions are made about the nature of the interactions, the equivalence can also be established. However, the principle as stated is clearly applicable to contexts far beyond those which can be developed by such comparisons. To apply the principle one just needs to identify the appropriate potential energy V and kinetic energy T for the system of interest. But does the principle actually work for such situations? The reality is that predictions based on the variational formulation have been confirmed in a broad variety of situations and no discrepancy between predictions and observation has ever been shown to be associated with the invocation of the variational formulation. In fact the variational formulation appears to have a far more fundamental position in the structure of physics than Newton's Laws. For a detailed account of the standard deductive approaches to the variational formulations of dynamics the reader is referred to Goldstein (1950).

Comment: One might well ask: Why bother with an alternative formulation if the theoretical predictions are the same? The answer is that from a *practical* point of view the formulations are *very* different. One formulation is not better than the other. Under certain circumstances it's best to use one formulation; under other circumstances the other is best. *Both* formulations are quite basic in the armoury of the mathematical modeller. One of the aims of this chapter is to make comparisons; however, it's as well at this stage to point out an obvious difference. The variational formulation works with scalars (T, V); the Newtonian approach with vectors $(\mathbf{F}, \ddot{\mathbf{x}})$. Scalars are of course normally easier to work with. Thus, for example, with the Newtonian formulation one pays a heavy price if one chooses to work with a non-standard co-ordinate system. Witness the struggle needed to determine the acceleration of a particle in polars. No such difficulty arises with the other approach, and often one works with convenient *generalized* coordinates, specific to the problem. All that is necessary is that one can relate (T, V) to the chosen coordinates—usually a simple matter. Of course the advantage of avoiding the need to deal with forces in the variational approach is forfeited if one is actually interested in forces. Often, for example, we'd like to know if some mechanical system will fail, so of necessity in such circumstances we must work with forces at some stage.

A general subjective comment based on the experience of the authors is that for mechanical systems the Newtonian approach is normally much to be preferred for the simple problems often set up to gain understanding of a situation, basically because intuitionally and analytically the approach is more tractable. For complex circumstances and computational circumstances the variational approach is normally much to be preferred. The table fable problem in Chapter 3 is in this sense very typical. Perhaps the reader should keep this in mind when reading the material that follows.

Example 4.4.1 *The Heavy Spring*

We illustrate the use of the variational ideas by applying them to a heavy spring, with a mass attached, oscillating in a vertical line, see Fig. 4.8. Let l be the natural length of the spring when it is unloaded and let x be the co-ordinate specifying position along the spring measured from the point of suspension in this unloaded state. Let the density per unit length of the spring be σ in that unloaded state. With the weight attached and the whole system moving, let the bottom end of the spring be at $l + Z(t)$ at time t. If we assume uniform stretching along the string then the part of the spring

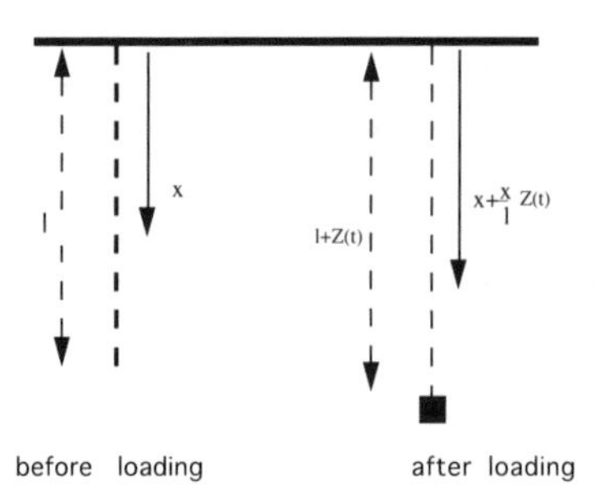

Figure 4.8: The heavy spring.

which was at x in the unstretched state will be at $x + (x/l)Z(t)$ at time t. Thus the spring element lying between x and $x + dx$ in the unloaded state (with mass σdx) is moving with velocity $(x/l)Z'(t)$ at time t. Let M be the mass of the body on the end of the spring. Then the kinetic energy of the spring and mass system is given by

$$T = \frac{1}{2}\left[MZ'^2 + \int_0^l \sigma(x/l)^2 Z'^2\, dx\right] = \frac{1}{2}Z'^2[M + m/3],$$

where $m = \sigma l$ is the mass of the spring. Using the suspension point as a datum the potential energy of the system due to its presence in a uniform gravitational field is

$$-\int_0^l \sigma g(x + Z(x/l))\, dx - Mg(l + Z) = -mg[l + Z(t)]/2 - Mg(l + Z),$$

where g is the gravitational acceleration. The elastic energy of the spring is given by the expression (4.18) obtained earlier for the rod i.e.

$$\mathcal{E}_e = kZ^2/2,$$

where k is the stiffness of the spring. Thus the Lagrangian L for this system, see (4.20), is given by

$$L = \{[M + m/3]Z'^2/2 + g[(M + m/2)(Z + l)] - kZ^2/2\},$$

and the corresponding Euler-Lagrange equation for an extremum is given by, see (4.21),

$$[M + m/3]Z'' - [M + m/2]g + kZ = 0.$$

This gives us a second order linear constant coefficient ordinary differential equation for $Z(t)$, which you'll no doubt recognize as being the forced harmonic equation. The solution corresponds to a simple sinusoidal oscillation

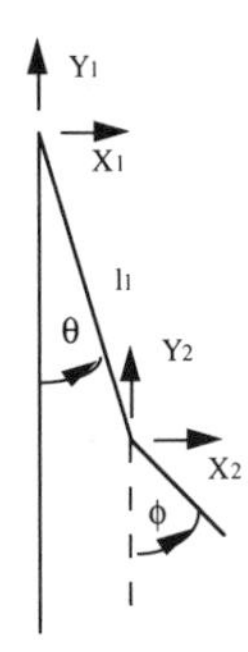

Figure 4.9: A compound pendulum.

about the point $Z = [M + m/2]g/k$ with period $2\pi\sqrt{k/[M+m/3]}$. The dependence of the frequency of oscillation on the non-dimensional mass ratio parameter m/M is the most interesting physical feature. Clearly it's a trivial matter to extend this work to cover the case of many connected springs, and the natural frequencies of motion can be extracted using an algebraic package.

The above procedure for determining the describing equations is much simpler than the Newton approach, and the Newtonian approach becomes clumsy rapidly with coupled oscillators. The Newtonian approach is also very problem specific (i.e. each problem has to be treated on its own merits) and so ingenuity is always required to set up the required equations. By comparison little ingenuity is required for the variational approach, and furthermore one can call on available results for the kinetic and potential energies of various sub-components of complex systems so that it's simply a matter of plugging them into the expression for L. We'll do this in the next example.

It is perhaps worth remarking that the result obtained above for the spring is sometimes questionable because of the failure of the assumption that the spring extension is linear in x and proportional to $Z(t)$ at any time. Waves can in fact be generated in the spring and that can cause quite interesting effects.

Example 4.4.2 *A Compound Pendulum*

One further example will serve to illustrate the comparative advantage of the Euler-Lagrange approach. We consider a compound pendulum consisting of two rods, not necessarily uniform, hinged so that they may swing independently in a plane. Let the position of the rods be specified by the angles θ and ϕ as shown in Fig. 4.9.

For the plane motions considered here the only non-zero angular velocity component is the one perpendicular to the plane of rotation. We shall use $(M_1k_1^2, M_2k_2^2)$ for the corresponding moments of inertia of the two pendulum components about their centres of mass. The parameters k_i are the radii of gyration and have the dimensions of length. We shall use the notation l_1 and l_2 for the lengths of the upper and lower rods respectively. Let the centre of mass of each rod be located at a distance $\alpha_j l_j$ from its upper end. The velocity of the centre of mass of the upper rod is $\alpha_1 l_1 \theta'$ perpendicular to the axis of the rod and the velocity of the centre of mass of the lower rod is the vector sum of the velocity of the upper end $l_1\theta'$ perpendicular to the first rod, and $\alpha_2 l_2 \phi'$ perpendicular to the second. Thus the translational kinetic energy of the system is

$$\mathcal{E}_t = \frac{1}{2}[M_1(\alpha_1 l_1 \theta')^2 + M_2\{(l_1\theta + \alpha_2 l_2 \phi' \cos(\phi - \theta))^2 + (\alpha_2 l_2 \phi' \sin(\phi - \theta))^2\}].$$

The rotational kinetic energy is

$$\mathcal{E}_r = \frac{1}{2}[M_1 k_1^2 \theta'^2 + M_2 k_2^2 \phi'^2].$$

The potential energy of the system is

$$V = -g[M_1 \alpha_1 l_1 \cos\theta + M_2\{l_1 \cos\theta + \alpha_2 l_2 \cos\phi\}].$$

Thus the Euler-Lagrange equations for the system can be readily calculated, but the algebraic complexity is such that if one wants the equations exactly it is better to use an algebraic package to carry out the analysis, particularly since the resulting equations will need processing. The questions of physical interest here are:

1. As a function of the parameters of the problem what are the natural vibrational frequencies for the system for small amplitude motions? Knowing this one could avoid possible resonances that could occur when an external force acts on the system.

2. How does the frequency of the motion change with the amplitude of vibration?

3. What types of evolving behaviour are possible?

4. Are the hinge forces likely to be of sufficient magnitude to cause breakage?

The process of using algebraic packages to answer these questions is still tedious, and the results are not of great interest in the present context so the Maple code will not be presented; however, we'll list the Maple steps required to produce useful output:

- Identify the dimensionless groups; a sensible choice for the Lagrangian scale would be $(M_1 g l_1)$, so one should input the scaled dimensionless Lagrangian $L/(M_1 g l_1)$.

- Use Maple to generate the two Euler-Lagrange equations (use **diff** with care).

- To answer question (1) above we need to use **subs** to replace $\sin(\phi-\theta)$ by $\phi - \theta$ etc. The resulting two equations will be second order linear constant coefficient equations in (ϕ, θ). Given that the system is conservative we expect purely oscillatory solutions, so the characteristic equations will have eigenvalues of the form $\pm i\omega_1, \pm i\omega_2$,where ω_1, ω_2 are the required natural frequencies. Maple can be used to extract these values, and to determine the relative amplitude of the oscillations.

- To obtain small amplitude solutions corresponding to different initial conditions one could use **dsolve** and **plot** the results. However such results are not likely to be very informative—we need to know what to look for!

- To answer (2) and (3) again **dsolve** may be used but analytic solutions won't be possible so the **numeric** option would be necessary. Maple is not at this stage very useful for such numeric work, so it would be sensible to output the Euler-Lagrange equations in Fortran or Pascal and use one of the standard numeric packages (eg. the NAG routines) to produce solution graphs. In the near future links between algebraic packages and numeric programs should be standard. Note, however, that such simulations are not likely to lead to any *real* understanding—again we need to know what to look for.

- To determine the hinge forces we would need to invoke Newton's Laws.

The important thing to notice about the above is that algebraic complexity is no longer an issue if one uses algebraic packages. In the past, traditional courses in dynamics went to all sorts of idealizations to produce problems as student exercises which would lead to much simpler equations, but the practical reality is that situations of interest are likely to involve *much* greater complexity than the present problem. The real issue that remains is the understanding. What are the possibilities for such systems? For example one might well ask if it's possible for all the energy to eventually be concentrated in just one of the pendulum's components so that failure could result. Some of these issues will be taken up in the Vibrations Chapter.

Is the situation more complicated if one uses the Newtonian approach to this problem? For our present problem, see Fig. 4.9, the unknowns are the forces at the two hinges with components $(X_1(t), Y_1(t))$ and $(X_2(t), Y_2(t))$, and the angular displacements $(\theta(t), \phi(t))$—in all six unknowns. For each of the two sections of the pendulum there are two independent non-trivial equations describing the translational motion of the centre of mass of the section (see (4.1)); and one non-trivial equation describing the rotational motion of that section (see (4.2))—in all six equations in the above six unknowns. Thus we can expect to generate enough equations to determine all the unknowns in principle. Note the extra work we have to perform in order to get the equations we want:

- We have to calculate accelerations of the centres of mass in terms of the co-ordinates, whereas previously we needed only velocities.
- We write down more equations (6 rather than 2) involving the forces and then eliminate the forces from the equations. Note also that the equations are all second order differential equations. How much work this involves depends upon how clever we are in writing down the best set of equations by choice of special features of the system. Of course the more complex a system gets, the less likely it is that a clever way will be found.

Of course if the objective were to predict the forces at the hinges then the advantages of the Euler-Lagrange approach would be largely illusory. In such cases one again should expect to perform the calculations using a symbol manipulative software package in any real problem, for otherwise the chances of obtaining correct equations (let alone correct solutions) would be low. The following Maple code could be used to calculate the accelerations of the centre of mass of the lower body

```
Y2:=l1*cos(theta(t))+a2*l2*cos(phi(t));     %theta ≡ θ, phi ≡ φ
X2:=l1*sin(theta(t))+a2*l2*sin(phi(t));
AY2:=diff(Y2,t,t);
AX2:=diff(X2,t,t);,
```

and the hinge forces can then be determined.

Whichever approach is used the code used for building up the equations required is more meaningful than the actual complex expressions detailing these equations—the equations are unreadable!

4.4.2 Static continuous configurations

When a system is at rest its kinetic energy is zero so Hamilton's Principle (see (4.19)) simply requires that the potential energy be an extremum. This can be a powerful tool for analysing continuous static systems. The moorings problem is a case in point.

The cable stretches from the mother ship to the sea bed. Now the cable can be modelled as an elastic system that can stretch under the influence of the forces acting. Thus energy can be stored in the cable in the form of elastic potential energy. Any such stretching will cause the cable to sag and thus cause a decrease in the gravitational potential energy of the

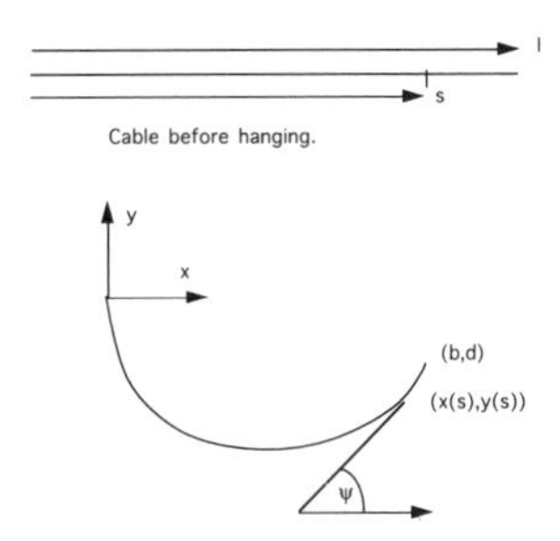

Figure 4.10: The hanging cable

system. Some balance will be reached between these opposing tendencies under equilibrium conditions. Hamilton's principle tells us that the balance that results is the one that minimizes the total potential energy of the system. Of course it makes sense to investigate this situation first for the uniform cable case, and we'll do this. Now one might expect intuitively that the cable will stretch relatively little, so that ignoring the stretching may produce accurate results. The standard textbook accounts of this problem do just this and the results obtained are accurate. We shall not make such an assumption, and will allow the mathematics to determine the relative importance of elastic and gravitational effects. The work shows some of the features which arise regularly in energy methods and serves to illustrate how approximations are made in applying such methods.

4.4.3 The hanging cable.

Let m be the mass per unit length and l the length of the cable in its unloaded state. Typical lengths in the moorings context might be several kilometres. The analysis to follow will not be restricted to the moorings situation; it will cover the gamut of possible hanging cable situations. We'll work with lengths that are scaled relative to the unloaded length of the cable l, and we'll use the scaled arc length s along the cable in the *unloaded* state as a means of specifying position along the curve C describing the cable's position in the actual static equilibrium hanging configuration, see Fig. 4.10. By making this choice we ensure we know the domain of definition; it's $0 \leq s \leq 1$. (If for example we used arc length along the *stretched* cable the domain wouldn't be known a priori.) Let $x(s)$ be the scaled horizontal co-ordinate measured from the left hand end support and $y(s)$ the scaled vertical upwards co-ordinate measured from the left hand support. Thus $(x(s), y(s)), 0 \leq s \leq 1$, provides us with a parametric representation of C. The scaled length of the element of cable lying between s and $s + ds$ will thus become $\sqrt{x'^2 + y'^2} \cdot ds$ in the hanging position, where x' denotes dx/ds

etc. Thus the local increase in length per unit length of the cable is given by $[\sqrt{x'^2+y'^2}-1]$. The elastic energy stored in the stretched element corresponding to ds (of real length lds) is thus

$$\frac{EA}{2}[\sqrt{x'^2+y'^2}-1]^2\cdot(l\,ds),$$

see equations (4.16,4.18), where E is the Young's modulus of the material and A is the initial cross-sectional area of the cable. Now, because of the way we've parameterized the problem, the mass of cable corresponding to our cable element remains $(ml\cdot ds)$ in the hanging position. This element is located at $(x(s),y(s))$ in the hanging position, so its gravitational potential is given by $(ml\,ds)g\,l\,y(s)$. Thus the total potential energy of the cable is given by

$$V=\int_0^1\left\{\frac{EA}{2}[\sqrt{x'^2+y'^2}-1]^2+mg\,ly(s)\right\}lds,$$

and if we choose to use the potential or energy scale mgl, so that $\mathcal{V}=V/mgl$ we obtain the expression

$$\mathcal{V}=\int_0^1(\mathcal{E}_e+\mathcal{E}_g)\,ds, \tag{4.22}$$

where

$$\mathcal{E}_e(s)=\frac{1}{2}[\sqrt{x'^2+y'^2}-1]^2,\quad\text{and}\quad\mathcal{E}_g(s)=\epsilon y, \tag{4.23}$$

represent the (local) scaled elastic energy and gravitational energy per unit scaled cable length stored in the hanging cable, and where

$$\epsilon=\frac{mgl}{EA}=\frac{\rho gl}{E},$$

where ρ is the density of the cable material. For the equilibrium hanging cable configuration this integral has to be stationary amongst all admissible functions taking prescribed values at $s=0$ and $s=1$.

Apart from the geometry, it's the dimensionless parameter ϵ which specifies the problem. This parameter provides us with a measure of the relative importance of gravitational and elastic effects. For steel the value of $\rho g/E$ is about 1.5×10^{-7} per metre, so that for cables of the length (several kilometres) of interest ϵ is of order 10^{-4}—very small! In fact most cables used for engineering purposes are hard to stretch so the relevant ϵ is small. Simply interpreted this indicates that the amount of elastic energy stored in the cable as a result of an *arbitrary* displacement will be *much greater* than the amount of gravitational energy stored. Given the smallness of ϵ it seems sensible to ignore the gravitational component of $\mathcal{V}$; i.e. put

$\epsilon = 0$. The problem then reduces to the one of determining the curve C that minimizes

$$\int_0^1 [\sqrt{x'^2 + y'^2} - 1]^2\, ds.$$

We know the answer to this problem. The minimum is obviously 0 and is reached if $x'^2 + y'^2 \equiv 1$, i.e. if the cable doesn't stretch anywhere along its length (recall that this expression represents the square of the increase in length per unit original length of the cable). If you're disturbed by this result don't worry—you should be! The point is that there is an infinite set of curves that correspond to this situation in which the cable is nowhere stretched. Now the real world knows what it's doing (it certainly selects a unique solution) so there's something wrong with our formulation. Hamilton's principle is well tested, so we expect that the original variational problem is well posed. We must therefore conclude that the apparently small gravitational term that we removed in the above work must be retained; even though it looks to be minute! The situation is even worse than this—to obtain a *robust* formulation we should *only* retain the gravitational term in the functional to be minimized. This type of paradoxical situation occurs very often in variational problems, and so merits careful attention. In the present case the solution of the exact variational problem can be obtained (a luxury normally denied to us) so we'll obtain this solution to see what's gone wrong.

The Euler-Lagrange equations (4.14) for the functional (4.22) are given by

$$\frac{d}{ds}[x'(1 - \frac{1}{\sqrt{x'^2 + y'^2}})] = 0,$$

$$\frac{d}{ds}[y'(1 - \frac{1}{\sqrt{x'^2 + y'^2}})] - \epsilon = 0.$$

The first of these integrates to give

$$x'[1 - 1/\sqrt{x'^2 + y'^2}] = C,$$

for some constant of integration C. Using this result the second equation then yields

$$C\frac{d}{ds}[y'/x'] = \epsilon,$$

which integrates to give

$$Cy'/x' = D + \epsilon s,$$

for some integration constant D.

We use this equation to eliminate y' in favour of x' in the first equation to give

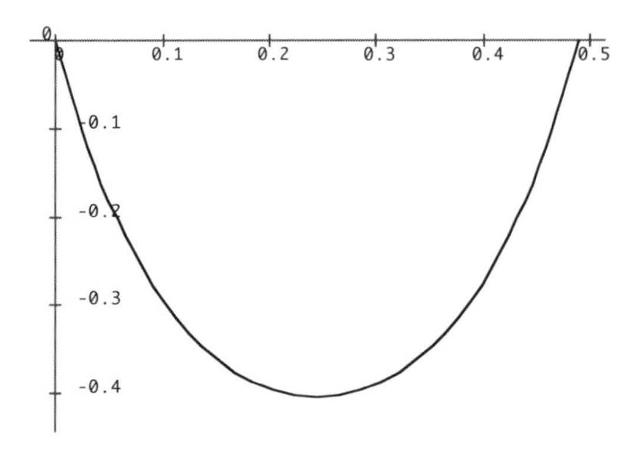

Figure 4.11: The hanging cable solution.

$$x' = C + 1/[1 + (\frac{\epsilon s + D}{C})^2]^{1/2}.$$

An integration from $s = 0$ (with $x(0) = 0$) gives $x(s)$. The equation for y' can then be integrated from $s = 0$ (with $y = 0$) to give $y(s)$. The two constants (C, D) are then determined so that $(x(1), y(1)) = (b, d)$, the prescribed location of the end of the cable. Numerical solutions for these two equations for (C, D) can be obtained, and the solutions plotted. The Maple code for doing this in the symmetric case in which $x(1) = .5, y(1) = 0$, and $\epsilon = 0.1$ is:

```
xp:=C+1/sqrt(1+((e*s+D)/C)^2);                         %xp ≡ x',e ≡ ε
x:=int(xp,s=0..S);                                      % x(S) ≡ x(s)
yp:=(D+e*s)*xp/C;                                            %yp ≡ y'
y:=int(yp,s=0..S);
x1:=subs(S=1,x);y1:=subs(S=1,y);            % (x1,y1) ≡ (x(1),y(1))
sol1:=fsolve(subs(e=.1,x1=.5,y1=0),C,D);        % solve for (C,D)
sol1[1];sol1[2];                         % we identify the solutions
plot(subs(C=sol1[2],D=sol1[2],e=.1,[x,y,S=0..1]));       % plot C,
```

and the solution curve is shown in Fig. 4.11. We're concerned with the very small ϵ case, but we know that the $\epsilon = 0$ case is ill-posed, so it's sensible to commence with a moderately small value of ϵ and then reduce its value, keeping a look out for difficulties.

Aside: Often it's useful to "creep up" on difficult situations.

Table 4.1 summarizes the results for (C, D), and typical values for the scaled elastic and potential energy $(\mathcal{E}_e, \mathcal{E}_g)$ stored in the cable (see (4.23)) for various ϵ values.

It should be noted:

- The scheme is not homing in on solution values for (C, D) as $\epsilon \to 0$!

Table 4.1: Table of cable solution values

ϵ	C	D	$\mathcal{E}_e$	$\mathcal{E}_g$
.1	.11 10^{-1}	-.5 10^{-1}	.6 10^{-4}	-.4 10^{-1}
.01	.11 10^{-2}	-.5 10^{-2}	.6 10^{-6}	-.4 10^{-2}
.001	.11 10^{-3}	- .5 10^{-3}	.6 10^{-8}	-.4 10^{-3}
.0001	.11 10^{-4}	-.5 10^{-4}	.6 10^{-10}	-.4 10^{-4}

All that's happening is that values of C and D are becoming smaller and smaller as ϵ gets smaller. Thus we suspect that our numerical scheme has problems. (Also it's likely that round-off errors would swamp any correct solution for the small values of ϵ of interest.)

- $\mathcal{E}_e \ll \mathcal{E}_g$, and the ratio becomes smaller as ϵ gets smaller. You'll recall that earlier we presented scaling arguments that suggested that this inequality should be reversed! In fact, based on these earlier arguments, we argued that the approximation $\epsilon = 0$ was sensible—we can see now that this approximation retains precisely the *wrong* term!

You'll recall that the values of ϵ of interest are of order 10^{-4}, so the above calculation scheme is suspect for the parameter range of interest. The reason the scheme has difficulties for very small ϵ is not that the solution doesn't exist (a unique solution exists for any $\epsilon > 0$) but that, because $\epsilon = 0$ is a singularity of the system, numerical difficulties arise for small ϵ. It's the same type of difficulty that arises when one tries to invert numerically a set of non-homogeneous linear algebraic equations when the determinant of the defining matrix is almost zero. The numerical procedure is said to be *ill – conditioned*. In practical terms one needs to approach the problem differently; if one understands the nature of the difficulty it's usually possible to devise a sensible robust scheme.

Comment: In our present circumstances we're lucky because we know the solution in the $\epsilon = 0$ case, and thus have anticipated difficulties. Normally we do not have such knowledge. If, for example, we had thrown out the small ϵ term and then proceeded with a numerical evaluation we would have been attempting to compute a unique solution that didn't exist! The computer would either produce confusing results or, worse still, might even find a solution! (As it is the numerical evaluation of the "exact" functional is still all but useless.)

Comment: From a basic mathematical point of view the above result is quite disturbing. The variational problems corresponding to small values of ϵ and $\epsilon = 0$ are almost identical, and yet in one case we end up with a unique solution (that's hard to obtain computationally); and in the other case there is *no unique* solution. There's nothing about the form of the functionals that indicates such difficulties may arise! The obvious question is : How can we recognize if a variational problem is properly posed? Unfortunately the answer is that there are no known rules. This is one of the real difficulties besetting this variational method; it's often difficult to know (by observing its form) if a variational problem is well posed, and experience of the above type suggests that small terms can make all the difference. Compare this with ordinary differential equations where difficulties are usually obvious. Of course this difficulty should not deter us from using variational methods, and experience (usually arising out of the physical context, rather than the mathematics) can often enable us to anticipate difficulties. The power of variational methods, especially for numerical work, is such that they can't be ignored. *Singular* variational problems of the above type are an active research area.

To summarize the above: The suggested approximation ($\epsilon = 0$) failed because the assessment of the relative size of the energy terms was incorrect, and the exact equations led to an unsatisfactory numerical scheme. Two obvious questions arise:

A For this particular problem, and for variational problems in general, how do we obtain sensible term estimates?

B How do we obtain a *robust* numerical scheme ?

Let's address these questions:

A In the above work we've stuck strictly to energy ideas and thus have avoided any reference to the forces generated in the cable. Now, although all information concerning the required solution must be contained in the variational formulation (including the size of terms), it's not contained in an *accessible* form. Hooke's Law provides us with the relevant information concerning the stretching of the cable but the variational approach, in eliminating reference to forces, has concealed this important information. (Newton's approach is in this sense much more revealing.) To obtain the required estimates simply note that the total weight of cable is mgl. The cable supports must bear this load and so the tension in the cable must be of the same order. The proportional stretch resulting from such a load is given by Hooke's Law to be of order mgl/EA i.e. ϵ. Thus the elastic energy stored in the cable will be of order $EA\epsilon^2$—much smaller than suggested by the scaling used in (4.22)! In fact the scaled elastic energy term in the functional is of order ϵ^2 whereas the gravitational term is only of order ϵ. The numerical results contained in Table 4.1 thus make sense; on the above basis we'd expect the elastic energy component in the functional to be relatively smaller than the gravitational component for solutions close to optimal. In energy terms the apparent paradox can be rationalized: the energy expense associated with any stretching of the cable is so great that all near contenders for the optimal solution will involve no cable stretching, leaving the gravitational potential energy component to sort out the "real" contenders. There is a lesson to be learned here concerning other situations in which terms of disparate size occur in a functional. One might expect in such cases that the large terms may vanish identically; however, *in modelling there are no concrete results—just useful lessons.* If if weren't this way modelling would not be challenging.

B In a minor way we could improve the above numerical scheme by just getting the scales correct. The C's and D's obtained when solving the exact variational problem are very small as a result of incorrect scaling. However, this does not get to grips with the real problem; the gravitational and elastic energy terms are of different orders of magnitude, so that numerical errors will enter any calculation that treats them on a par. Given the above discussion it's clear what is needed to obtain a robust numerical scheme. We need to:

- Ensure the cable length is 1 for curves $\mathcal{C}$ of interest.
- Pick out from such constant length contenders the one that minimizes the gravitational potential. A little thought about the physics should convince the reader that there should be just one solution.
- Calculate the extension that would be induced in such a cable and thus evaluate the elastic energy associated with this solution and *verify* that it is indeed small in comparison with the gravitational energy.

Interestingly enough the above procedure transforms the nature of the mathematical problem strikingly; from an ordinary optimization problem to a *constrained* optimization problem. We now discuss in detail this (standard) procedure.

The standard hanging cable calculation

We could continue to work with $y(s)$ as our unknown function and with x being obtained by integrating $x' = \sqrt{1 - y'^2}$; in this way we would ensure that curves being considered involved no stretching. An alternative approach using Lagrange Multipliers has been found, however, to be *much* simpler for such *constrained optimization* problems. Also we'll use x as our independent variable and thus we'll use $Y(x)$ to describe $\mathcal{C}$. Reorganizations of the mathematics of this type are typical; often a seemingly impossible problem becomes trivial when viewed from an equivalent but different perspective. It's no exaggeration to say that the very simple Lagrange procedure described in this section in this particular context has proved to be a *major* mathematical breakthrough, relevant to many areas of mathematics. The technique is particularly valuable for optimization problems. The length of the cable will still be retained as the length scale, and the potential scale mgl will again be adopted. The scaled potential energy of the system is thus the same as before, see (4.22), with the elastic component $\mathcal{E}_e$ removed i.e.

$$\mathcal{V} = \int_0^1 \epsilon y(s)\, ds,$$

so, dropping the irrelevant multiplicative factor, and changing from variable s to variable x, our objective is to determine the stationary conditions for the functional

$$J(Y) = \int_0^b Y(x)\sqrt{1 + [Y'(x)]^2}\, dx. \tag{4.24}$$

Two further statements are needed to complete the mathematical formulation of the problem. We need to ensure the length of the curve is 1, so that the functional

$$S(Y) = \int_0^b \sqrt{1 + [Y'(x)]^2}\, dx = 1. \tag{4.25}$$

We of course also need to specify the locations of the supports, namely

$$Y(0) = 0, \quad Y(b) = d.$$

Our problem is now in the form of a variational problem, but with the added complication that only curves with the prescribed length are acceptable.

The Lagrange multiplier method simply recognizes that the function $Y(x)$ that minimizes the expression $K = J - \lambda(S - 1)$ with respect to $Y(x)$ *and* λ satisfies our requirements. For clearly $\partial K/\partial \lambda = 0$ recovers the required condition on the cable length, and with $S = 1$ it's clear that stationary values for K and J are realized for the same $Y(x)$. This simple device avoids the problem of eliminating from contention functions that don't satisfy the integral constraint. Notice that the problem we now have to deal with is just an ordinary *unconstrained* optimization problem.

Thus we seek to find the extremum of

$$\int_0^b [Y - \lambda]\sqrt{1 + Y'^2}\, dx.$$

We note further that the integrand does not explicitly involve x so we can use the corresponding integrated form of the Euler-Lagrange equations, see (4.10), available in such cases. Thus we require,

$$[Y - \lambda]\sqrt{1 + Y'^2} - [Y - \lambda]Y'^2/\sqrt{1 + Y'^2} = C$$

for some constant C. From this we obtain

$$Y'^2 = [(Y - \lambda)^2 - C^2]/C^2,$$

which may be arranged as

$$\int^{Y} \frac{dy}{\sqrt{(y-\lambda)^2 - C^2}} = \pm(D + x/C),$$

where D is a constant of integration. Thus

$$Y = \lambda + C\cosh(D + x/C).$$

This result involves three unknown constants (C, D, λ) which have to be chosen so as to satisfy the conditions that the points $(0,0)$ and (b,d) lie on the curve and that the total length is 1 i.e.

$$\int_0^b \sqrt{1 + Y'^2}\, dx = 1.$$

In general the resulting equations require a numerical solution for particular values of the parameters. The symmetric solution corresponding to the boundary condition $Y(\frac{1}{2}) = 0$ is more easily extracted and is given by

$$Y = \lambda[1 - \cosh(\frac{x - 1/4}{\lambda})/\cosh(\frac{1}{4\lambda})],$$

where the value of λ needs to be determined (numerically) so that the curve length is 1. Maple can be used and comparisons can be readily made with the stretched cable solution obtained earlier. These show that, providing one doesn't go to very small values of ϵ, the results closely coincide. As we've seen, for small ϵ values numerical errors render the earlier stretched cable approach useless; by comparison the above approach is both accurate and robust.

Important Comment: The interesting feature is that the *exact* model used earlier produces *incorrect* results for small ϵ cases of interest, whereas the *approximation* based on cable inextensibility produces *accurate* results. *Thus the approximate model is in a practical sense better than the exact model!* This brings into question what one means by a *good* model. *It is often true that exact models are useless in practice.* Thus, for example, the aim of most meteorological models is to *remove* real effects from *accurate* fluid dynamics equations so that one can see and compute the weather patterns of interest.

The above results determine the shape in which the cable hangs but almost certainly one would be interested in the loads required to support the cable and these loads are related to the values of the tension T in the cable at the ends. Let ψ be the angle the cable makes with the positive x axis, see Fig. 4.10, so that $Y' = \tan\psi$. The tensions at the two ends have to support the weight of the cable and also must be such as to give a net zero horizontal force component. Thus

$$T(b)\sin\psi(b) + T(0)\sin\psi(0) = mgl \quad \text{and} \quad T(b)\cos\psi(b) + T(0)\cos\psi(0) = 0,$$

provide two equations from which the tensions at the two ends may be determined, given $\psi(0), \psi(b)$. In the symmetrically hanging cable case we get $T(0) = mgl/(2\sin\psi(0))$.

All that remains to do is to check that the assumption that the cable doesn't stretch is valid. A simple way of checking whether this assumption is reasonable is to calculate the extension which results as the result of the tension calculated from the above results. An accurate expression for the local cable tension $T(s)$ could be obtained using the above results and thus the local extension could be determined using Maple, but unless this information is of special interest, it's hardly worth the effort. It suffices to note that the maximum tension is realized at the cable ends and is given by $mgl/(2\sin\psi(0))$ in the symmetric case. Hooke's Law thus tells us that the proportional length change will therefore be less than $mgl/(2EA\sin\psi(0)) = \epsilon/2\sin\psi(0)$ i.e. of order 10^{-4} for our cable, unless the cable is taut. Thus, although this doesn't *ensure* that our result is accurate, it's very reassuring. In more doubtful cases the calculations should be more thorough.

4.5 A Cheaper Mooring

You'll recall that the mother ship is to be restrained by a series of chains that are to be attached to the sea bed, and that our problem is to design the chains so that they are strong enough to carry the loads needed to moor the ship in water of given depth, and so that the capital cost is least. By "design the chains" we mean determine their length, where they should be anchored, and the weight distribution along their length. The work of the previous Section tells us that elastic stretching of the chains can be ignored.

4.5.1 Setting up a model.

In this problem the chain length is unknown but the water depth is known so we'll find it convenient to parameterize the chain curve in terms of its height (measured in this case) above the sea bed, see Fig.4.12. Note that, since we don't expect the chain to be horizontal in any location, this specification will be unique. The angle ψ that the tangent to the chain makes with the positive x axis is also useful to work with. Arc length s will be measured from the chain anchor.

The form of model

There are a number of decisions to be made about the form of this model:

- We could examine what is happening in each of the individual chain links and then, by considering how a system of such links would behave, determine the behaviour of the complete chain. There will be,

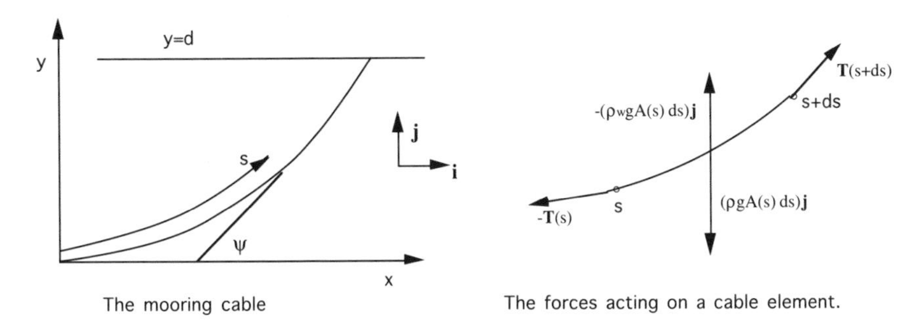

Figure 4.12: The mooring cable.

however, an extremely large number of links in the chain and conditions will differ but little between neighbouring links, so that it makes sense to model the chain as a continuous system. This reduces the complexity enormously and makes available all the mathematical power associated with continuous systems which is not available for discrete systems. Thus we will model the chain as a cable with varying cross-sectional area $A(s)$; where $A(s)$ represents the total sectional area across the two parallel sides of a link. The area will vary with length s along the cable. There may be special (high stress) problems associated with the links close to the ship; these would have to be handled in a sub-model.

- In its design configuration there will be a variable tension $T(s)$ along the cable. The force per unit area will therefore be $T(s)/A(s)$ and if this exceeds the specified (experimentally determined) value for steel the cable will draw under the load. Thus there is a maximum value of this ratio which cannot be exceeded with safety. There are, however, bound to be variations in the properties of the material in the chain, the accuracy with which it is manufactured, and the accuracy of the analysis used in the design. Thus normal engineering practice is to build in a safety factor which makes it improbable that the design will fail in any foreseen circumstances. Often this safety factor is specified by a government imposed design code but in circumstances like this one the safety factor is more likely to be imposed by requirements to be met in obtaining insurance cover for the arrangement. A failure of a mooring which severed the connection of an oil well with the ship and left it discharging oil into the ocean could prove very expensive and so is normally insured against. We merely assume that a maximum figure α is set for this ratio T/A. Clearly, since the total weight is to be minimized, there should be no excess weight at *any* location along

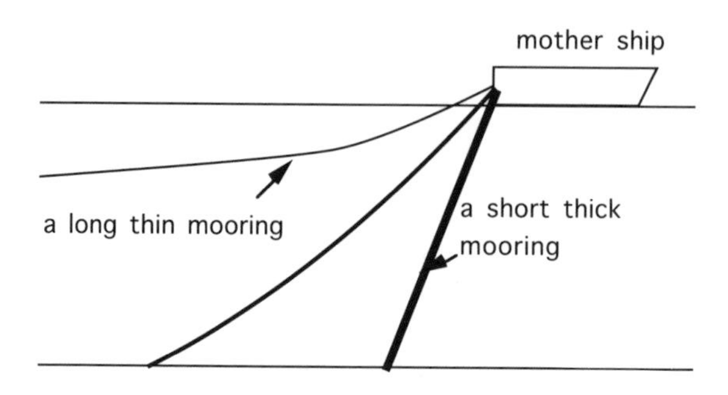

Figure 4.13: Alternative moorings.

the chain, so we require the critical stress condition

$$T(s)/A(s) = \alpha \tag{4.26}$$

to be satisfied at all points along the cable.

- The chains are required to exert a horizontal force on the ship which will be sufficient to restrain it. At this stage we assume that the size of this horizontal force is specified to be H. Thus we require, see Fig. 4.12

  $$T \cos \psi = H \quad at\ y = d. \tag{4.27}$$

  Equivalently, using the critical stress condition (4.26), this imposes the area verses angle constraint

  $$A_d \cos \psi_d = H/\alpha, \tag{4.28}$$

  on the chain at the point of attachment to the mother ship, where $A_d = A(d), \psi_d = \psi(d)$.

The plan of attack

Clearly at the point of attachment of the chain to the mother ship there is a range of possible values of chain area that fulfill both the strength and restraining condition requirements. The area verses angle relationship (4.28) encapsulates both requirements. Thus for *any* given ψ_d in the range $0 \leq \psi_d \leq \pi/2$, A_d can be selected to satisfy this relationship. A_d will lie in the range $H/\alpha \leq A_d \leq \infty$ with the large areas corresponding to almost vertical attachment angles, see Fig. 4.13. Presumably there will be physically realistic solutions satisfying the full range of values of A_d.

Thus a plan of attack is to determine the optimal solutions corresponding to fixed values of A_d. The chain masses corresponding with all such solutions can then be compared and the optimal of these selected. We should first check to see if intuitively we'd expect an optimal solution to exist. Note that as $\psi_d \to \pi/2$, $A_d \to \infty$, so short (almost vertical) moorings (see Fig.4.13) will not be optimal. Also as $\psi_d \to 0$, $A_d \to H/\alpha$, but the length of such chains is likely to be great, so that such long thin chains (see Fig. 4.13) are also likely to be expensive. Thus one might feel confident that there will be an optimal configuration with an A_d somewhere between these extremes.

4.5.2 The model

The mathematics turns out to be simpler if one works with a Newtonian approach rather than the variational approach, basically because the tension in the chain is closely connected to the determination of the required shape. There is no way of knowing this in advance, and the choice can only be sensibly made by examining the appearance of different formulations. The chain will be in static equilibrium provided the forces on elements of the cable balance.

Consider the element of cable of length ds lying between s and $s+ds$. Let $\psi(s)$ be the angle the tangent to the cable makes at the positive x axis, see Fig. 4.12, $\mathbf{i}$ and $\mathbf{j}$ are unit vectors in the x and y directions. The external forces acting on this element (see Fig. 4.12), are $\mathbf{T}(s+ds)$ and $-\mathbf{T}(s)$, where $\mathbf{T}$(s) is the vector tension in the chain; the body force $(-\rho g A(s)ds)\mathbf{j}$, where ρ is the density of the chain material; and the buoyancy force due to the pressure exerted by water on the cable $(\rho_w g A(s)ds)\mathbf{j}$, where ρ_w is the density of water. Thus static equilibrium for this element requires

$$[\mathbf{T}(s+ds) - \mathbf{T}(s)] + gA(s)(\rho_w - \rho)ds\mathbf{j} = \mathbf{0}.$$

Now Taylor's expansion gives

$$\mathbf{T}(s+ds) = \mathbf{T}(s) + \frac{d\mathbf{T}(s)}{ds}ds + O(ds^2),$$

so in the limit as $ds \to 0$ we get the static equilibrium equation

$$\frac{d\mathbf{T}(s)}{ds} = gA(s)(\rho - \rho_w)\mathbf{j}.$$

Now $\mathbf{T} = T(\cos\psi\mathbf{i} + \sin\psi\mathbf{j})$ so the $\mathbf{i}$ and $\mathbf{j}$ component equations are given by

$$\frac{d}{ds}[T(s)\cos\psi(s)] = 0,$$

and

$$\frac{d}{ds}[T(s)\sin\psi(s)] = (\rho - \rho_w)gA(s).$$

We thus have two first order differential equations and one algebraic equation in $(T(s), \psi(s), A(s))$, see (4.26), so we'd expect two boundary conditions. Evidently we need to ensure that the cable provides the right restraining force at the mother ship (see (4.27)), and that the cable is anchored to the sea bed. This completes the specification of the problem. The manipulations to follow are straightforward but a little tedious. Maple is not up to the task.

After integration the first of the above equations gives

$$T(s) \cos \psi(s) = H, \tag{4.29}$$

where use has been made of the horizontal restraining force condition (4.27). Using this result $T(s)$ can be eliminated from the vertical equilibrium equation giving

$$\frac{d}{ds}[H \tan \psi(s)] = H \sec^2 \psi(s) \frac{d\psi(s)}{ds} = (\rho - \rho_w) g A(s). \tag{4.30}$$

The safety requirement equation (4.26) and the horizontal equilibrium condition (4.29) give

$$A(s) = H \sec \psi(s)/\alpha, \tag{4.31}$$

which can be used to eliminate $A(s)$ in favour of $\psi(s)$ in (4.30) to give

$$\sec \psi \frac{d\psi}{ds} = (\rho - \rho_w) g/\alpha, \tag{4.32}$$

a differential equation involving the single dependent variable $\psi(s)$.

The boundary conditions are imposed at $y = 0, d$, so it's best to change the independent variable from s to y. Note that for the monotonic solutions $y(s)$ of interest there's just one s for each y, so the transformation will be one to one. Now

$$\frac{d}{ds} = \frac{dy}{ds} \frac{d}{dy} = \sin \psi \frac{d}{dy},$$

using standard differential geometry results. Thus we obtain

$$\tan \psi \frac{d\psi}{dy} = (\rho - \rho_w) g/\alpha.$$

This integrates to give

$$\cos \psi(y) = \cos \psi(0) e^{-Jy/d},$$

with

$$J = (\rho - \rho_w) g d/\alpha, \tag{4.33}$$

and where $\cos \psi(0)$ is yet to be determined. Thus

$$\cos \psi(d) = \cos \psi(0) e^{-J}, \tag{4.34}$$

and if we choose $\cos\psi(0)$ so that ψ_d is consistent with the assumed A_d value, see (4.28), i.e. if we choose

$$\cos\psi(0) = \frac{H}{\alpha A_d}e^J,$$

then the solution corresponding to A_d is completely determined. Thus the required sectional area is given by, using (4.26,4.29)

$$A(y) = A_d\frac{\cos\psi(d)}{\cos\psi(y)} = A_d e^{-J(1-\frac{y}{d})},$$

and the other chain characteristics follow. The dimensionless parameter J may be regarded as the dimensionless depth of water.

We're now in a position to make comparisons between chains with different attachment areas A_d. The total mass M of the chain with attachment area A_d is given by, using equation (4.32)

$$\begin{aligned} M &= \rho\int A\frac{ds}{d\psi}\,d\psi \\ &= \frac{\rho A_d\alpha\cos\psi_d}{g(\rho-\rho_w)}\int_{\psi(0)}^{\psi_d}\sec^2\psi\,d\psi, \\ &= \frac{\rho A_d\alpha\cos\psi_d}{g(\rho-\rho_w)}(\tan\psi_d-\tan\psi(0)). \end{aligned}$$

Now, using the area verses angle attachment condition (4.28) and equation (4.34), we get

$$M = \frac{\rho}{\rho-\rho_w}\frac{H}{g}[\sqrt{\mathcal{A}^2e^{2J}-1}-\sqrt{\mathcal{A}^2-1}],$$

where the dimensionless parameter $\mathcal{A} = \alpha A_d/H$ may be regarded as the dimensionless sectional area of the chain at the point of attachment.

The parameter J is fixed by the material of the chain and the water depth and so for a given design may be regarded as fixed. The selection of the cheapest chain thus involves finding the value of $\mathcal{A}$ which minimizes M. Using Maple it can be shown that M achieves a minimum value M_{opt} for $\mathcal{A}^2 = \exp(2J)+1$ and the required optimal mass is given by

$$M_{opt} = \mathcal{M}[e^J - e^{-J}], \tag{4.35}$$

where

$$\mathcal{M} = \frac{\rho}{\rho-\rho_w}\frac{H}{g}. \tag{4.36}$$

The value of $\mathcal{A}$ at which the minimum is achieved serves to determine the cross-sectional area at the ship for the optimal solution, and thus the area

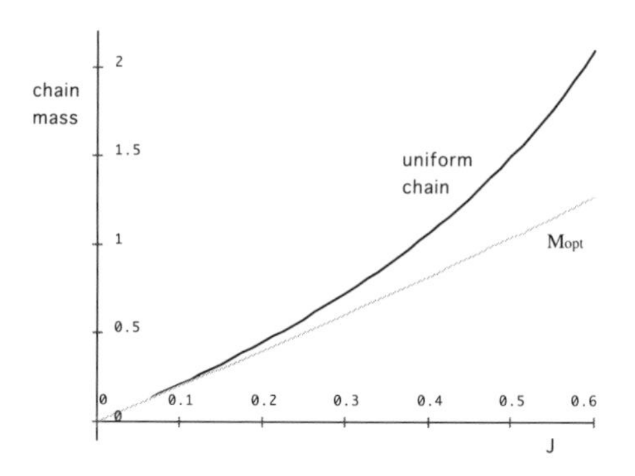

Figure 4.14: Optimal and uniform chain comparisons.

as a function of y. Using this it's a simple matter to determine the chain length and distance the anchor should be from the mother ship. Explicitly

$$s(y) = \int_0^y \frac{dy}{\sin\psi(y)} = \int_0^y \frac{dy}{\sqrt{1-\cos^2\psi(d)e^{-2J(1-y/d)}}},$$

and $s(d)$ is the required chain length, and

$$\int_0^d \frac{dy}{\tan\psi(y)} = \int_0^d \frac{dy}{\sqrt{\sec^2\psi(d)e^{-2J(1-y/d)}-1}},$$

determines the anchoring location.

Comparisons For comparison purposes we need the corresponding results for a uniform chain. The pattern of calculation is very similar, see Exercise 4.13. If we assume that the cross-sectional area is chosen so that the critical stress condition (4.26) is satisfied at the point of attachment to the ship where stress levels are largest we obtain the result:

$$M_{unf} = \mathcal{M}[1/(1-J) - (1-J)]. \tag{4.37}$$

As one would expect, the two chain mass expressions, (4.35) and (4.37), for M_{opt} and M_{unf} share the same scale $\mathcal{M}$, but the dimensionless expression multiplying this scale is different. The scale factor, see (4.36), simply indicates that the required mass increases in proportion to the required restraining force H. The dependence on the dimensionless parameter J is where the interest lies.

The required chain mass is of course greater for the uniform chain case, and the saving becomes large as $J \to 1$, see Fig. 4.14. The value of J, see (4.33), is proportional to the depth and otherwise is fixed largely by the density, the failure stress levels for the material of the chain, and the safety factors that are applied. For a material such as steel, failure results under the application of a steady load when the elastic limit is exceeded; under these circumstances the steel deforms plastically i.e. it stretches and obtains a permanent set. For steel, assuming this is the limiting feature and ignoring an additional safety factor, J evaluates to 2.5×10^{-4} per metre of depth. On this basis for a depth of 200 metres the proportional saving is small; however, the following additional factors need to be considered:

- Each link will be passing on to the next a force T but the maximum stress in the link will be in excess of this value. To compute this requires a detailed calculation of the behaviour of stress in each link; such calculations are certain to be available in the literature.

- Each link will have been welded and this will have weakened the material. The design must be based on the worst case since the chain is no stronger than its weakest link.

- Over the life of a chain the metal is likely to develop weakness as a result of metal fatigue, so an allowance must be made for this.

- In its role as a dynamic restraint for the ship, the chain will carry loads greater than the static design load.

- During the motion there is the possibility of impact loadings leading to clanking of the chain. Impacts can have a greater adverse effect than constant loads and so the allowable stress must be further reduced.

All these effects are not necessarily additive and many call for judgment based on practical experience. An insurance company may also call for substantial margins to cover doubt and so the safety factor applied to accommodate all the above is likely to be large. If a safety factor of 10 covering all such effects is assumed (not an uncommon figure in much safer circumstances), then for a depth of water of 200 m the saving of mass in using the variable area chain is substantial (about 35%). Since the uniform chain has an infinite mass for $J = 1$ the savings grow quite markedly with the value of J. It suffices to say that variable thickness chains have saved hundreds of thousands of dollars for mooring a single ship.
Aside: Some attempt is now being made to understand the failure mechanisms described above. Clearly this is an interesting and challenging experimental and theoretical area.

4.6 ⋆⋆The Dynamics of a Moored Ship

A complete discussion of the dynamics of a moored ship will not be undertaken. Our aim is to indicate how one can build up to quite complex

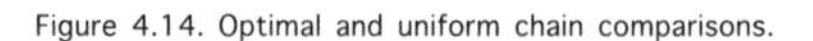
Figure 4.14. Optimal and uniform chain comparisons.

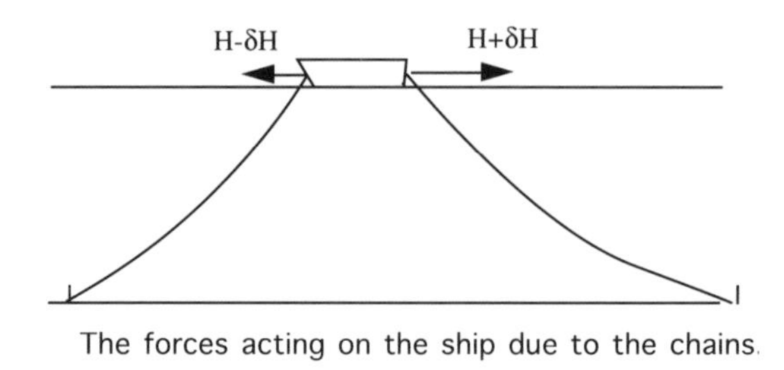

The forces acting on the ship due to the chains.

Figure 4.15: Ship dynamics.

problems from simpler component problems. We'll simply consider the translatory movement of a ship in one direction under a prescribed force in that direction when moored by two cables organized so as to act in the line of motion, see Fig. 4.15.

When the ship moves, because the cables are of fixed length, one will slacken and thus will exert less force on the ship, and the other will become more taut and thus exert more force. The changed situation will thus be one in which both cables will exert additional restoring forces on the ship. An alternative way to look at the problem is that in the disturbed state there is a net increase in potential energy of the cables, and thus of the whole system, so that the system is no longer in a viable static equilibrium state and will change in a way described by Hamilton's principle.

Now a cable only exerts a force on the ship via the point of attachment so if we could obtain a relationship connecting the movement of the attachment point to the force exerted by the cable, then we'd be in a position to write down the equation of the ship under the influence of all the cables and the other forces, and solve for the ship's motion. To carry out the required cable analysis we need to express mathematically the following physical requirements:

1. The ensuing cable's motion is described by Newtonian or Hamiltonian dynamics.

2. The chain remains of fixed length l.

3. The attachment point moves horizontally in a prescribed way.

4. The horizontal component H of chain tension will act on the ship.

Item 1 Now the movement of the attachment point will change the potential energy of the chain and increase its kinetic energy, and one would expect the relative importance of these effects to be as displayed in the Lagrangian of the cable (see (4.19)),

$$\int_0^l \rho A[(y_t^2 + x_t^2)/2 - gy]\, ds,$$

where $x(s,t), y(s,t)$ specifies the location of points (identified as before by the arc length parameter s) on the curve $\mathcal{C}$ defining the chain at time t. Now, since the ship is being restrained, we'd expect the scale of the movement of the attachment point (a say) to be small compared with l. If the time scale of the attachment point motion is τ then $\rho A g a$ provides an estimate for the potential energy change, and $\rho A(a/\tau)^2$ the kinetic energy change of the chain (per unit cable length) induced by the movement. We'd thus expect the ratio of these two terms (i.e. $a/[g\tau^2]$) to provide us with an estimate of the relative importance of kinetic energy effects to gravitation effects as far as the chain is concerned. Now the observed ship accelerations are very small compared with the gravitational acceleration g, so the kinetic energy term should contribute very little to the Lagrangian. The upshot is that the (minimum potential) static cable solutions obtained in the last Section can be strung together (in time) to generate an accurate *quasi-static* approximation to the dynamic cable problem of interest here. (Since $L(t) \approx -V(t)$ *for all* t, it follows that the Hamiltonian ($\int_0^t L(t)dt$) will be (almost) rendered stationary over *any* time interval by the minimum $V(t)$ solutions.) Arguments of this type can fail (as we saw when considering cable stretching effects in Section 4.4) but generally we'd expect success, and should proceed optimistically. After completing calculations based on these assumptions checks on consistency can be made.

The calculations are very much simplified by the above *quasi-static* approximation. Thus if we had an explicit expression for the horizontal component of the force exerted by the chain at its attachment point, as a function of the distance between the attachment point and the anchor under equilibrium conditions, then we could simply substitute this expression in the ship's equation of motion, and determine its motion.

Unfortunately the results of the previous Section are not immediately applicable, and it turns out to be easier (and more instructive) to work directly with the static equilibrium equations, rather than adapt those results. Let's therefore proceed with the items listed above. Our aim is to determine the changes brought about by a relatively small change in position of the ship's attachment point from location (h_0, d_0) to $(h_0 + \eta, d_0)$ say, where $\eta \ll l$ and h_0. The analysis to follow is a *perturbation* analysis similar to that introduced in Chapter 2. Variables with subscripts $_0$ will refer to their values before the ship is disturbed from its equilibrium location.

Item 2 As before (see (4.30)) we have

$$\sec^2\psi(s)\frac{d\psi(s)}{ds} = \frac{(\rho-\rho_w)gA_0(s)}{H}.$$

The present situation is one in which the cross-sectional area $A_0(s)$ is the *prescribed* optimum associated with (H_0, h_0, d_0) obtained in the previous Section. This value of $A_0(s)$ will *not* be also optimum for the new disturbed conditions, so the substitution $A_0 = H_0\sec\psi(s)/\alpha$ employed in the optimization calculation (see (4.31)) is not exact. It is, however, accurate to the order of subsequent calculations and, if this substitution is made, the above equation reduces to the more convenient (and equivalent) form

$$\sec\psi(s)\frac{d\psi(s)}{ds} = \beta\frac{H_0}{H}, \tag{4.38}$$

where

$$\beta = \frac{(\rho-\rho_w)g}{\alpha}.$$

Integrating we obtain

$$\int_{\psi(0)}^{\psi(l)} \sec\psi\, d\psi - \beta l\frac{H_0}{H} = 0,$$

which provides us with a mathematical expression of the fact that the chain remains of length l. This equation holds both for the initial equilibrium state with $(\psi(0),\psi(l)) = (U_0, V_0)$ (say) with $H = H_0$; and for the new state with $(\psi(0),\psi(l)) = (U_0+\delta U, V_0+\delta V)$ and $H = H_0+\delta H$. If we subtract the equations corresponding to these states and use Taylor's expansion, we obtain to first order the constant cable length requirement

$$\sec V_0\delta V - \sec U_0\delta U = -\beta\frac{\delta H}{H_0}.$$

Note that the values U_0 and V_0 are known from the initial equilibrium configuration calculations, so the above equation represents a constant coefficient linear algebraic equation relating the changes in attachment angles at both cable ends to the change in H that is consistent with the constant length chain requirement.

Item 3 From the geometric cable relationship

$$\frac{dy}{ds} \equiv \frac{dy}{d\psi}\frac{d\psi}{ds} = \sin\psi$$

for y, one obtains

$$\tan\psi\frac{d\psi}{dy} = \frac{H_0}{H}\beta,$$

after using (4.38). An integration over the y domain gives

$$\int_{\psi(0)}^{\psi(l)} \tan\psi \, d\psi = \beta d_0 \frac{H_0}{H},$$

and thus we obtain the approximate result

$$\tan V_0 \delta V - \tan U_0 \delta U = -\beta d_0 \frac{\delta H}{H_0}.$$

Similarly the equation for x gives

$$\delta V + \delta U = -\beta h_0 (\frac{\delta H}{H_0} - \eta).$$

Thus, in summary, the mooring attachment conditions have led us to the set of linear algebraic equations

$$\begin{pmatrix} \sec V_0 & -\sec U_0 & \beta l \\ \tan V_0 & -\tan U_0 & \beta d_0 \\ 1 & 1 & \beta h_0 \end{pmatrix} \begin{pmatrix} \delta V_0 \\ \delta U_0 \\ \delta H/H_0 \end{pmatrix} = \begin{pmatrix} 0 \\ 0 \\ \beta\eta \end{pmatrix},$$

which can be inverted to determine the changes brought about by a displacement η of the ship from its equilibrium location. In particular the relative change in the horizontal force exerted on the ship by the cable is given by $\delta H/H_0 = -K_0\eta$, where K_0 can be explicitly determined from the above equations, and depends only on the equilibrium cable configuration. Thus the movement of the ship induces a restoring cable force that's proportional to the ship's displacement η.

We are now in the position of being able to consider the derivation of the equation of motion of the ship. The complete details of this depend on using results for the effect of the motion of the water which is carried along with the ship. This can be included by adding an additional mass to that of the ship. It may also be necessary to model viscous water drag effects and drag due to wave making; these effects will dampen any motion that is set up (an expression of the form $D\dot{\eta}$ would normally be used). Thus the equation of motion for the ship is of the form

$$M\ddot{\eta} + D\dot{\eta} + 2K_0H_0\eta = F(t),$$

where $F(t)$ represents the disturbing force acting on the ship. The factor 2 in the chain term comes from the two chains, both of which contribute the unbalanced restoring force $K_0H_0\eta$. It can be seen from this equation that under the influence of a steady force F^0 (say) the ship will be displaced by an amount $F^0/(2K_0H_0)$. Also after such a force is removed the equation tells us that the ship's motion will be of the damped harmonic type with a time scale of order $\sqrt{M/(2H_0K_0)}$. Thus from the point of view of the dynamics, larger H_0's are better, so heavier (more expensive) chains are preferable, see (4.35). Obviously there is a decision concerning capital outlay vs. insurance expenses that must be made. Such questions are important, but not particularly interesting from a mathematical viewpoint; they're in the decision makers' area of interest.

This example serves to illustrate how one can build up the solution to a major problem in a series of relatively simple steps. The simple equation derived above includes the effect of a moving chain in the dynamics to a quite reasonable accuracy, and represents a useful starting point for further investigations. Having obtained a first estimate for the ship's behaviour, additional modifications can easily be achieved using energy techniques to derive the appropriate equations, and approximation techniques can then be used to obtain solutions to the desired accuracy.

4.7 Summary

The Newtonian and Variational approaches to understanding dynamical systems represent parallel and complementary streams that occur throughout physical science. In the chapters that follow both streams will be evident. For the simple situations discussed in this chapter the Newtonian approach has lead to a clearer understanding of the issues involved, and the variational approach was found to be more useful for handling complexity.

Optimization problems are of interest in their own right throughout engineering and economics.

The ideas of approximation have been developed further in this chapter.

4.8 Exercises

Exercise 4.1 *Vertical Hanging Chain*

A manufacturer wishes to determine the minimum mass chain of prescribed length l_0 to suspend safely a large load. The chain is to be manufactured from a material of density ρ. The sectional area A may vary along the chain; in fact the aim is to determine the sectional area variation that minimizes the mass. The chain hangs vertically; the arrangement is as indicated in Figure 4.16. The maximum stress the chain material can sustain without plastically deforming is known (α say) so that $T(y)/A(y) \leq \alpha$. Assuming the chain doesn't stretch significantly, write down the static equilibrium conditions for the length of chain lying between y and l_0, and thus determine the $A(y)$ that minimizes the chain mass. Comment on features of the solution that you think are significant.

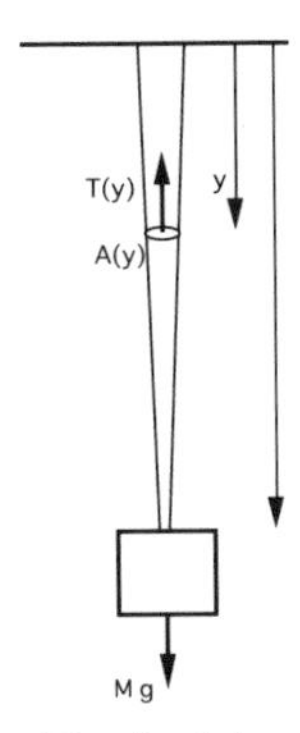

Figure 4.16: Vertical hanging chain.

Exercise 4.2 *The Brachistochrone Problem*

Using Maple complete the analysis of the brachistochrone problem of Section 4.3.1, and draw solution graphs for a range of b/d values. Make any observations you think appropriate about the path shape or the dependence of the travel time on (b, d). Especially observe the fixed d, large b *asymptotic* behaviour.
Comment: It should be noted that algebraic packages are often unreliable when faced with a choice of possible solutions (for example when roots or other multivalued functions are involved), so one should always proceed with care and carefully check to see that the algebraic package chooses the correct branch in context.

Exercise 4.3 *Brachistochrone Extensions*

The problems to follow are close relatives of the brachistochrone problem of Section 4.3.1.

(a) Consider first the situation depicted in Fig. 4.4 in which one end A of the required minimum time path is located at a fixed point ($(0,0)$ say), but the other end B can be located *anywhere* along the vertical line $x = b$ (at $y = \ell$ (say)). In this case ℓ, as well as the path, has to be determined.

  (i) It's as well to first check to see if the problem is likely to have a well behaved solution. By choosing simple explicit paths show that both shallow (i.e. small ℓ paths) and deep paths, have large transit times; so that an interior minimum is to be expected.

  (ii) $\star$ This problem is of the free boundary conditions type discussed in Section 3.2. Working from first principles (start from equation (4.5)) show that the Euler-Lagrange equation for the optimal solution remains unaltered but that the boundary conditions required to achieve a zero first order variation are

$$y(0) = 0, \ y'(\ell) = 0.$$

Do these boundary conditions make physical sense? Explain.

(iii) $\star$ Use Maple to determine the solution of the resulting equations, and plot the results.

Aside: Note that an alternative (much less efficient) procedure for obtaining the solution is to examine brachistochrone solutions corresponding to fixed b and variable d values. The required solution is of course the one that has the least associated transit time.

(b) $\star\star$ Consider now the situation in which the end points A and B are fixed, but the allowed paths are restricted to be not too steep (so that $|y'(x)| < \alpha$ (say)). This problem derives from the problem of designing a safe but exciting slippery slide.

A skier would offer the solution: "Accelerate immediately as much as possible and then coast home." Based on this intuitive guess, construct a solution (use Maple) and show that this does in fact lead to the desired minimum time path, at least in some cases.

Hint: You'll need to consider the cases $d/b \leq \alpha$ and $d/b > \alpha$ separately.

Comment: This situation is typical in that often the Euler-Lagrange equations lead to results that are incompatible with the prescribed boundary conditions or constraints. In such cases the optimal solution is often obtained by patching together Euler-Lagrange equation curves, and curves arising out of the additional constraints. It's also often the case that an understanding of the physics of the situation helps greatly.

Exercise 4.4 *Shortest Distance*

The shortest distance between two points is of course a straight line, and it's easy to show this without using the calculus of variations. It's instructive, however, to also see how this result can be obtained from variational principles.

(a) Show directly from the integral representing the length of an arbitrary curve that the straight line joining the two points $(0, 0)$ and (a, b) in the plane minimizes the curve length.

(b) By solving the appropriate Euler-Lagrange equation determine the shortest length curve.

(c) A sensible question to ask is: of all curves $Y(x)$ passing through $(0,0)$ and (a,b) and with the additional requirement that these curves have zero slope at $x = a$ (so $Y'(a) = 0$) which is of minimum length? Determine the Euler-Lagrange equation for this problem and show that, except under very special circumstances, there is no solution to this equation satisfying the required boundary conditions. From the defining equations we can see why difficulties arise; the Euler-Lagrange equation is of second order and there are three boundary conditions to be satisfied. Although this is an interesting observation it doesn't indicate the nature of the difficulty, or how to overcome it.

(i) The nature of the difficulty is best understood by examining the geometry directly. It can be shown that, although the length of an acceptable curve can be made as close to straight line value $\sqrt{a^2+b^2}$ as one wishes, this limit length cannot be reached. To show this result examine the kinked curve consisting of straight lines joining $(0,0)$ to $(a-\epsilon, b)$, and $(a-\epsilon, b)$ to (a,b).

(ii) Even if one insists that acceptable solutions must be smooth (eg. have continuous second order derivatives) it's still possible to determine solutions with lengths arbitrarily close to $\sqrt{a^2+b^2}$. This is evident geometrically. To demonstrate its truth analytically consider acceptable functions of the form $(b/a)x + g(a-x, \epsilon)$, and choose the function g so that it's small except within ϵ of a and contributes at most a term of order ϵ to the curve length. *Hint:* Consider exponential functions g of $(a-x)/\epsilon$.

Comment: If the curve sought above corresponds to a wire that is to be stretched between A and B, but needs to enter a horizontal housing at B , then bending stresses would be introduced because of the kink near $x = a$, and these would play a role in determining the physically achievable solution. A modified formulation taking into account such additional constraints would then be necessary to deal with the situation. (For example in this case excessive wire bending may be unacceptable, so admissible curves may be required to have less than a prescribed curvature.) It's in fact often the case that mathematical difficulties encountered in variational problems arising out of real world situations arise because of modelling inadequacies.

Exercise 4.5 *The Compound Pendulum*

⋆⋆ We wish to use Maple to determine the Euler-Lagrange equations for the compound pendulum example considered in Section 4.4.1. In such a complicated algebraic situation it's usually sensible to examine first a simpler related problem for which the answers are known (to provide a check on the code). The pendulum with just one rod is the obvious choice.

(a) Write down the Lagrangian for the pendulum consisting of just one rod. Use Maple to perform the calculations to determine the Euler-Lagrange equations for this situation. Determine the natural frequency for the pendulum, and plot out a few solutions.

(b) ⋆⋆⋆ Extend the above calculations to the compound pendulum case.

Exercise 4.6 *Elastic String Vibrations*

An elastic string (of unstretched length l_0 and mass per unit length m) is stretched between two supports distance $l > l_0$ apart. The tension in the string in this configuration is T_0. The string is set into vibration by displacing it laterally from this equilibrium and then releasing it. Let $u(x,t), 0 \leq x \leq l$, denote the lateral displacement of the string at time t, see Figure 4.17. We'll use the Hamilton's principle to determine the equations governing the string's vibration.

(a) Show that the total change in length of the string as a result of (i) the initial stretching and (ii) the additional lateral displacement $u(x,t)$ at time t, is given by

$$\int_0^l \sqrt{1+u_x^2}\; dx - l_0.$$

(b) Using Hooke's Law (4.16) and the expression (4.18) for the elastic energy stored in the string, show that for relatively small displacements $u(x,t)$, compared with $(l-l_0)$, the increase in elastic energy in the string at time t over that in its equilibrium situation is given approximately by (use Taylor's expansion)

$$\frac{1}{2}T_0 \int_0^l u_x^2 dx.$$

(c) Thus show that the Hamiltonian integral for the string is given by

$$J = \int_0^t \int_0^l [\frac{1}{2}mu_t^2 - \frac{1}{2}T_0 u_x^2] dx dt.$$

(d) Using the Euler-Lagrange equation for such two dimensional problems, see (4.15), show that the motion is governed by the *string equation*

$$mu_{tt} - T_0 u_{xx} = 0.$$

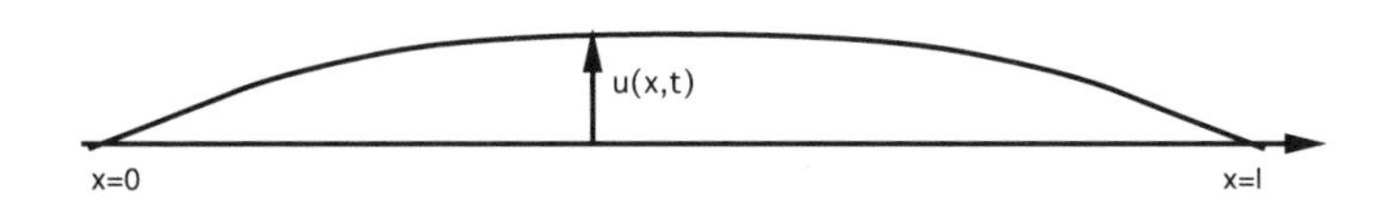

Figure 4.17: String vibrations.

(e) Use scaling arguments to determine the time scale of the motion.

Exercise 4.7 *The Flagpole Revisited*

⋆⋆ The elastic energy stored in a beam of length l as a result of bending is given by

$$\frac{1}{2}\int_0^l EIu_{xx}^2 dx,$$

where $u(x,t)$ is the lateral displacement of the beam, E is the Young's modulus, and I the second moment of area of its cross-section (see Exercise 4.8) .

(a) Determine the Lagrangian for the beam.

(b) For the flagpole being shaken by earthquake waves considered in Chapter 2, the appropriate boundary and initial conditions are:

$$u(0,t) = a\sin\omega t, \quad \frac{\partial u}{\partial x}(0,t) = 0,$$
$$\frac{\partial^2 u}{\partial x^2}(l,t) = 0, \quad \frac{\partial^3 u}{\partial x^3}(l,t) = 0$$
$$u(x,0) = 0 \quad \frac{\partial u}{\partial t}(x,0) = 0.$$

Working from first principles show that the first order variation of the pole's Hamiltonian vanishes if the Euler-Lagrange equation given by

$$\rho\frac{\partial^2 u}{\partial t^2} + EI\frac{\partial^4 u}{\partial x^4} = 0 \tag{4.39}$$

is satisfied, and the above boundary and initial conditions are imposed.

(c) ⋆⋆ Determine the *natural* boundary conditions for the beam problem. Recall that the *natural* boundary conditions for the variational problem are the conditions that need to be imposed to ensure a zero first order variation if there are no imposed constraints.

Exercise 4.8 *Maximum Strength Beams*

Beams used for construction are most likely to fail by buckling under applied longitudinal load or by bending excessively under a load at right angles to the beam. It pays to use the minimum weight beam of given length that can withstand a prescribed bending or buckling load, and significant savings can be realised if the shape of the beam's section is appropriately chosen. Our aim is to determine the best sectional shape for all such beams.

The equation governing the transverse deflection u of a beam under the action of a force per unit length of f (including the beam's weight) is given by

$$EI\frac{d^4u}{dx^4} = f,$$

and, when supplemented with appropriate boundary conditions, determines the deflection of the beam. Here E is the Young's modulus of the material of the beam and

$$I = \int_A z^2 dA,$$

with y measured from the centroid of the beam's section, is called the second moment of the sectional area of the beam about the y axis, see Fig. 4.18; $dA = dzY(z)$ is the sectional area element corresponding to z. $Y(z)$ specifies the shape of the section. It suffices for our purposes to note from the deflection equation that the minimum deflection for a beam with prescribed sectional area $\int_A dA$ and length will be realized by the beam with greatest second moment of area I. To determine the shape required to achieve this optimal we thus need to consider the constrained variational problem:

Find $Y(z)$ such that the integral

$$I = \int_A z^2 Y(z) dz$$

is minimized with the integral $\int_A Y(z)dz$ fixed.

(a) Using Lagrange multipliers write down the appropriate Euler-Lagrange equation and show that this leads to a vacuous result. This suggests that the solution is likely to have non-smooth properties. (The Euler-Lagrange equation derivation fails in such cases.) Usually this means that we should look for a solution with an "exaggerated" behaviour.

(b) Show that I is increased by having as much area as possible as far as possible from the y axis. Why are eye beams and hollow structures (eg. pipe flagpoles) used by builders? Note that biological structures that are required to withstand large bending stresses are also often hollow (eg. bamboo).

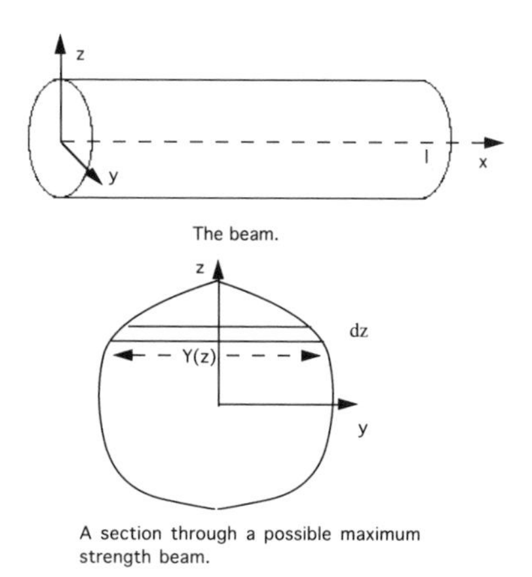

Figure 4.18: The eye beam problem.

Exercise 4.9 *Large Area Figures*

Using the following arguments show that the plane figure of maximum area enclosed by a curve of prescribed length l_0 is the circle.

(a) Show that the greatest area will be achieved by a curve that's symmetric about a line joining any two points distant $l_0/2$ apart along the curve.

(b) If this symmetric figure is not convex, show (by a simple construction) that the area can be increased without increasing the curve length. Under such circumstances it's possible to define a single valued function $y(x)$ for the boundary on one side of the line of symmetry.

(c) Using the Euler-Lagrange equations or otherwise show that the required figure is a circle.

Comment: Geometric arguments are often valuable for producing solutions of variational problems.

Exercise 4.10 *Minimum Length Path on a Surface*

It is desired to find the paths of shortest length between two points on the surface of the ellipsoid

$$\frac{x^2}{a^2} + \frac{y^2}{b^2} + \frac{z^2}{c^2} = 1.$$

(a) Verify that the parameters θ and ϕ defined by

$$x = a\cos\theta, \quad y = b\sin\theta\cos\phi \quad \text{and} \quad z = c\sin\theta\sin\phi$$

serve to specify the location of points on the surface.

(b) Assuming that the path of least distance along the surface is given by $\phi = \Phi(\theta)$, determine an ordinary differential equation for $\Phi(\theta)$. Use Maple to calculate the arc length along this curve. Examine special cases to check your solution.

Exercise 4.11 *Orienteering anyone?*

The objective is to come up with a procedure for determining the shortest time path for a runner going from a location A to a location B in circumstances in which the terrain (and thus the speed of movement ($v(x, y)$ (say)) varies with position. Obviously this problem is of relevance in many (more important) contexts.

Comment: This problem is typical of the type of "open ended" problem that is often presented to modellers. What is required here is a reasonably *robust* and quick computational procedure for determining a useful approximation of the best path for v's that are likely to be met in practical circumstances. (It's almost always possible to defeat any computational procedure by carefully selecting awkward examples; our aim isn't to search for such curiosities.)

(a) Choose the origin at A and let the terminus B be at (c, d). Show that the functional to be minimized is

$$\mathcal{T} = \int_0^c \frac{\sqrt{1 + y'^2}}{v(x, y)} dx,$$

where $\mathcal{T}$ is the time taken if $y(x)$ is the adopted path.

(b) Write down the Euler-Lagrange equation and boundary conditions for the problem.

The equation is a nonlinear horrible mess! Clearly a computational procedure will eventually be required.

(c) One might be tempted at this stage to throw the problem at a computer, but it's as well to think carefully about what the solution is likely to look like, and so it's useful to examine some special $v(x, y)$ cases for which exact analysis is possible. Such cases will also provide checks on any computational solution procedures that may be later set up. Examine a number of such cases.

Suggestions: Try $v(x, y) = const$, $v(x, y) = const$ in patches, $v(x, y) = Y(y) \cdot X(x)$. You should ask:

- Is there always a unique solution?
- Are there commonly occurring situations that need to be specially handled?

Hint: To obtain solutions, use physical intuition as well as mathematics, and recall that one often needs to patch together solutions of the Euler-Lagrange equation to obtain the minimizer.

(d) ⋆ To obtain a numerical solution one can either solve the Euler-Lagrange differential equation numerically or use the variational formulation directly. *Usually the variational formulation is to be preferred for numerical work, and the differential equation approach is best for any analytic work.*

A simple and very effective numerical technique is to determine the particular member of a narrow class of admissible functions that minimizes the defining functional. In our present case the quadratics

$$Y(x) = \frac{d}{c}x + \gamma x(x - c),$$

pass through the required points for any value of γ. Thus we can substitute this expression directly into the defining functional and determine the value of γ that minimizes it. *Providing these quadratics are of the right general shape* for the $v(x, y)$'s of interest one might expect the outcome to provide a good estimate for the minimum path. One could improve the approximation by using higher order polynomials. Experiment with this idea using Maple (use `int, fsolve`), making use of the known exact solutions you've obtained earlier to provide a check. *This procedure is especially useful when used in conjunction with an algebraic package*, basically because it's a trivial matter to improve the approximation (by extending the class of admissible functions), until the desired accuracy is obtained.

(e) ⋆ The speed of travel on hilly terrain is unlikely to be handled by the above model. Why? How would you modify the above model to deal with such situations?

Exercise 4.12 *Elastic Cables*

In Section 4.4, we considered hanging cables that stretched very little under the influence of the driving gravitational force. Situations involving easily stretched elastic cables are also of interest. Set up the Maple code of the text and explore such situations. For example, for unit order ϵ values, see what happens as the distance between the cable suspension points increases.

Hint: One might expect the character of the solution to change when the spacing between the supports exceeds the unstretched length of the cable.

Exercise 4.13 *The Uniform Mooring Cable*

Determine the minimum cost uniform sectional area mooring cable. The calculations are very similar to those already carried out for the optimal sectional area mooring cable in Section 4.5.

Exercise 4.14 *Moorings analogues*

Often to get a feel for a problem it's useful to examine simpler analogous situations. The problem of determining the dynamics of the mother ship moored by a cable addressed in Secion 4.6 is sufficiently complex to warrent the examination of an analogue. Two simple analogues are represented in Fig.4.19: (Perhaps the reader can think of others.)

- In the first analogue a trolley, subject to external forces, is stabilized by horizontal stretched elastic springs attached to fixed supports.
- In the second a trolley is stabilized by weights attached to cords as shown.

The questions of interest to us are:

1. Under the application of a steady force the trolley will be permanently displaced from its equilibrium position. By how much?
2. If, because of the application of an external force no longer present, the trolley is displaced from its equilibrium situation, how long will it take to return to the equilibrium location? In particular we'd like to know the dependence of the time scale on the appropriate spring and mass parameters specifying these situations. Note that the trolley will not remain at the static quilibrium location but will oscillate about this location until dissipative mechanisms remove the energy introduced by the external force.

(a) Using either Newtonian or variational methods determine the equation of motion of the trolley (assume frictionless wheels) for either of the above arrangements, and examine the questions listed above. Discuss the results i.e. determine time scales etc., and make comparisons with the results obtained in the text.

(b) ⋆ By avoiding the difficulties of variable mass distribution along the chain and shape changes of the curve describing the moorings in the water, the above moorings analogues lead to a simple analysis, without changing major features of the problem. One would expect only minor complications to arise if the mass/cord arrangement in the second analogue were replaced by a symmetrical, uniform chain arrangement. Using the results obtained in the text for the hanging cable, analyse this situation and make appropriate comparisons.

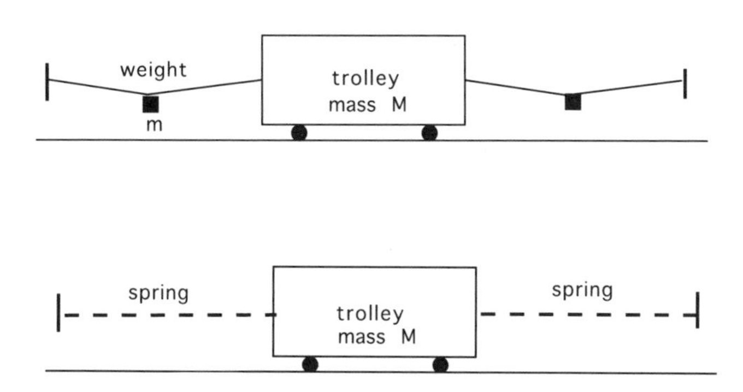

Figure 4.19: Moorings analogues.

It should be noted that the above simple analogues represent not only devices introduced to get to grips with the moorings problem, but also represent an ideal framework for presenting the results associated with the real problem. Referring to the second analogue, the aim of the complete moorings analysis should be to determine (if possible) an expression for the *effective* (m, M) values as a function of the moorings parameters; the results of the text have gone a long way towards doing just this.

Bibliography

Akhiezer, N.I. (1962). *The Calculus of Variations,* Blaisdell Publishing Company, New York.

Courant, R. and Hilbert, D.(1953). *Methods of Mathematical Physics. Vol 1*, New York:Interscience Publishers. a division of John Wiley & Sons, Inc.

Fowles, Grant R. (1970). *Analytical Mechanics,* Holt Rinehart and Winston Inc., New York.

Goldstein, H. (1950). *Classical Mechanics,* Addison-Wesley, Reading, Mass.

Strang, Gilbert (1986). *Introduction to Applied Mathematics.* p720, Wellesle Cambridge Press, Mass.

Chapter 5

The table fable revisited

Energy techniques
Scaling
Applied linear algebra
Organizing numerical calculations

5.1 Introduction

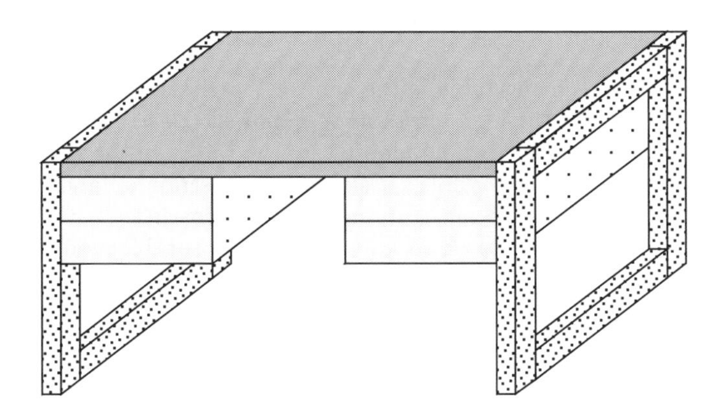

Figure 5.1: A real table.

We return to the table fable. You'll recall that in our earlier work we highlighted the need for an alternative approach to handle the complexities arising when dealing with manufacturing imperfections. The energy minimization methods discussed in the last chapter will now be brought to bear

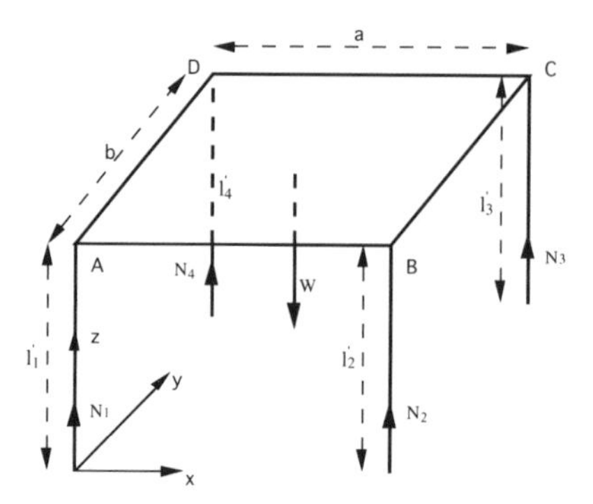

Figure 5.2: The simple table model.

on this problem, in Section 5.2.

A major deficiency of the models discussed to this stage has been the assumption that the table top is rigid. In order to see how to handle a bending table top we'll first examine a closely related (but simpler) problem involving a beam supported on three posts in Section 5.3. We'll then return to the table problem in Section 5.4.

5.2 Energy Methods

The energy stored in the table is the sum of the elastic and gravitational potential energies in the legs and table top, and the elastic energies stored in the joints. The configuration adopted by the table will be the one that minimizes the total energy while satisfying the constraints imposed by the geometry and loading. To illustrate how this approach leads to solutions for the table problem we'll first re-examine the simple table model, see Fig. 5.2.

The only energy stored in the *rigid* top is gravitational potential energy given by the product of its weight and the height l'_{cm} of its centre of mass above the floor. The weightless legs of the model only store elastic energy given by $N_i^2/(2k)$ (see (4.17)) where, using the notation adopted in Chapter 3, N_i is the load in the ith leg of length l'_i with the table on the floor, see Fig. 5.2. The energy function, $\mathcal{E}(N_i)$, which has to be minimized as a function of N_i is thus

$$\mathcal{E} = \sum_1^4 N_i^2/(2k) + W l'_{cm}(N_i).$$

If all legs make floor contact then

$$l'_{cm}(N_i) = \frac{1}{4}\sum_1^4 l'_i. \tag{5.1}$$

The leg length l'_i is related to the unstressed leg length l_i and N_i via. Hooke's Law

$$l'_i = l_i - N_i/k. \tag{5.2}$$

The constraints which have to be met are:

- the reactions must be non-negative, i.e.

$$N_i \geq 0, \quad \text{for } i = 1\ldots 4,.$$

- the table top must be planar, which has been shown previously (3.12) to require that

$$G = (l'_1 + l'_3) - (l'_2 + l'_4) \geq 0, \tag{5.3}$$

so that in terms of N_i this condition becomes

$$G = [(l_1 + l_3) - (l_2 + l_4)] - [(N_1 + N_3) - (N_2 + N_4)]/k \geq 0. \tag{5.4}$$

In this form we have what is known as a *quadratic programming* problem which can be attacked by straightforward numerical means, using standard packages if particular values of the parameters are substituted into the equations. As indicated in our earlier table work the numerical results obtained in this way would be virtually uninterpretable, so it's essential to process the system further and if possible obtain understandable analytic results.

It's the inequalities in the above equations that make the situation analytically awkward; basically because we can't tell which of the inequalities are *active* in determining the solution. If we choose to follow the path used in our earlier work of assuming that all four legs are in contact with the floor, the inequality constraint conditions no longer have significance, and the problem reduces to the much simpler one of minimizing $\mathcal{E}(N_i)$ subject to the equality constraint $G(N_i) = 0$. This problem is best attacked using Lagrange multipliers, see Chapter 4 Section 4.2. Thus we introduce a parameter λ and look for values of the loads N_i which minimize $\mathcal{E} - \lambda G$. The minimization with respect to N_i leads to the set of linear equations

$$N_i/k - W/(4k) \pm \lambda = 0, \quad \text{for } \; i = \ldots 4,$$

where the plus sign applies for $i = 1, 3$ and the minus sign otherwise. Thus

$$N_i = W/4 \mp k\Delta,$$

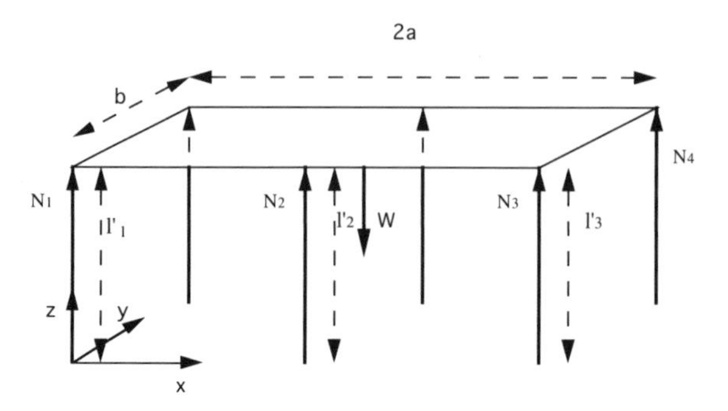

Figure 5.3: The table with six legs.

and the substitution of this solution into the constraint condition $G = 0$ leads to

$$\lambda = -\Delta/4, \text{ so } N_i = W/4 \mp k\lambda \tag{5.5}$$

where

$$\Delta = (l_1 + l_3) - (l_2 + l_4).$$

Thus we've been able to extract the same analytic result using energy arguments as we obtained earlier using Newton's Laws, see (3.15,3.16). The great advantage of the energy method is that it's a relatively simple matter to extend the approach to complicated situations; all that's required is a knowledge of the internal energy associated with the various components. To see this we'll very briefly examine a situation with rather more complex geometry; namely that of a six legged table, again with a rigid top. This work can be skipped without losing continuity.

⋆ The six legged table

The geometric arrangement is as displayed in Fig. 5.3; the rectangular top, with sides of length $(2a, b)$ is rigid with line legs on all four corners and in the middle of the sides of length 2a. Again we first consider the case in which all legs are in contact with the floor. The legs will be indexed by an integer i which takes the value 1 for the leg situated $(0,0)$ and with the index increasing by one as the legs are encountered anticlockwise starting with that at $(a, 0)$. The elastic energy $\mathcal{E}_e$ is obviously

$$\mathcal{E}_e = \sum_{i=1}^{6} N_i^2/(2k),$$

while the potential energy V measured relative to the floor is given by

$$V = \frac{W}{6}\sum_{i=1}^{6} l'_i, \equiv \frac{W}{4}(l'_1 + l'_3 + l'_4 + l'_6) \tag{5.6}$$

where (as before) $l'_i = l_i - N_i/k$ are the lengths of the legs with the table on the floor.

The constraint conditions again result from the fact that the tops of the six legs are coplanar, and hence the associated points must satisfy the equation of a plane. The coefficient matrix generated in this way is

$$\begin{pmatrix} 0 & 0 & l'_1 & 1 \\ a & 0 & l'_2 & 1 \\ 2a & 0 & l'_3 & 1 \\ 2a & b & l'_4 & 1 \\ a & b & l'_5 & 1 \\ 0 & b & l'_6 & 1 \end{pmatrix}.$$

The formation of this matrix should become clear if one examines the previous derivation of the 4×4 matrix for the four legged table, see equation (3.11). In order that the corresponding six homogeneous equations should admit of non-trivial solutions for the four unknowns (the coefficients of the equation of the plane) one expects three conditions to be met. These consistency equations may be generated by performing a Gaussian reduction on the columns not containing the l'_i's. The manner of performing this is not unique but one set of equations is as follows:

$$G_1 \stackrel{\text{def}}{=} l'_1 + l'_5 - l'_2 - l'_6 = 0,$$

$$G_2 \stackrel{\text{def}}{=} l'_1 + l'_4 - l'_3 - l'_6 = 0,$$

$$G_3 \stackrel{\text{def}}{=} l'_1 + l'_3 - 2l'_2 = 0, \tag{5.7}$$

which evidently are generalizations of the constraint (3.12) obtained in the four legged table case.

After eliminating the l'_i's in favour of the N_i's and the l_i's using Hooke's Law (5.2), the problem is thus reduced to the determination of the six unknown forces N_i's and three unknown Lagrange multipliers λ_i, such that

$$H = \mathcal{E}_e + V - \sum_1^3 \lambda_i G_i$$

is a minimum with respect to the N_i and λ_i. The nine equations in N_i, λ_i thus generated are sufficient to determine the nine unknowns. It would be a great advantage for interpretation and further processing to have an analytic solution in a useful form and it's clear one exists. It's also clearly a job for an algebraic package; not a human! The appropriate mathematical setting is matrix algebra and it is an interesting application of linear algebra ideas and Maple to obtain a solution *in a useful form*. This work, described in the Appendix, gives

$$N_i/W = 1/6 + (k/W)M \sum_{j=1}^{3} [\mathbf{l} \cdot \mathbf{e}_j] \mathbf{e}_j(i), \tag{5.8}$$

where $\mathbf{e}_i, i = 1,2,3$ is an appropriate basis for the problem, and M is a 6×6 matrix; both are defined in the Appendix, see (5.37,5.36). Also $\mathbf{l}$ is the vector made up of components l_i and $\mathbf{e}_j(i)$ denotes the ith component of the jth basis vector. All legs make floor contact if $N_i > 0$ so it's possible, using this result, to establish conditions under which the table will be well supported.

The specifics of the above solution are not important for our purposes, but it's interesting to compare the results with those obtained for the four legged table, see (5.5). Note that unfortunately there is not a simple *stability parameter,* see (3.18), defining the problem in this case—basically because there are many possible rocking modes. Mathematically this result is pretty, because it very neatly isolates those components of $\mathbf{l}$ that can give rise to rocking from those that can't (see the Appendix). However, it's not a result that the average manufacturer would find interesting or useful. If one assumes that the leg lengths are described by a random distribution and that they are randomly placed around the table, then one can use (5.8) to determine the probability of rocking as a function of k/W and the standard deviation σ of the leg lengths. Thus one can obtain the required *stability parameter*

$$r = k\Delta^\star/W \text{ where } \Delta^\star = \Delta^\star(\sigma)$$

for the six legged table. To do this one simply needs to average over the sample space, as we did in Section 3.3.2. This would provide information that the manufacturer may get excited about.

Comment: Our main purpose here has been to observe the use of the energy procedure in a reasonably complex situation. The feature you should notice is that there was no real difficulty in including two additional legs. Other factors could be dealt with in the same way. Flexible joints, for example, (that we've seen are really difficult to handle with a Newtonian approach) could be trivially dealt with in this way. The internal energy stored in a joint (for the small angular displacements of interest) is quadratically dependent on the small angular displacement (the Hooke's Law analysis of Chapter 4 , see Section 4.4, is readily extended to cover this case), so the expression for $\mathcal{E}_e$ which is required to include joint effects is

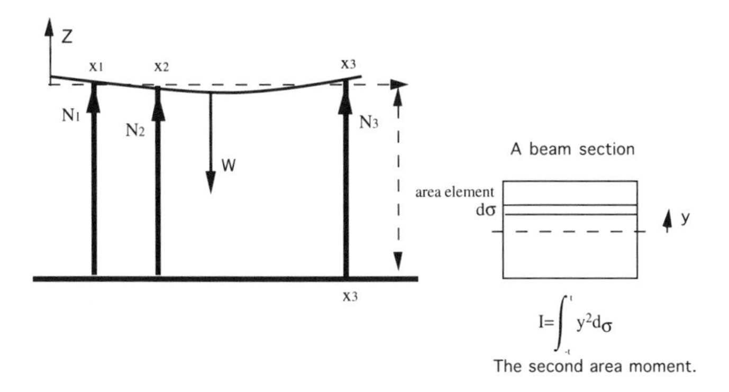

Figure 5.4: The supported beam.

the same one as above with the summation index ranging over more values, and with the l_i''s being either angles or displacements. A larger set of linear equations needs to be solved and, as indicated above, Maple is effective and *essential* for producing an output in a *usable* form. In fact the code described in the Appendix requires little adaptation to handle this case.

One feature of the problem cannot be so easily handled; the table top. We now turn our attention to this problem.

5.3 The Supported Beam

Under the influence of the forces due to the four legs and its own weight the table top bends into a new shape. The legs of course also compress, so the problem is to determine the shape of the top and the change in length of the legs that results in a minimum energy arrangement. As we've seen in the moorings Chapter it's a non-trivial matter to determine a minimizing *shape* (i.e. a complete function), so this is why we can't immediately extend the above work. The one-dimensional analogue of this problem is the problem of determining the deflection of a supported beam, see Fig. 5.4. This is obviously a sensible starting point for the investigation.

A beam rests upon three vertical posts of almost the same length. The objective is to calculate the loads N_1, N_2, and N_3 borne by the posts. There are three unknown forces acting on the system as a whole and only two non-trivial equilibrium conditions (no vertical translational motion and no horizontal rotation) so, as in the rigid table situation, statics considerations alone don't determine the loads carried by the posts. All four elements of the structure deform slightly under the contact and gravitational loads they're subjected to and the experience we've gained suggests that such considerations determine the situation as a whole, and in particular determine the post loads of interest. Let the beam be of weight W, length L. We'll assume the posts are of negligible weight and are at distances x_1, x_2 and x_3 from the left hand end of the beam. Our first task is to determine the shape of the beam under arbitrary post loading.

5.3.1 The beam

The vertical displacement, $Z(x)$, of the bottom of a beam subject to a distributed vertically upwards load per unit length of $P(x)$ is accurately given by

$$E_b I_b \, \mathbf{D}^4 Z = P(x), \tag{5.9}$$

for small displacements, see Segel (1977), where $\mathbf{D}$ denotes the operation of differentiation with respect to x, E_b is the Young's Modulus of the material of the beam, and I_b is the second moment of area of the beam about the centroid of the cross-section i.e.

$$I_b = \int_A y^2 dA,$$

see Fig. 5.4. Thus I_b is strongly dependent on the beam width (varies roughly in proportion to width cubed).

We'll measure x from the left end of the beam, and it's convenient for later purposes to measure the beam displacement $Z(x)$ from a datum of height l_0 above the horizontal floor, where l_0 is the average unstressed post length, see Fig. 5.4. ($Z(x)$ is thus negative for most x). Assuming the ends of the beam are "free" (i.e. unsupported) the boundary conditions

$$Z''(0) = Z''(L) = 0, \quad \text{and} \quad Z'''(0) = Z'''(L) = 0,$$

need to be satisfied. The beam is subjected to load per unit length $-W/L$ due to its own weight, and also of course there are the forces due to the posts acting on it. Now if N_i is the total force exerted on the beam by the ith post (of thickness $2\epsilon \ll L$) then the force per unit length due to beam i will be of the form

$$N_i f_i(x - x_i, \epsilon)$$

where the functions f_i are such that

$$f_i(x - x_i, \epsilon) = 0 \ \text{ for } |x - x_i| > \epsilon$$

and

$$\int_{x_i-\epsilon}^{x_i+\epsilon} f_i(x - x_i, \epsilon) dx = 1;$$

so we can write the equation for the deflection of the beam in the form

$$E_b I_b \, \mathbf{D}^4 Z = -W/L + \sum_{i=1}^{3} N_i f_i(x - x_i, \epsilon) \ \text{ for } \ 0 < x < L. \tag{5.10}$$

One might expect the beam displacement to be insensitive to the shape of the functions f_i except in the contact zones; we'll see that this is the case. Under such circumstances it's sensible (and usual) to introduce a notation that avoids entirely the introduction of such shape functions. Using this notation the above equation is written

$$EI\,\mathbf{D}^4 Z = -W/L + \sum_{i=1}^{3} N_i \delta(x - x_i) \quad \text{for} \quad 0 < x < L, \tag{5.11}$$

where $\delta(x)$ is referred to as the **Dirac delta function** and is defined in a formal sense by the rules

$$\delta(x) = 0 \text{ for } x \neq 0$$

and

$$\int_{-\epsilon}^{\epsilon} \delta(x)dx = 1,$$

and is not defined at $x = 0$; no definition is required. The notation makes sense providing the sums work; and (with care) they do in many circumstances. Such *generalised functions* prove to be generally useful. Here we'll just show the notation is convenient and produces correct results for the problem being considered. Generalised functions will be discussed in greater detail in Part II (Section 7.7).

Scaling the problem

It is apparent that the loads on the posts depend upon the weight of the beam and this suggests that a natural scale for the forces N_i is W. Thus we introduce scaled variables, R_i, defined by

$$R_i = N_i/W.$$

The length of the beam defines a natural length scale for the x co-ordinate, so that

$$\xi = x/L, \quad \text{with} \quad \xi_i = x_i/L,$$

is a sensible scaling. The scale of the vertical displacement is not so obvious so we'll select it by introducing a new variable, Y, defined by

$$Y = Z/c,$$

where c is some unknown constant which contains the units of length. If the differential equation is now stated in terms of these new variables it becomes, after a trivial rearrangement,

$$D^4 Y = \frac{WL^3}{E_b I_b c}\left[-1 + \sum_{i=1}^{3} R_i f_i(\xi - \xi_i)\right] \quad \text{for } 0 < \xi < 1, \tag{5.12}$$

where D now denotes differentiation with respect to ξ. The choice of c given by

$$c = \frac{WL^3}{E_b I_b}, \tag{5.13}$$

reduces the equation to its simplest non-dimensional form

$$D^4 Y = \left[-1 + \sum_{i=1}^{3} R_i f_i(\xi - \xi_i, \bar{\epsilon}) \right] \quad \text{for} \quad 0 < \xi < 1, \tag{5.14}$$

where $\bar{\epsilon} = \epsilon / L$.

Our expectation is as usual that this choice determines the appropriate scale for the deflection, and so provides a rough estimate of the actual deflection.

In terms of the scaled variables the conditions at the ends of the beam become

$$Y''(0) = Y'''(0) = Y''(1) = Y'''(1) = 0, \tag{5.15}$$

where the primes now denote differentiation with respect to ξ.

The solution process

If equation (5.14) is integrated with respect to ξ from 0 to ξ and use made of the value of $Y'''(0)$ the result obtained is

$$Y'''(\xi) = -\xi + \int_0^{\xi} \sum_{i=1}^{3} R_i f_i(\xi' - \xi_i, \bar{\epsilon}) d\xi'. \tag{5.16}$$

Now for regions outside the post contact zones we have

$$\int_0^{\xi} \sum_{i=1}^{3} R_i f_i d\xi' = \begin{cases} 0 & \text{if } \ 0 < \xi < \xi_1 - \bar{\epsilon} \\ R_1 & \text{if } \ \xi_1 + \bar{\epsilon} < \xi < \xi_2 - \bar{\epsilon} \\ R_1 + R_2 & \text{if } \ \xi_2 + \bar{\epsilon} < \xi < \xi_3 - \bar{\epsilon} \\ R_1 + R_2 + R_3 & \text{if } \ \xi_3 + \bar{\epsilon} < \xi < 1 \end{cases},$$

and in the post contact ranges $\xi_i - \bar{\epsilon} < \xi < \xi_i + \bar{\epsilon}$ the integrals of the f_i's need to be explicitly determined. Notice, however, that unless one focuses one's attention on the regions of contact between the posts and the beam, there's no need to evaluate the integrals, because the actual expressions obtained outside the contact regions depend only on the *integrals* of the f_i's *across the contact zones* (which are all unity, by definition). This justifies the intuitive result that the deflection of the beam is insensitive to the shape functions f_i, and also justifies the use of the Dirac delta function notation. In order to avoid the awkwardness of always breaking up the ξ domain in the above way it's convenient to introduce another generalised function, the **Heaviside step function,** $H(x)$, defined by the rules

$$H(x) \stackrel{\text{def}}{=} \begin{cases} 0 & \text{if } x < 0 \\ 1 & \text{if } x > 0. \end{cases}, \tag{5.17}$$

in terms of which the appearance of the integral improves and equation (5.16) can be written in the form

$$Y'''(\xi) = -\xi + \sum_{i=1}^{3} R_i H(\xi - \xi_i).$$

Again, providing one's not interested in detailing displacements in the region of contact, this description is correct.

Aside: The behaviour of the system as a whole could be strongly affected by the contact regions if the posts and the beam were jointed. In such cases a more careful analysis would be necessary, because clearly the joints could exert a torque on the beam. Our present concern is not with such issues, so we have avoided the necessity of dealing with such complications by assuming the beam is balanced on the posts.

Substituting $\xi = 1$ in the expression for Y''' we see that the boundary condition $Y'''(1) = 0$, will be satisfied providing

$$0 = -1 + \sum_{i=1}^{3} R_i, \tag{5.18}$$

which, simply interpreted, states the Newtonian result that the sum of all the vertical forces acting on the beam must vanish for it to remain at rest.

A second integration of the beam equation yields

$$Y''(\xi) = -\xi^2/2 + \sum_{i=1}^{3} R_i H(\xi - \xi_i)(\xi - \xi_i)$$

so that the boundary condition $Y''(1) = 0$ requires

$$0 = -1/2 + \sum_{i=1}^{3} R_i(1 - \xi_i). \tag{5.19}$$

The interpretation of this result is that the net torque on the beam about the end $\xi = 1$ must vanish for equilibrium; which again provides a comforting check on our calculations. All the boundary conditions have now been utilized so that further integrations to obtain Y will introduce arbitrary constants of integration A and B. The resulting expression for Y is

$$Y = A + B\xi - \xi^4/24 + \sum_{i=1}^{3} R_i H(\xi - \xi_i)(\xi - \xi_i)^3/6. \tag{5.20}$$

Note that the solution consists of a particular solution and a complementary function. The particular solution term $-\xi^4/24$ represents the displacement due to the beam's weight (scaled to unity), and the other particular solution terms are due to the post forces. All these effects are purely additive (because the beam equation is linear). The complementary function (satisfying the homogeneous equation and boundary conditions) is linear in ξ. Physically the complementary function can be interpreted as being associated with a vertical translation and horizontal rotation of the beam as a whole. (Clearly such rigid body beam displacements can't affect stress levels in the beam.) Since both the vertical location of the centre of the beam and the orientation of the beam can't be determined until the beam is "fitted" to the posts, this result seems sensible. Although there's no need to "identify" terms physically in this way (one can simply trust the mathematics) it can be most beneficial; particularly when the situation is complex. We'll see this when we extend the present work to deal with the table top.

This completes our analysis of the beam. We now need to ensure that the beam deflections match those of the posts.

5.3.2 Beam/posts matching

The 5 unknowns R_i, A, B in the above general expression (5.20) for the beam deflection need to be chosen so that the displacements match those at the 3 posts, and the 2 static equilibrium conditions (5.18,5.19) are satisfied.

Let's assume the unstressed post lengths are l_i and in situ they are l_i'. If we adopt the same datum (l_0 above the floor) and the same length scale $c = WL^3/E_bI_b$ (see (5.13)) as used for the beam, then the scaled displacements of the post tops are given by

$$\ell_i' = (l_i' - l_0)/c,$$

and the beam and post displacements will match providing

$$\ell_i' = Y(\xi_i) \quad \text{for } i = 1, 2, 3.$$

The complete set of matching requirements is thus given by

$$\begin{pmatrix} 0 & 0 & 0 & 1 & \xi_1 \\ \lambda_{12} & 0 & 0 & 1 & \xi_2 \\ \lambda_{13} & \lambda_{23} & 0 & 1 & \xi_3 \\ 1 & 1 & 1 & 0 & 0 \\ (1-\xi_1) & (1-\xi_2) & (1-\xi_3) & 0 & 0 \end{pmatrix} \begin{pmatrix} R_1 \\ R_2 \\ R_3 \\ A \\ B \end{pmatrix} = \begin{pmatrix} \ell_1' + \mu_1 \\ \ell_2' + \mu_2 \\ \ell_3' + \mu_3 \\ 1 \\ 1/2 \end{pmatrix},$$

where the coefficients λ_{ij} and μ_i are defined by

$$\lambda_{ij} = (\xi_j - \xi_i)^3/6 \quad \text{and} \quad \mu_i = \xi_i^4/24.$$

The l_i''s are determined by the N_i's via the Hooke's law, explicitly

$$l_i' = l_i - N_i/k,$$

which in scaled terms becomes

$$\ell_i' = \ell_i - (W/ck)R_i,$$

where $\ell_i = (l_i - l_0)/c$.

Using this result, transferring the additional R_i terms to the left hand side of the equations, and substituting for c, we arrive at the following *explicit* equations for R_i, A, B:

$$\begin{pmatrix} \gamma & 0 & 0 & 1 & \xi_1 \\ \lambda_{12} & \gamma & 0 & 1 & \xi_2 \\ \lambda_{13} & \lambda_{23} & \gamma & 1 & \xi_3 \\ 1 & 1 & 1 & 0 & 0 \\ (1-\xi_1) & (1-\xi_2) & (1-\xi_3) & 0 & 0 \end{pmatrix} \begin{pmatrix} R_1 \\ R_2 \\ R_3 \\ A \\ B \end{pmatrix} = \begin{pmatrix} \ell_1 + \mu_1 \\ \ell_2 + \mu_2 \\ \ell_3 + \mu_3 \\ 1 \\ 1/2 \end{pmatrix},$$

where

$$\gamma = E_b I_b/(kL^3). \tag{5.21}$$

At this stage it is apparent that the completion of the calculation is best performed numerically because the algebraic complexity of the answers would make them meaningless anyway.

Our principal interest in this particular problem is to shed light on the question as to whether the bending of the beam or the compression of the posts is the more significant factor in allowing posts of different height to carry some of the load. The dimensionless number γ defines the situation and provides a measure of the relative significance of two effects. Here γ is a rough measure of the ratio of the contraction of the legs to the deflection of the beam. If γ is small, then the posts are effectively rigid and the beam (if it can) will bend to fit the posts. On the other hand if γ is large, then the posts will deform much more than the beam. If γ is neither large nor small then both effects contribute significantly to the determination of the loads.

Comment: Simple scaling observations may have enabled us to avoid the above detailed matching analysis in particular cases. We observed earlier from the defining equations that the scale of deflection of the posts under forces typified by W is W/k, and the beam equations gave a deflection scale of order $WL^3/(E_b I_b)$, so we are led to the above ratio γ for the relative size of these deformations from simple scaling arguments. If γ is either large or small then half the analysis can be avoided and matching becomes irrelevant. In the present problem exact solutions to the complete problem are available so it's no great issue; however, usually an all inclusive model is impractical and scaling observations are then *essential* to arrive at a working model.

5.4 The Table

To extend the above analysis to deal with the table we need the equations describing the small deflections of a plate. We can easily look up the required equations in a standard reference, eg. Landau and Lifshitz (1989), or Timoshenko and Goodier (1970).

Aside: As indicated on earlier occasions it's not absolutely necessary to know the detailed derivation of the equations describing the physical situation of interest. It's useful, however, to have a physical interpretation of the various terms, and to be aware of the approximations that have been used to derive the equations.

The equation governing the deflection $Z(x, y)$ of a uniform rectangular plate of weight W with sides of length a, b, supported by posts at each of the four corners exerting forces N_i on the plate, as in Fig. 5.2, is given by "the plate equation"

$$\mathcal{D}\, \nabla^4 Z(x,y) = -W/(a \cdot b) + \sum N \text{ for } 0 < x < a, 0 < y < b, \tag{5.22}$$

where

$$\sum N = N_1\delta(x)\delta(y) + N_2\delta(x-a)\delta(y) + N_3\delta(x-a)\delta(y-b) + N_4\delta(x)\delta(y-b),$$

$$\nabla^4 Z = Z_{xxxx} + 2Z_{xxyy} + Z_{yyyy},$$

and

$$\mathcal{D} = \frac{E_t t^3}{12(1-\sigma^2)}, \tag{5.23}$$

is referred to as the *flectural rigidity* of the plate of thickness t with Young's modulus E_t, and Poisson's ratio σ ($\sigma \approx 0.3$ for most materials). The above plate equation is accurate for describing situations in which the deflections are small compared with the plate thickness (certainly true for most table tops of interest). The similarity to the beam equation (5.11) is apparent. Note that the flectural rigidity is strongly dependent on plate thickness. The (homogeneous) boundary conditions for the "free edges" of the plate are given by

$$Z_{xx} + \sigma Z_{yy} = 0, \text{ and } Z_{xxx} + (2-\sigma)Z_{xyy} = 0 \text{ on } x = 0, a, \tag{5.24}$$

with corresponding homogeneous conditions on $y = 0, b$. All this looks daunting, but we don't have to solve the equations to determine the size of the deflection, so we'll do this first. Scaling the equations we see that the deflection scale is given by

$$\mathcal{Z}_t = \frac{Wa^4}{a \cdot b\mathcal{D}} = \frac{12W(1-\sigma^2)a^2}{(a/b)E_t t^3}. \tag{5.25}$$

Earlier we've seen that the scale of the length change of the legs is

$$\mathcal{Z}_l = W/k = Wl_0/(E_lA_l) \approx Wl_0/(E_lr^2),$$

where the individual legs are of average length l_0, with sectional area A_l and typical sectional length r, and E_l is the Young's modulus for the material of the legs.

The ratio of the deflection of the top to that in the legs is therefore

$$\mathcal{F} = \frac{\mathcal{Z}_t}{\mathcal{Z}_l} = \left[\left(\frac{(1-\sigma^2)}{a/b}\right)\left(\frac{a}{l_0}\right)\right]\left[\frac{E_l}{E_t}\right]\left[12\left(\frac{r}{t}\right)^2\left(\frac{a}{t}\right)\right], \tag{5.26}$$

where various dimensionless combinations have been bracketed to enable us to see more clearly the relevance of various parameters. The first factor in square brackets is of magnitude 1 for most tables; the second factor would often be 1 but could be small if materials with considerably different properties were used for top and legs; the last factor would normally be large because table tops are usually relatively thin. One needs to examine the specific data for a method of construction of a table to be entirely sure, but the likely outcome is that the top will deflect rather more than the legs will compress. Assuming this is the case we'll proceed with the analysis of the table top. We'll use the beam work to guide us; thus we'll first determine the displacement of the table top (treated as a plate) under arbitrary leg loading and then ensure that the displacements match those of the legs.

5.4.1 Plate solutions

Unfortunately, unlike in the beam case, simple exact solutions are not available, so approximate answers are necessary. This is not particularly concerning, especially since standard computer packages for determining plate solutions are readily available—even on PC's! It is, however, necessary to carefully organize the calculations.

Organizing Numerical Calculations

If a numerical calculation takes many hours on a machine but only has to be done once, then that's fine. If, however, the calculation takes only minutes but has to be continually repeated, then that's more of a problem. In terms of our table problem this is very relevant. There are two features of our plate bending problem that make it non-trivial:

- We do require extremely accurate results. We need answers for deflections accurate to within fractions of millimetres (typical for leg length differences) for tables that may be 2 or 3 metres long! It should be added that available programs for plate problems are not particularly good at determining displacements (as opposed to forces).

- In order to solve the table problem, (as in the beam case) we need the solution to the plate problem under *arbitrary* post loading. Thus particular numerical evaluations are not of *any* use to us! One could, of course, imagine setting up an iterative scheme to search for the required solution. Thus one could input estimates for l_i''s, calculate corresponding N_i's, determine the plate deflections using a package, adjust the l_i''s accordingly etc., and repeat the calculation until the observed changes in the results obtained from successive iterates were small. Such an approach would be computationally intensive (probably wouldn't work) as well as being ugly! The following linearity ideas reduce the general problem to *just 2* evaluations of the plate equations and thus *in essence* solve the problem. Even if the computer needs to work hard to get the required accuracy *it only has to make the calculations once*!

Comment: For a modeller any analytic *or numerical procedure* is regarded as a *solution* to a problem providing it's *interpretable in context* and it can be *usefully* and quickly *evaluated.* In this sense an (exact) convergent expression for a solution is a solution in a modelling sense *only if* it converges rapidly, whereas an integral (without singular behaviour) that requires numerical evaluation is normally an acceptable solution, as is also any procedure that can be effectively handled by a standard package.

We return to the problem in hand and examine the square plate case (sides of length a). Using the scales $\mathcal{Z}_t$, (see (5.25), a and W, so that

$$Z = \mathcal{Z}_t Y(\xi,\eta) \text{ where } x = a\xi, y = a\eta \text{ and } N_i = WR_i,$$

the plate equation (5.22) becomes

$$\nabla^4 Y(\xi,\eta) = -1 + \sum R \text{ for } 0 < \xi < 1, 0 < \eta < 1, \tag{5.27}$$

where

$$\sum R = R_1\delta(\xi)\delta(\eta) + R_2\delta(\xi-1)\delta(\eta) + R_3\delta(\xi-1)\delta(\eta-1) + R_4\delta(\xi)\delta(\eta-1), \tag{5.28}$$

with *homogeneous* boundary conditions

$$Y_{\xi\xi} + \sigma Y_{\eta\eta} = 0, \text{ and } Y_{\xi\xi\xi} + (2-\sigma)Y_{\xi\eta\eta} = 0 \text{ on } \xi = 0,1, \tag{5.29}$$

$$Y_{\eta\eta} + \sigma Y_{\xi\xi} = 0, \text{ and } Y_{\eta\eta\eta} + (2-\sigma)Y_{\eta\xi\xi} = 0 \text{ on } \eta = 0,1. \tag{5.30}$$

Since both the plate equation and the boundary conditions are linear, one can obtain solutions by superposition. Thus, if $Y^{(1)} = Y(\xi,\eta,R_i^{(1)},W^{(1)})$ is a solution to the plate equation associated with a plate weight $W^{(1)}$*and* leg forces $R_i^{(1)}$ and with edges free (so that the above boundary conditions are satisfied); and $Y^{(2)} = Y(\xi,\eta,R_i^{(2)},W_1^{(2)})$ is a second such solution with different forces acting, then $cY^{(1)} + dY^{(2)}$ is a plate solution corresponding to the forcing $(cR_i^{(1)} + dR_i^{(2)}, cW^{(1)} + dW^{(2)})$. We can use this result to build up general solutions to the plate problem from particular basis solutions. There are alternative basis solutions, but we should keep in mind that a computer package will be needed to produce results, so physically sensible, easily evaluated, basis solutions should be sought. We'll use the solutions corresponding to the situations illustrated in Fig. 5.5 as our basis. Explicitly:

- The first (symmetric) solution denoted by $Y^a(\xi, \eta)$ corresponds to the situation in which the plate of unit weight is supported under all four corners (so that Y = 0 at the four corners). Of particular interest to us are the forces and displacements at the corners. It's clear from the physics of the situation that the forces exerted by the posts (numbered anticlockwise) are given by $(1/4, 1/4, 1/4, 1/4)$ and the displacements at the four corners are of course $(0, 0, 0, 0)$, as prescribed.

- The second solution $Y^b(\xi, \eta)$ corresponds to the situation in which a unit weight plate is pinned on three of its corners (#1,#2,#4) and the other corner (#3) is free. The free corner will deflect downwards by an amount which we'll denote by y^b (to be determined computationally); so the corner displacements will be $(0, 0, -y^b, 0)$. Static equilibrium conditions (no net force or torque) require that the forces exerted by the posts are given by $(0, 1/2, 0, 1/2)$.

By adding together combinations of the symmetric solution and the four possible free corner solutions (that can be generated using the above free corner solution by a suitable change in the variables), one might expect to be able to generate a solution corresponding to the required (R_1, R_2, R_3, R_4) corner force distribution. However, given that global equilibrium considerations for our symmetric plate require $R_2 = R_4$, $R_1 = R_3$, and $R_2 = 1/2 - R_1$, we only need to ensure that the combination has the correct associated plate weight and the correct force on leg #1. The force requirements at the other posts will be automatically satisfied. Thus we expect the two basis solutions described above to suffice, and it can be seen that the required combination, see Fig. 5.5, is given by

$$Y_p(\xi, \eta) = 4R_1 Y^a(\xi, \eta) + (1 - 4R_1) Y^b(\xi, \eta).$$

For matching purposes we need the corresponding displacements at the corners which are given by $(0, 0, -y^b(1 - 4R_1), 0)$ (see Fig. 5.5).

Now the solution Y_p represents a *particular solution* to the plate problem satisfying the free edge conditions and with the prescribed post forces acting. To obtain the *general* solution we need to append to this the *complementary function* i.e. the solution to the corresponding homogeneous

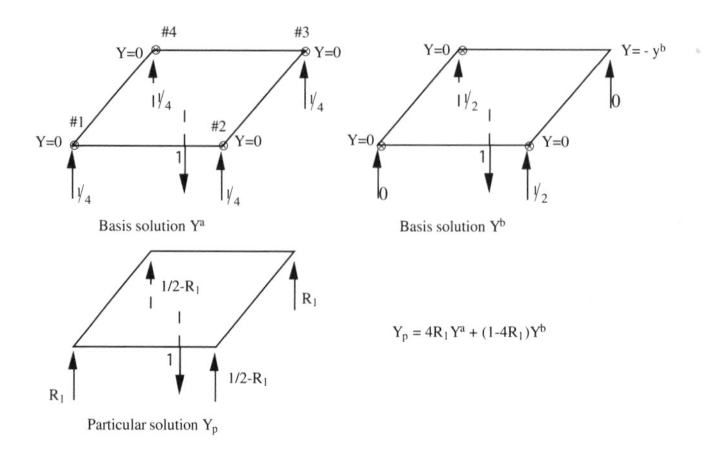

Figure 5.5: The generation of a particular plate solution.

problem. In the beam problem we found the complementary function was linear function of ξ, (see (5.20)) and corresponded to a rigid body translation and rotation of the beam. In our present case also a rigid body translation and rotation would not affect the stress distribution in the plate, and again a linear function corresponds to such a displacement, so one might anticipate this to be the required homogeneous equation solution. You can check that such a solution satisfies the homogeneous plate equation with the required free boundary conditions. The general solution of our plate problem is thus given by:

$$Y(\xi, \eta) = 4R_1 Y^a(\xi, \eta) + (1 - 4R_1) Y^b(\xi, \eta) + A\xi + B\eta + C. \qquad (5.31)$$

For matching purposes we need the corresponding displacements at the four corners which are given by

$$(C, A + C, A + B + C - y^b(1 - 4R_1), B + C) \qquad (5.32)$$

The plate analysis is now complete.

5.4.2 Matching

The 4 unknowns R_1, A, B, C in the general expression (5.32) for the plate deflection need to be determined so that the displacements match those of the legs. It's necessary to adopt a common datum (l_0 above the ground) and length scale ($\mathcal{Z}_t$). The matching conditions are given by, using (5.32,)

$$\begin{pmatrix} 0 & 0 & 0 & 1 \\ 0 & 1 & 0 & 1 \\ 4y^b & 1 & 1 & 1 \\ 0 & 0 & 1 & 1 \end{pmatrix} \begin{pmatrix} R_1 \\ A \\ B \\ C \end{pmatrix} = \begin{pmatrix} \ell_1' \\ \ell_2' \\ \ell_3' + y^b \\ \ell_4' \end{pmatrix},$$

where

$$\ell_i' = (l_i' - l_0)/k\mathcal{Z}_t.$$

Again, as in the beam case, the ℓ_i''s are determined by the R_i's and the unstressed leg lengths ℓ_i using Hooke's law, which leads to the following explicit equations for R_1, A, B, C:

$$\begin{pmatrix} \frac{1}{\mathcal{F}} & 0 & 0 & 1 \\ \frac{1}{\mathcal{F}} & 1 & 0 & 1 \\ \frac{1}{\mathcal{F}} + 4y^b & 1 & 1 & 1 \\ \frac{1}{\mathcal{F}} & 0 & 1 & 1 \end{pmatrix} \begin{pmatrix} R_1 \\ A \\ B \\ C \end{pmatrix} = \begin{pmatrix} \ell_1 \\ \ell_2 \\ \ell_3 + y^b \\ \ell_4 \end{pmatrix},$$

where

$$\mathcal{F} \equiv \mathcal{Z}_t/\mathcal{Z}_l = \frac{12k(1-\sigma^2)a^2}{E_t t^3},$$

is the dimensionless group, arrived at earlier when we scaled the problem, see (5.26).

The legs will maintain floor contact if $0 \le R_1 \le 1/2$ so that, as a function of $\mathcal{F}$ the stability regime for the table can be mapped out. Maple can be used to do this. The only factor that isn't explicitly known in the expression for R_1 is the displacement y^b which has to be determined numerically. Note that this is the *only* information that we need from the numerical plate package to feed into our solution!

Comment: Almost assuredly the package used to determine this displacement would be based on the **finite-element** technique, which we'll come across later in the book in other contexts. This extremely powerful technique was initially introduced to deal with precisely complex elasticity problems of the type considered here but its use has now extended to most areas of classical applied mathematics, as we'll see. The technique is a straightforward application of variational principles. The elastic energy stored in a plate is given by the expression

$$\mathcal{E}_e = \frac{\mathcal{D}}{2} \int\!\!\int [(\frac{\partial^2 Z}{\partial x^2} + \frac{\partial^2 Z}{\partial y^2})^2 + 2(1-\sigma)\{(\frac{\partial^2 Z}{\partial x \partial y})^2 - \frac{\partial^2 Z}{\partial x^2}\frac{\partial^2 Z}{\partial y^2}\}]dxdy,$$

a horrible expression! The displacement $Z(x,y)$ actually realized is the function that minimizes the total energy of the system subject to the relevant constraints. Now, if one could search through all acceptable functions and choose the minimizer then we'd obtain this exact solution; but this is not usually possible. One might expect to obtain a reasonable approximation, however, if one focused one's attention on a sub-class of all functions (eg. polynomials of 4) and selected the minimizer. This is the type of procedure used by finite-element and related techniques, with different techniques using different approximating functions. Later we'll return to discuss this matter in a simpler context.

Advice to the manufacturer

We're now in a position to provide reasonably complete advice to our manufacturer. For table designs of interest:

1. If $\mathcal{F} \ll 1$ then the analysis of Chapter 3 is relevant and, as indicated there, the sectional area of the legs should be made sufficiently small to accommodate any leg length differences resulting from manufacture. The probability analysis of Chapter 3 Section 3.2 provides more precise information.

2. If, as anticipated by our scaling arguments, $\mathcal{F} \gg 1$, then the table thickness t should be chosen so that $\mathcal{Z}_t \approx \delta l$, i.e. thin enough so that the deflection of the top is sufficient to counteract expected differences in leg lengths.

3. If $\mathcal{F} \approx 1$ then both legs and top influence the table's behaviour and the solution obtained above could be used. A probability analysis could be undertaken to arrive at a practically useful result for the manufacturer.

Comment: We've used a Newtonian type approach to extract the solution of the beam equation and to discuss the plate equation. It is a simple matter to write down the elastic and gravitational energy associated with such solutions and in this way include deflection terms in a general energy formulation of the problem (i.e. one that includes leg compression, flexible joints etc., as well as table top effects). If many imperfections are significant, and the application warrants the effort, this is the sensible thing to do. For the table application such effort is probably unwarranted but in special applications (eg. large structures that need to be accurately designed) such work may be necessary.

5.5 Summary

It's been a long (and hopefully instructive) fable, of interest because of the modelling and mathematical ideas that arise—the application is more amusing than important. It's interesting to note the modelling pattern. Initially the model considered was simple, but this simple model not only enabled us to see the important issues, but provided a framework for follow up work (the stability parameter observations etc.). The models then became more complicated because the physics examined (coupled plate deflection and leg compression etc.) is more complex. Eventually, however, after the important parameters were identified, the model simplified again, and in fact (although we didn't carry it this far) a generalised stability parameter could be identified. This is a very common pattern. Whether it's because Nature has designed a simple physical world, or man chooses to see it this way is a matter of conjecture.

In conclusion it is suggested that if you encounter tables which rock you might like to look at their method of construction to see whether you can see the reason. In the University of Western Australia there is a set of tables which are particularly prone to rocking, which in fact motivated the development of this as a student exercise. They have steel legs and substantial wooden rails supporting a wooden top. For this situation one has to allow for the deflection of the rails under the loads from the legs and the distributed load from the plate along the contact zone between the plate and the rails. The rails are there, as in many tables, to add rigidity to the top, and the obvious advice to the manufacturer would be to make the top less rigid. To obtain precise solutions in this case it would be necessary to examine the deflection of the beam/plate combination that is the table top. It's not at all clear how one could arrive at a Newtonian description for such a complex set-up. Such complex situations are, however, easily handled using energy methods and appropriate software is available.

5.6 Exercises

Exercise 5.1 *Flexible Ladders*

⋆ The three static equilibrium equations are insufficient to determine the four (frictional and normal) surface forces acting on a ladder resting against a wall, see Exercise 3.6. Our earlier illustrations of statically indeterminate systems suggest that this lack of uniqueness may be resolved by allowing for the elastic deformation of the ladder. Our aim is to find out which form of deformation controls the determination of the solution, and then to determine the solution. To do this we need to find out the relative scales of the elastic energy stored due to the compression of the ladder along its length and that due to the lateral deformation of the ladder.

(a) From the Hooke's Law result

$$E\frac{du}{dx} = T/A,$$

which relates the longitudinal displacement, $u(x)$ (where x is the location along the ladder), to the longitudinal force T, the cross sectional area of the ladder A, and other physical parameters, estimate the typical longitudinal displacement of a ladder of length l subjected to T, and calculate the elastic energy stored due to the longitudinal force.

Hint: The energy stored due to compression is well estimated by the product of the force and the overall displacement.

(b) As far as the elastic bending energy is concerned the ladder may be regarded as two beams tied together by the steps, so that they deflect laterally in the same way, but otherwise their deflection is unaffected by the presence of the steps. (The steps just act to connect the two bending beams). From the equation for the bending of a beam

$$EI\frac{d^4w}{dx^4} = N/l,$$

which relates the lateral displacement, $w(x)$, to the *lateral* force per unit length, N, and other physical parameters of the beam, obtain an estimate of the lateral displacement and the elastic energy associated with bending. The energy associated with the bending is well estimated by the product of force and the maximum displacement.

(c) Using the above results show that the ratio of the elastic energy stored as a result of compression to that associated with bending is of the order A/l^2. This suggests that one may neglect the effects of compression for most circumstances unless very precise answers are required.

(d) ⋆⋆ A man walks up the ladder. Compute the transverse displacement of the ladder for each value of the distance the man is along the ladder in terms of the undetermined vertical friction force. The elastic energy stored in the ladder is precisely half that indicated by the sum, over all applied forces, of the product of the final displacement of the ladder at the point of application of the force and the force. Hence determine the values of the forces on the ladder. Comment on the physical implications.

5.7 Appendix 1: The 6 legged table solution

The practical implementation of this solution is more easily achieved by writing the system in matrix form. Also the generalizations are easier to perceive and implement. Thus we introduce vectors $\mathbf{N}^T, \mathbf{l}', \mathbf{l}$ representing the leg force components etc., and also we introduce a Lagrange multiplier vector

$$\boldsymbol{\lambda}^T = (\lambda_1, \lambda_2, \lambda_3).$$

In order to specify the gravitational potential energy we introduce the vector

$$\mathbf{U}^T = (1, 0, 1, 1, 0, 1),$$

in terms of which the gravitational potential is given by $(W/4)\mathbf{U}^T\mathbf{l}'$, see (5.6). To specify the geometric constraints (the table top is flat) we introduce the (3×6) matrix G so that, see (5.7),

$$G\mathbf{l}' = \mathbf{0}.$$

The objective is to find the stationary values of

$$\mathbf{N}\mathbf{N}^T/(2k) + (1/4)W\mathbf{U}^T(\mathbf{1} - \mathbf{N}/k) - \boldsymbol{\lambda}^T G(\mathbf{1} - \mathbf{N}/k)$$

subject to the restraints

$$G(\mathbf{1} - \mathbf{N}/k) = 0. \tag{5.33}$$

The conditions for a stationary value are

$$\mathbf{N} - (W/4)\mathbf{U} + G^T\boldsymbol{\lambda} = \mathbf{0}, \tag{5.34}$$

so that the system of equations to be solved for $\mathbf{N}, \lambda$ is

$$\begin{pmatrix} \mathbf{I} & G^T \\ G & 0 \end{pmatrix} \begin{pmatrix} \mathbf{N} \\ \boldsymbol{\lambda} \end{pmatrix} = \begin{pmatrix} W\mathbf{U} \\ kG\mathbf{1} \end{pmatrix},$$

where $\mathbf{I}$ is the sixth order unit matrix. In formal terms (5.34) can be used to eliminate $\mathbf{N}$ from (5.33) yielding the solution for λ

$$\boldsymbol{\lambda} = (GG^T)^{-1}G[(W/4)\mathbf{U} - k\mathbf{1})].$$

Substituting this expression back into equation (5.34) for $\mathbf{N}$ gives

$$\mathbf{N} = (W/4)(\mathbf{I} - M)\mathbf{U} + kM\mathbf{1}, \tag{5.35}$$

where

$$M = G^T(GG^T)^{-1}G.$$

The solution is thus complete once we have M.

Now all the above manipulations are purely formal. To actually carry out the envisaged calculations is a major task on paper (finding inverses using Gaussian elimination etc.); here is where we can really appreciate an algebraic package! Assuming that G has been entered, the Maple commands

```
multiply(G,transpose(G));
inverse(");
multiply(",G);
M:=multiply(transpose(G),");
```

lead to the result

$$12M = \begin{pmatrix} 5 & -4 & -1 & 3 & 0 & -3 \\ -4 & 8 & -4 & 0 & 0 & 0 \\ -1 & -4 & 5 & -3 & 0 & 3 \\ 3 & 0 & -3 & 5 & -4 & -1 \\ 0 & 0 & 0 & -4 & 8 & -4 \\ -3 & 0 & 3 & -1 & -4 & 5 \end{pmatrix}, \tag{5.36}$$

for M. It isn't easy to understand and interpret this result in a practical way, so further processing is required. Some reduction in the complexity may be achieved if we understand the effect M has when acting on an arbitrary vector. This is the stuff of Linear Algebra; the information is contained in the eigenvalues and eigenvectors of M, which can be obtained using

```
eigenvectors (M);
```

The eigenvalues (μ say) turn out to be 0 and 1; both with multiplicity three.

The $\mu = 0$ eigenvalue. That $\mu = 0$ should be a threefold eigenvalue is a consequence of the fact that M is constructed using the matrix GG^T which is of rank 3. That it is no more than threefold implies that the constraint conditions are linearly independent, which is a consequence of the geometric fact that six arbitrary points being coplanar produces three linearly independent conditions. An orthogonal basis for the null space of the matrix can then be generated by using the `Gram Schmidt` instruction. One such basis for the null space generated in this way is

$$(1,1,1,0,0,0)^T, \quad (0,0,0,1,1,1)^T, \quad \text{and} \quad (1,0,-1,-1,0,1)^T.$$

Note that, see (5.35), for any vector $\mathbf{l}$ lying in the subspace generated by this basis, the weight of the table will be carried equally by all six legs; since M annihilates such a vector. This is an interesting result with physical implications. For example, since $(1,1,1,0,0,0)$ is one such vector, tables which have all three legs on one side equal in length will have the weight carried evenly by all legs, regardless of any differences in the length of the other three legs (providing all make floor contact).

The $\mu = 1$ eigenvalue. Given the above result, the length differences which give rise to the possibility of rocking are to be found in the subspace corresponding to the $\mu = 1$ eigenvalue. An orthonormal basis $\mathbf{e}_i, i = 1,2,3$ associated with this eigenvalue is obtained using the same procedure as above. This gives

$$\mathbf{e}_1 = \frac{1}{6}(-3,0,3,-1,-4,1)^T,$$

$$\mathbf{e}_2 = \frac{1}{2}(0,1,-1,1,-1,0)^T, \text{ and}$$

$$\mathbf{e}_3 = \frac{1}{2\sqrt{15}}(4,-5,1,3,-3,0)^T. \tag{5.37}$$

In terms of this basis the loads carried by the legs are given by (again use Maple)

$$N_i = W/6 + kM\sum_{j=1}^{3}[\mathbf{l}\cdot\mathbf{e}_j]\mathbf{e}_j(i),$$

where $e_j(i)$ denotes the ith component of the jth basis vector. Checking that the normal reactions are not negative thus involves seeking the minimum of these six loads, a process which is computationally simple; the analysis has achieved our aim.

Bibliography

Landau,L.D. and Lifshitz, E.M.(1989).*Theory of Elasticity*, p 42, Maxwell MacMillan International Editions.

Segel, Lee A.(1977). *Mathematics Applied to Continuum Mechanics*, p 207, Macmillan Publishing Co., Inc.

Timoshenko, S.P. and Goodier, J.N.(1970). *Theory of Elasticity*, McGraw-Hill Book Co.. Inc., New York.

Part II

Diffusion

Chapter 6

Preliminaries

Basics of heat conduction
Elementary solution techniques
Uniqueness theorems

6.1 Situations of Interest

- How can one increase the efficiency of an industrial cooker?
- Radiation from the sun falls on the Earth's surface and is partially absorbed, the remainder is re-radiated into space. The gaseous products of combustion produced by human activity influence the amount of absorption. What temperature rise is to be expected?
- Iron sheets can be cast by pouring molten steel on to a cooled rotating drum. Is this procedure an economic alternative to the procedures presently used?
- Radiative heat from a fire falls on a combustible material. Will the material spontaneously burst into flame?

These are some of the situations that will be discussed in this Part of the book.

Inexperienced cooks, when attempting to barbecue a steak, can easily manage to convert the outside of the meat to charcoal while leaving the middle virtually frozen. A chef who has to cook steaks to order from rare to well-done knows how to ensure that the cooking takes place at just the right rate to achieve the customer's requirements. The need to control heating processes so as to achieve a suitable spatial and temporal distribution of temperature within a body arises in a wide variety of industrial contexts involving metal processing. The tempering and quenching of steel is an example. The appropriate mix of the desired properties, usually toughness and hardness, can only be realized if suitable heating and cooling treatments are used. Much of present day practice rests heavily on knowledge obtained originally by trial and error, that has been passed on from generation to generation. The modern drive for greater efficiency in industrial processes and the desire to avoid the wastage of energy has led to the search for optimal ways of achieving the desired heat treatment.

When heat is applied to the surface of an object, some of the heat goes into the heating of the surface material, some is conducted inside to warm the interior, and the remainder is lost to the surroundings by a mixture of radiation, convection, and diffusion. Any chemical reactions that occur at the surface (combustion is such an example) or within the material (eg. crystallization) will affect the heat balance. In order to determine the optimal heat treatment of a body it is thus necessary to understand how the various heat components are determined by the physical mechanisms.

Greenhouse effect investigations represent another area in which heat partitioning is the central issue. The observed atmospheric temperature changes over the past 50 years are sufficiently large, and the time span for these changes sufficiently short, to be of major concern. In fact, although opinion is divided, the vast majority of researchers in the area believe the danger to be real. Given the importance of work on this problem and the requirements that models will only be effective if they are convincing to policy makers (who could be adversely affected by suggested remedies), the challenge is real and daunting.

In general terms the physics of the situation is well understood and is as follows: Solar radiation, in the form of high frequency light radiation, is intercepted and partially absorbed by the Earth with its atmosphere, causing a rise in temperature. The Earth re-radiates heat into space in the form of low frequency heat radiation. A balance can be realized between these two competing heat transfer processes for sufficiently high earth temperatures, and this equilibrium situation will sustain itself providing the heat transfer processes are not altered—presumably this was the existing situation before the use of fossil fuels became widespread. The additional absorption of infra-red radiation by carbon dioxide and other gases emitted by mankind into the atmosphere has altered this balance and thus has led to the retention of a greater fraction of solar radiation reaching the Earth. Two obvious questions are of concern:

- If present or projected gas production levels are sustained, what temperature conditions on the Earth can be expected when a new equilibrium state is reached?

- How long will it take for this state to be reached?

Although the qualitative picture presented above is sensible it is difficult to quantify the processes described and so answer the questions posed.

The difficulty is partially observational and statistical—the great variability with both position and time of the thermal parameters of interest makes interpretation of observations difficult, particularly since our concern is with global changes over a time span that's large compared with the observed time scale of fluctuations. Even if data collection etc. can be successfully managed, there are major questions concerning the mechanisms that need to be answered before sense can be made of collected data so that reliable predictions can be made. It is in this area that the modeller can play a major role. Thus, for example, it's necessary to ask "What process primarily determines the response time of the system ?" Is it the time taken for heat to build up in the atmosphere, or the time for this heat to penetrate the oceans or the earth, or is the situation much more complicated? In fact the situation is complicated. The large transports of energy by both sea and air, and the effects of clouds on the reflection of the sun's rays and the infra-red radiation emitted by the Earth, make the situation both complex and difficult to predict. Modern computing facilities fall far short of the capacity required to include all such effects directly in a mathematical model and even if one could do this the output of such a model would be doubtful. The state of the art/science models that are used to make greenhouse predictions attempt to include the difficult processes "on average", with different models using different averages and producing different results. Work in this area has barely commenced and the models are not adequate. A disturbing feature, however, is that runs on such models all seem to lead to dire predictions. This is why the vast majority of the scientific community believes that urgent action is necessary. In this book some simple models will be considered which provide information concerning the significance of some of the various heat transfer processes.

Many heat transfer problems, typified by the above examples, are currently under serious investigation and there is likely to be a continuing need for such investigations. Furthermore the mathematical equations governing the diffusion of chemical species are similar to the heat equations so that the material of this chapter has many applications beyond the particular contexts raised here. Thus, for example, equations of the same type can be used to understand the diffusion of pollutants in the environment, of oxygen into metals (rusting etc.), or the movement of elements within metals, which is, for example, relevant to the understanding of the premature aging and consequent failure of metals used in nuclear reactor cooling systems. Some of these applications will be described in the exercises.

In the next Section relevant information on the experimental basis for heat flow studies and also some mathematical preliminaries are presented. The **heat equation,** that's fundamental to all heat flow studies, is derived in Section 6.3. In the remainder of this Section insights concerning the behaviour of diffusive systems are obtained using scaling arguments and by extracting special solutions. Following this, some mathematical results associated with the heat equation are presented for completeness; these results do not form an essential part of the text. This chapter is followed by two chapters devoted to development of mathematical techniques of major significance to the modeller. These techniques will be developed in the context of determining the temperature distribution in a body due to surface heating. There are then three chapters describing the modelling of particular situations.

6.2 Background

6.2.1 The experimental basis for heat flow studies

The following observations and concepts underlie our understanding of heat flow processes:

1. **Temperature** T: dimensions $[\Theta]$, units °C or K.

   The temperature of a body is a measure of how hot a body is in the sense that if two bodies at different temperatures are brought into contact, heat will flow from the body with the larger temperature to that with the lower. No heat transfer will occur if the bodies are at the same temperature. The commonly used unit is the Celsius degree but in many scientific applications, for example in radiation studies, the absolute or Kelvin scale is more natural. The difference between the two is purely one of choice of datum; the Kelvin temperature may be obtained from the Celsius temperature by adding 273.16.

2. **Quantity of heat** H: dimensions $[H] = [M][L]^2[T]^{-2}$, units joules, or equivalently watts secs.

   Heat is merely one form of energy which is frequently convenient to consider in isolation from other forms. The dimensions are thus simply those of energy. Sometimes for historical reasons calories rather than joules are used. To convert from one to the other simply use 1 cal= 4.182 joules.

   Usually we're concerned with changes in heat content rather than absolute values relative to an appropriate datum. The amount of heat δH required to change the temperature T of a body by an amount δT is given by

   $$\delta H = mc\delta T, \tag{6.1}$$

   where m is the mass of the body and c is a property of the material the body is comprised of, known as its **specific heat**. The specific heat is usually found to be (weakly) dependent upon the temperature of the body, so a better way of expressing the above result is given

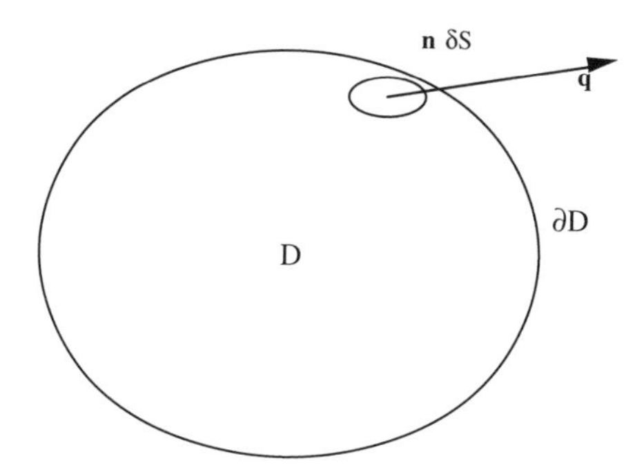

Figure 6.1: The flux through a surface element.

by its integral equivalent

$$\mathcal{H}(T) - \mathcal{H}(T_0) = \int_{T_0}^{T} mc(T')\, dT', \tag{6.2}$$

where $\mathcal{H}$ is called the **Enthalpy** or heat content function. For many cases of practical interest the variation of the specific heat with temperature is sufficiently slight, or the temperature range sufficiently small, to ignore this variation. In the work that follows, unless specifically stated to the contrary, it will be assumed that the specific heat is independent of the temperature.

3. **Heat Flux q**: a vector, dimensions $[H][L]^{-2}[T]^{-1}$, units watts/m^2.

   When the temperature in a body varies with position there is a flow of heat energy from hotter parts to cooler parts. The heat flux specifies the direction of heat flow and the magnitude of the rate of flow across a unit area at right angles to the flow direction. The flow rate across a surface element $\delta S\mathbf{n}$ is thus given by

$$\delta q = \mathbf{q} \cdot \mathbf{n}\delta S, \tag{6.3}$$

   where the usual convention of using the *external* surface normal, see Fig. 6.1, is adopted.

4. **Thermal Conductivity** k: dimensions $[H][T]^{-1}[L]^{-1}[\Theta]^{-1}$, units watts/m °C.

   Experiments show that for many solid materials the heat flux, $\mathbf{q}$, is given by

$$\mathbf{q} = -k\nabla T, \tag{6.4}$$

   so that heat flows in the opposite direction to the temperature gradient and flow rate is proportional to the magnitude of this gradient. The parameter k is a property of the material known as the *thermal conductivity*. The thermal conductivity is temperature dependent but frequently varies so little for cases of interest that it is reasonable to neglect this variation; we will do this here.

Diffusion processes also transport heat in fluids, and the thermal conductivity as defined above provides the required heat transfer information *providing* the fluid isn't moving. In most circumstances, however, the fluid will be moving either as a result of external stirring and/or as a result of the buoyancy induced *natural convection* motions generated by heating. The heat transport is greatly enhanced by such motions and coupled fluid dynamics/heat flow models are often necessary to determine the resulting heat transport. In certain circumstances simple empirical relationships of the Newtonian type (see item 8. below) can be used to describe this process.

5. **Diffusitivity** κ: dimensions $[L]^2[T]^{-1}$, units m^2/s.

   This derived property of a material is defined by

$$\kappa = \frac{k}{\rho c}, \tag{6.5}$$

   where ρ is the material density, and is of importance because the thermal response time of a material depends on this particular combination of physical parameters, as will be seen.

6. **Latent Heat** L, and **Heat of Reaction** $\mathcal{R}$: dimensions $[H][M]^{-1}$, units joules/kg.

   There are circumstances when heat energy is put into a body and no change in temperature is observed. If this occurs it's a good indication that the heat flow is resulting in a change in internal chemical or physical structure; for example when there is a change of state from solid to liquid. Often such changes occur at specific temperatures. It is observed that the amount of heat absorbed is proportional to the mass of material δm which changes state, so that

$$\delta H = L\delta m, \tag{6.6}$$

   where L is known as the **latent heat** associated with the state change. More generally, any physical or chemical structural change within a body will result in either the release or absorption of heat described by

$$\delta H = \mathcal{R}\delta m, \tag{6.7}$$

   where $\mathcal{R}$ is called the **heat of reaction**.

7. **Stefan's Law of Radiation**

   The amount of radiative heat energy per unit area per unit time lost from a surface is given by the Stefan-Boltzmann law

$$R = \epsilon\sigma T^4, \tag{6.8}$$

   where the Stefan-Boltzmann constant $\sigma = 5.669 \times 10^{-8}$ watts/m^{2}K^4. T is measured in degrees Kelvin and the constant ϵ, referred to as the **emissivity**, depends on the nature of the surface. The emissivity ranges over $0 \leq \epsilon \leq 1$ with the theoretical limit $\epsilon = 1$ realized for a "perfect black body". For real surfaces ϵ has to be experimentally determined.

8. **Newton's Law of Cooling**

   This is a semi-empirical relation which is often used as a means of calculating small corrections in applications like calorimetry. It uses the idea that the heat loss from the surface of a body is dependent mainly on the temperature drop $T_b - T_0$ from the surface (at T_b) to the surroundings (at T_0). Under such circumstances one might expect the linear Newton's "Law" of cooling

$$\text{Heat loss/area} = \beta(T_b - T_0), \tag{6.9}$$

   to provide a useful approximation, if T_0 is appropriately measured. The *heat transfer coefficient* β, with units watts/m^2/°C, depends on heat transfer processes in the surroundings and is normally measured in the laboratory under circumstances comparable to those being investigated. The usefulness of this description depends on whether the laboratory and application situations are sufficiently similar over the required temperature range. It represents quite a good approximation to the radiative loss from a body when the temperature difference between the surface and the surrounds is a very small fraction of the absolute temperature of the surface. It is also often a good approximation for well designed heat exchangers where fluid is brought into close contact with the surface and allowed to absorb the available heat before it is removed. It does not apply to the transfer of heat to another material by conduction, and is of limited use if natural convection dominates the heat transport process.

For a detailed description of this background material see Carslaw and Jaeger (1959), the classic book in the area. For an excellent account of more recent work on heat conduction see Hill and DeWynne (1987).

9. **Thermal Parameters for Common Substances**

Substance	Density ρ kg/m^3	Specific heat c kjoule/kg °C	Conductivity k watts/m °C
Aluminium	2.707	0.896	204
Iron	7897	0.452	73
Steel	7833	0.465	54
Copper	8954	0.383	386
Brick	1600	0.84	0.69
Concrete	2000	0.88	1.37
Granite	2600	0.88	2.5
Soil (average)	2540	0.85	0.96
Wood (oak)	540	2.4	0.166
Water	999.8	4.225	0.566
Air	1.177	1.006	0.026

Table 6.1: Thermal Properties. (1 kjoule=10^3 joules.)

6.2.2 Mathematical background

Theorem 6.1 (Gauss's Divergence Theorem) *Gauss's Theorem states that if* $\mathbf{F}(\mathbf{x})$ *is a vector valued function of position* $\mathbf{x}$ *defined over a domain* D *with boundary* ∂D*, then*

$$\int_D \nabla \cdot \mathbf{F} dV = \int_{\partial D} \mathbf{F} \cdot \mathbf{n} dS, \tag{6.10}$$

where dV *is a volume element in* D*, and* $dS\mathbf{n}$ *is a surface element of* ∂D*, as in Fig. 6.1.*

This theorem is central to vector calculus and is useful for determining mathematical expressions of the conservation laws. The result holds under *all* circumstances for which meaning can be given to the terms of the equation; even when the terms are not defined in an "ordinary function" sense. Intuitive accounts of the proof of Gauss's theorem can be found in any second year calculus text, eg. Kreysig (1983).

6.3 The Heat Equation

We examine the movement of heat within a body that may be receiving heat from an *external* supply at a given rate. By an *external* supply we mean any supply other than body conduction, such as the heat produced by the passage of a current through the body or by a chemical reaction, or the heat produced by the absorption of radiation.

Now energy conservation requires that the rate of change of the heat content of any volume element dV within the body be equal to the heat being conducted into the domain through its boundaries plus that being absorbed from the external agents. We wish to convert this physical principle into a useful form of mathematical statement. Let the rate of supply of heat per unit volume from external processes be denoted by q, and consider the heat balance for an arbitrarily chosen domain D within the body, see Fig.6.1. Note that the heat energy content of D is given by

$$\int_D \rho c T \, dV,$$

where ρ is the density and c the specific heat of the body, so the rate of change of heat content of this domain is given by

$$\frac{\partial}{\partial t} \int_D \rho c T \, dV.$$

This heat content change occurs because external sources supply heat at a rate given by

$$\int_D q \, dV,$$

and also because heat is being transferred across the surface ∂D *into* the domain D by conduction, see (6.3,6.4,) and Fig. 6.1, at a rate given by

$$\int_{\partial D} k \, \nabla T \cdot \mathbf{n} \, dS,$$

where $\mathbf{n}$ denotes the outward normal to the surface ∂D. Thus heat conservation requires

$$\frac{\partial}{\partial t} \int_D \rho c T \, dV = \int_D q \, dV + \int_{\partial D} k \nabla T \cdot \mathbf{n} dS.$$

The surface integral can be converted to a volume integral using Gauss's Theorem (6.1), and the time derivative can be carried inside the integral on the left hand side to yield

$$\int_D \left[\rho c T_t - q - k\nabla^2 T\right] dV = 0.$$

This result holds for *any* arbitrarily selected domain D within the body (however small) and hence the integrand (provided that it is a continuous function of position) must vanish at all points. Dividing through by ρc we obtain the result that the temperature field satisfies the partial differential equation

$$T_t = \kappa \nabla^2 T + Q, \tag{6.11}$$

where $\kappa = k/\rho c$ is the *thermal diffusivity* of the material, and

$$Q = q/(\rho c) \tag{6.12}$$

is the heat supply rate per unit volume scaled relative to the *thermal mass* ρc per unit volume of the material. The 1D version of this equation is given by

$$\frac{\partial T}{\partial t} = \kappa \frac{\partial^2 T}{\partial x^2} + Q. \tag{6.13}$$

If the thermal properties of the material are temperature dependent then the enthalpy $\mathcal{H}$ is a more convenient variable to work with, see (6.2), and with minor adjustments the above approach leads to the more general equation

$$\mathcal{H}_t = \boldsymbol{\nabla} \cdot [\kappa \boldsymbol{\nabla} \mathcal{H}] + q. \tag{6.14}$$

Equation (6.11) is known as the **Heat** or **Diffusion equation** and arises in many contexts beside the conductive transfer of heat.

The heat equation quantifies the transport of heat *within* the body. Clearly, on physical grounds, to completely specify a problem it's necessary to append to this equation information concerning the initial thermal state of the body and the thermal conditions around the boundary of the body.

6.3.1 Some elementary observations

A typical question is: if a body is subjected to a new thermal environment what additional fluxes are introduced by the changed circumstances and how long will it take for the adjustments to occur? Although complete answers to questions of this type can only be obtained by extracting solutions to the defining heat equation with subsidiary conditions, scaling arguments lead to estimates for the sizes of interest.

Consider the following situation: A rod is initially at temperature T_0, and subsequently the temperatures of the rod ends are maintained at T_0 and T_1 with $T_1 > T_0$, see Fig. 6.2. How does the temperature evolve? As with all real problems it pays to think about what is likely to happen before proceeding. In this case, clearly heat will commence flowing from the hot boundary into the rod, causing its temperature to rise. This temperature rise will cause an outflow of heat from the rod into the cold boundary. Initially the inflow will exceed the outflow, however, as time goes on the inflow will reduce and the outflow will increase, because of the increase in the temperature of the rod. Eventually one might expect the two to balance

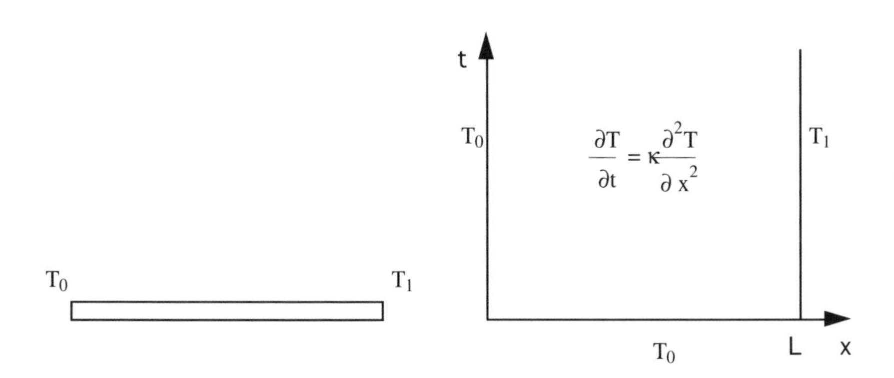

Figure 6.2: Heat flow in a rod.

resulting in a stable steady state situation. The equations describing the situation are:

$$T_t - \kappa T_{xx} = 0, \tag{6.15}$$

$$\text{with } T(0,t) = T_0, \;\; T(L,t) = T_1, \text{ and } T(x,0) = T_0. \tag{6.16}$$

Scaling the problem in the obvious way:

$$T(x,t) = T_0 + (T_1 - T_0)\theta(x', t')$$

$$\text{with } x = Lx', \text{ and } t = \tau t',$$

$$\text{where } \tau = \frac{L^2}{\kappa},$$

leads to the dimensionless formulation

$$\theta_{t'} - \theta_{x'x'} = 0, \text{ with} \tag{6.17}$$

$$\theta(0,t') = 0, \;\; \theta(1,t') = 1, \text{ and } \theta(x',0) = 0. \tag{6.18}$$

Note that, since there are no dimensionless groups remaining in the describing equations, *just one* real experiment on a convenient rod, or a single numerical evaluation could be used to determine the solution. Also note that the conduction time scale L^2/κ increases rapidly with rod length, and the combination of material parameters that matters as far as the transient response is concerned is the diffusivity $\kappa = k/\rho c$. To get a feel for the situation it's useful to see if the conduction time scale matches what you know from experience. Using the tabulated values for physical parameters, see Table 6.1, you might like to check if the scale is right for the heating a frying pan.

We'll proceed further with the analysis. The steady state solution of equation (6.17) with the prescribed boundary conditions (6.18) is given by $\Theta(x') = x'$. The interpretation is evident: after a time scale of order $\tau = L^2/\kappa$, the temperature settles down to the steady state linear profile given by (in unscaled variables)

$$T_s(x,t) = T_0 + (T_1 - T_0)\frac{x}{L}, \tag{6.19}$$

corresponding to a uniform flux of heat $-k(T_1 - T_0)/L$ along the rod. To determine an explicit expression for the time taken to "reach" this steady state one would need to solve the *transient* equations

$$\theta'_{t'} - \theta'_{x'x'} = 0, \tag{6.20}$$

$$\text{with } \theta'(0,t') = 0, \theta'(1,t') = 0, \text{ and } \theta'(x',0) = -x', \tag{6.21}$$

obtained by substituting $\theta(x',t') = \Theta(x') + \theta'(x',t')$ into the equations for θ. We'll examine these equations in the next section.

6.3.2 Simple solutions of the heat equation

In order to get an understanding of the behaviour of a physical/mathematical system defined by field equations and subsidiary conditions it's useful to first ignore the subsidiary conditions and extract particular solutions of the field equations. This can often be done by simply substituting a guessed solution form into the field equations. Sometimes this leads to an effective solution procedure for the complete problem.

Separation of variables

An elementary technique referred to as *separation of variables* is very often useful for extracting solutions to partial differential equations. One simply looks for solutions of a specific *separated* form. For linear partial differential equations this procedure leads to a full blooded solution method of broad applicability referred to as the *Fourier method* after its founder.

Thus for the homogeneous heat equation (6.20) solutions of the *separated* form $\theta'(x',t') = \mathcal{X}(x')\mathcal{T}(t')$, i.e. a function of x' times a function of t', are sought. Note that such solutions are very special in the sense that most functions of x' and t' can't be written in this form. For example $\sin(x't')$ can't be put into this form. Substituting the separated form into equation (6.20) gives

$$\mathcal{X}(x')\mathcal{T}_{t'}(t') - \mathcal{X}_{x'x'}(x')\mathcal{T}(t') = 0,$$

and after dividing through by $\mathcal{X}(x')\mathcal{T}(t')$ and rearranging, the equation *separates* into the form

$$\frac{\mathcal{T}_{t'}(t')}{\mathcal{T}(t')} = \frac{\mathcal{X}_{x'x'}(x')}{\mathcal{X}(x')}.$$

Now the left hand side of this expression is a function of t' only, while the right hand side is a function of x' alone. A little reflection will convince the reader that this situation can only arise if both expressions equate to a common constant (μ say). The implication is that the components of any separated solution of the above form must satisfy the coupled ordinary differential equations

$$\mathcal{X}_{x'x'}(x') - \mu\mathcal{X}(x') = 0, \quad \text{and} \tag{6.22}$$

$$\mathcal{T}_{t'}(t') - \mu\mathcal{T}(t') = 0.$$

Any value of the *separation constant* μ (real or complex) will generate solutions which may or may not be useful in a particular context. Some cases of special interest are now listed:

- If $\mu = 0$ the familiar steady state linear solution $\theta'(x',t') = Ax' + B$ is generated.

- For negative μ ($-\lambda^2$ say) the equation for $\mathcal{X}$ is the harmonic equation and the $\mathcal{T}$ equation is the exponential equation, so that the combinations

$$e^{-\lambda^2 t'} \times \begin{cases} \sin \lambda x' \\ \cos \lambda x' \end{cases},$$

  are solutions. Thus solutions that are "wavy" in the x' direction maintain the *same* wavy shape, but their amplitude decays exponentially in time with the shorter wave length waves (the wave length is $\ell = \frac{2\pi}{\lambda}$) decaying more rapidly than longer wave length waves.

Because the heat equation is linear, linear combinations of the above solutions are also solutions. Thus, for example, the combination

$$A_1 \sin(\lambda_1 x')e^{-\lambda_1^2 t'} + A_2 \sin(\lambda_2 x')e^{-\lambda_2^2 t'},$$

is a solution for arbitrary $\lambda_1, \lambda_2, A_1, A_2$. Note that this expression can't be written in a separable form, so that the solution class has been greatly extended by this superposition process. At any time t' this solution consists of two superimposed waves of wavelengths $2\pi/\lambda_1, 2\pi/\lambda_2$ but, because the amplitude of the shorter wave (with larger λ) decays much more rapidly than the other, the shape of the solution will change as time goes on. Thus for example if $\lambda_1 \gg \lambda_2$ so that there are ripples (with $\lambda = \lambda_1$) superimposed on long waves then, after a time t' of order $1/\lambda_1^2$, the ripples will disappear, leaving what appears to be simply a single long wave, which will in its turn disappear after sufficient time (t' of order $1/\lambda_2^2$)).

To illustrate how these ideas can be used to examine a specific situation we return to the problem of determining the transient thermal response of a rod. The relevant scaled equations to be satisfied are (6.20,6.21). The separable wavey solutions given above are appropriate, and if the cos terms are removed and the choices of λ given by

$$\lambda = n\pi, \text{ with } n \textit{ any } \text{integer}$$

are employed, then the required zero boundary conditions at $x' = 0, 1$ will be satisfied. We thus have many solutions, corresponding to different values of n (referred to as the *wave* or *mode* number) and also with arbitary multiplicative factors A_n, satisfying both the heat equation and the required boundary conditions. The problem is almost solved! All we need to do is satisfy the initial condition $\theta'(x', 0) = -x'$ (see (6.21)) and to accomplish this we have an infinite set of *modes* at our disposal (one for each n). Thus if we can find the A_n's so that

$$\sum A_n \sin n\pi x' = -x', \tag{6.23}$$

then the solution is given by

$$\theta'(x', t') = \sum A_n \sin n\pi x' e^{-(n\pi)^2 t'}. \tag{6.24}$$

The initial condition can be satisfied very closely by choosing the A's in various possible ways, however, a particular choice made by Fourier leads not only to a useful approximate answer but to an *exact* answer. We'll return to this question in Chapter 8. The exercises will illustrate these ideas.

For later purposes it should be noted that the separation of variables technique leads us to the consideration of an interesting and rather unusual question: the determination of the *solutions* $(\mathcal{X}(x'), \mu)$ of the homogeneous differential equation (6.22) *with* separation constant μ satisfying the boundary conditions $\mathcal{X} = 0$ at *two points* $x = 0$ and $x = 1$. Such a problem is an example of a class of problems referred to as *Sturm-Liouville problems* which are of major importance for Fourier work and other work. Again we'll return to this question later.

Comment: The physics of heat flow Although the details will vary with circumstances (and care is necessary) we might expect the above results to extrapolate to general 3D heat conduction problems in an obvious way. Thus we'd expect:

- The conduction process to cause a smoothing out of the temperature and flux profile with time.
- If L is the length scale associated with significant temperature variations, then the time scale for smoothing should be of order L^2/κ.
- If the heat supply rate is of scale q then over the same time scale we'd expect temperature changes of order $qL^2/(\rho c\kappa) \equiv qL^2/k$ to occur. To obtain this result scale the complete heat equation (6.13).

The exercises will reinforce these important ideas.

6.4 Mathematical Questions

- If the initial temperature state of a body is known, and the boundary conditions and heat supply rate are specified, then we would expect on physical grounds the temperature distribution at subsequent times to be determined. One might hope, therefore, that it would be possible to prove mathematically the existence of a unique solution under such circumstances.

- Since heat flows from hot to cold areas of a body then, in the absence of internal heat sources or sinks, one would expect the hotter parts of the body to get cooler and the cooler parts to get warmer. Thus, in the absence of any heat input, one would not expect hot and cold "spots" to spontaneously appear inside a body.

The above expectations serve to motivate the theorems in this section. Before proceeding with the proofs there are a number of general comments concerning the role of such theorems for modellers that are appropriate:

- It is noteworthy that it's the expectations based on physical intuition and simple observations that motivate the theorems, rather than the mathematical theorems leading to physical understanding. Although this is not always the case, and the two go hand in hand, the insight *usually* comes from the physics.

- The heat equation situation is special in that useful existence and uniqueness theorems can be proved—normally in modelling situations such theorems are just not on, even in cases in which one could bet such theorems are valid. Because such theorems are rarely available and usually the results (when available) are not in doubt in a practical sense, mathematical modellers tend to think of theorems as being irrelevant to their task. These results *are* of importance to the modeller for not so obvious reasons. Thus if one is assured by such a theorem that there exists just one solution to a given problem, then one can adopt imaginative, but not necessarily logical, arguments to discover the solution. (Often an ad hoc intuitive approach is necessary to determine the solution of the non-standard problems that arise out of real applications.) Having verified that the conditions of the problem are satisfied by the "guessed" solution, then one can be sure that it is *the* solution, and not just a mathematical curiosity irrelevant to the underlying problem.

- The approaches used to establish the theorems presented here are much more widely applicable than simply for the heat equation.

The theorems which follow provide a logical foundation to much of the work with specific models dealt with in later chapters. Having extracted a solution to a particular diffusion problem one should, if there's some doubt about the validity of the solution (or one is a purist), try to extend the following existence and uniqueness theorems to cover the non-standard circumstances of the model; often modellers don't bother with this step.

6.4.1 Uniqueness theorems for the heat equation

Theorem 6.2 *There is at most one solution, $T(\mathbf{x}, t)$, of the heat equation (6.11) in a fixed domain, D, which satisfies*

1. *the initial condition,*

$$T(\mathbf{x}, 0) = f(\mathbf{x}) \quad \mathbf{x} \in D,$$

where f is a given function,

2. *one of the three boundary conditions,*

$$T(\mathbf{x}, t) = g(\mathbf{x}, t) \quad \mathbf{x} \in \partial D, \ t > 0,$$

$$\text{or} \quad \mathbf{n}\cdot\boldsymbol{\nabla} T = h(\mathbf{x}, t) \quad \mathbf{x} \in \partial D, \ t > 0,$$

$$\text{or} \quad \mathbf{n}\cdot\boldsymbol{\nabla} T - \beta T = l(\mathbf{x}, t), \quad \mathbf{x} \in \partial D, \ t > 0,$$

where f, g, h and l are prescribed functions.

PROOF. Suppose there are two distinct solutions T_1 and T_2, satisfying the conditions of the theorem and consider their difference $u = T_1 - T_2$. We wish to show that the defining equations demand that $u \equiv 0$. Since all the equations defining the system are linear it is possible to proceed as follows: Taking the difference of the two defining field equations

$$T_{1t} = \kappa \nabla^2 T_1 + Q \text{ and}$$

$$T_{2t} = \kappa \nabla^2 T_2 + Q,$$

leads to the result that u satisfies the homogeneous heat equation,

$$u_t = \kappa \nabla^2 u. \tag{6.25}$$

By subtracting the corresponding subsidiary conditions we see that u satisfies the zero initial condition,

$$u(\mathbf{x}, 0) = 0,$$

and one of the homogeneous boundary conditions,

$$u = 0,$$

$$\text{or } \mathbf{n}\cdot\nabla u = 0.$$

$$\text{or } \quad \mathbf{n}\cdot\nabla u = \beta u.$$

Consider the positive definite functional $\mathcal{E}$ given by

$$\mathcal{E}(t) = \frac{1}{2}\int_D u^2 dV.$$

Now clearly $\mathcal{E}(t) \geq 0$ for any u because of the quadratic nature of the functional, and will take the value 0 *only if* $u(\mathbf{x},t) \equiv 0$. Now initially $u = 0$ so $\mathcal{E}(0) = 0$. Thus if we can show that $d\mathcal{E}/dt \leq 0$ is implied by the defining equations for u, then $\mathcal{E} \equiv 0$ is indeed the only viable option, and the uniqueness result follows. The following manipulations show that $d\mathcal{E}/dt \leq 0$:

$$\begin{aligned}
\frac{d\mathcal{E}}{dt} &= \int_D u u_t \, dV \\
&= \kappa \int_D u \, \nabla^2 u dV, \text{ using (6.25,)} \\
&= \kappa \int_D \left(\nabla\cdot[u\nabla u] - [\nabla u]^2\right) dV.
\end{aligned}$$

Now Gauss's Divergence Theorem (6.1) with $\mathbf{F} = u\nabla u$, can be used to convert the first volume integral to a surface integral, leading to the result

$$\frac{d\mathcal{E}}{dt} = \kappa\left(\int_{\partial D} u\,\mathbf{n}\cdot\nabla u\, dS - \int_D [\nabla u]^2 \, dV\right).$$

The surface integral vanishes if either of the first two boundary conditions applies, and becomes the non-positive expression

$$-\beta \int_{\partial D} u^2 dS,$$

if the third boundary condition applies. Thus in all cases it can be deduced that $d\mathcal{E}/dt \leq 0$, and the uniqueness result follows.

It's also clear from the above that the theorem still holds if the boundary conditions take the form of any of the alternative boundary conditions applying on portions of the boundary, but with just one of the three applying at all points of the boundary.

Pure mathematicians would spruce up the statement and proof of the theorem by invoking suitable regularity conditions. Any person who feels that the statement and proof are inadequate must surely be able to examine the proof to find what needs to be assumed, and to write their own statement of the theorem. It may also be remarked that the theorem can be generalized to deal with domains which are not fixed in time. The fact is that the theorem is true in even greater generality than the regularity assumptions required to make the statement and proof along the lines presented here entirely rigorous. To demonstrate this calls for the use of distribution theory and that calls for a more advanced level of mathematics than is assumed here. For our purposes, whenever it makes sense to think that the initial state is known and the external heating mechanisms and the boundary conditions are specified in a physically sensible way, we are justified in invoking the uniqueness theorem.

The functional $\mathcal{E}$ above is conventionally described by mathematicians as an energy functional, and the technique as an *energy technique* for proving uniqueness. In many other physical contexts, where a similar partial differential equation applies, the positive definite functional is closely related to the physical energy of the system. No such interpretation is available here; the energy content for the heat equation involves the spatial integral of the temperature (not its square). The procedure of looking for an energy integral or a positive definite form closely related to the problem of interest is a standard procedure for producing uniqueness proofs. It is often more difficult to establish the existence of a solution to a prescribed equation set; such proofs either rely on constructing convergent solution representations, or on using the *Fixed Point Theorem*. The Picard iteration procedure for ordinary differential equations is the prototype example of the first approach; and the normal proof of the *Intermediate Value Theorem* is the prototype example of the second type.

6.4.2 The maximum principle

Theorem 6.3 *If $T(\mathbf{x},t)$ is a solution of the* homogeneous *heat equation,*

$$T_t = \kappa\nabla^2 T, \quad \mathbf{x} \in D, \quad 0 < t \leq \tau,$$

for any positive value τ, then the maximum value for the solution T in the $(\mathbf{x},t)$ domain defined above occurs either at $t = 0$ or on the boundary ∂D, the union of which we'll call Γ, see Fig. 6.3.

This theorem is a consequence of the fact that the direction of heat flow, both within a body and to bodies in contact with it, is always from a higher to a lower temperature.

It is *almost* simple to construct a proof on the following trivial lines. Suppose there were such a maximum occurring at the interior point such as A in Figure 6.3 . Then at such an interior maximum

$$T_t = 0, \quad \boldsymbol{\nabla} T = \mathbf{0}, \quad \nabla^2 T \leq 0,$$

so if the condition $\nabla^2 T \leq 0$ were replaced by $\nabla^2 T < 0$, this would be in conflict with the heat equation. It could happen, however, that $\nabla^2 T = 0$

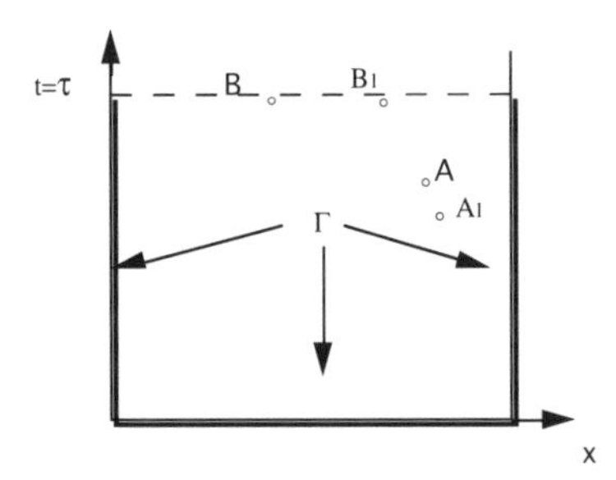

Figure 6.3: The maximum principle.

at the assumed interior maximum A; in which case no conflict would result. Similarly a maximum could occur at a point on the boundary $t = \tau$, for example B in Fig. 6.3, only if $T_t \geq 0$, and $\nabla^2 T = 0$, so the situation is again *almost* contradictory. Note that for the equation

$$T_t - \kappa \nabla^2 T = q, \quad \text{with} \quad q < 0 \tag{6.26}$$

there is no doubt—the heat extraction term q tips the balance and there cannot be an interior maximum *however small* q is! That is we cannot have an interior maximum if heat is being removed from all portions of the body. Note also that if $q > 0$, so that the body is heated, clearly hot spots could be created (any isolated source will produce such a hot spot); so the $q = 0$ case is marginal. It also happens that $q = 0$ is the most important case—such is life! The clue to the proof of our theorem lies in the small heat extraction result above and relies on the use of a very neat but technical mathematical argument.

PROOF. Suppose, contrary to theorem, T achieves its maximum outside Γ at the point $(\mathbf{x}_0, t_0)$; either in the interior of the domain at a point such as A, or on the boundary $t = \tau$ (point B say) and let its maximum value be T_0. Then, if such a situation exists, it is possible to construct another function v which is sufficiently close to T that it also has a maximum outside Γ; and yet it satisfies the heat equation with heat removal i.e. equation (6.26); the contradiction is thus exhibited. The v that does this is

$$v = T - \epsilon(t - t_0),$$

for sufficiently small positive ϵ. To see this, note:

- v satisfies (6.26) with $q = -\epsilon$,

- that $v(\mathbf{x}_0, t_0) = T_0$, and if ϵ is chosen so that $\epsilon t_0 < (T_0 - T_1)$ where T_1 is the least upper bound of values of T taken on Γ, then a little thought will convince the reader that the maximum value of v will be reached at some point not on Γ, eg. at B_1 or A_1, see Fig. 6.3.

The proof thus carries through.

It is also a simple matter to show that the solution minimum is reached on Γ; simply note that $(-T)$ also satisfies the homogeneous heat equation. This result permits an alternative proof of the uniqueness theorem for the case in which the temperature is given on the boundary. Its importance however lies in the fact that it generalizes to the steady state equation, with a modified trick function. In one such context (electrostatics) it has the remarkable interpretation that there can be no arrangement of particles, acting under an inverse square law of force, which can be stable. This result, in conjunction with the Rutherford scattering experiments confirming the applicability of the inverse square law, destroyed the possibility of building classical models of atomic structure and thus led to the development of quantum mechanics.

6.5 Summary

In this chapter the equations governing heat transport in stationary materials were discussed and theoretical results concerning the solutions to these equations were established. Scaling arguments were used to determine the time and temperature scales that are likely to occur in commonly occurring situations. Thus, for example, if L is the length scale over which significant temperature variations occur, then over a time scale of order L^2/κ these variations are likely to be smoothed out by conduction. Separation of variables ideas were used to extract simple exact solutions of the heat equation, which demonstrated diffusive smoothing.

6.6 Exercises

Exercise 6.1 *Linearity*

For the purpose of this exercise suppose we know very little about the physics of heat flow. Let's assume we simply suspect that a linear isotropic (i.e. direction independent) process governs the temperature distribution within a material. Thus we suspect that $T(\mathbf{x})$ is governed by an equation of the form

$$\mathcal{L}(T(\mathbf{x}, t)) = 0$$

where $\mathcal{L}$ is some linear operator.

(a) $\star$ Describe an experiment or experiments for both steady state and unsteady state situations that could be performed on a rod or plate of the material to test for linearity and isotropy. What materials would you expect to exhibit non-isotropic and/or nonlinear conduction behaviour?

(b) ⋆ Assuming that experiments confirm that the process is linear and isotropic, use linearity, everyday experience, and commonsense arguments to determine the steady state temperature at the centre of a square plate with three of its edges maintained at temperature T_0 and the remaining edge maintained at temperature T_1.

Hint: If all four edges are maintained at temperature T_0 what will be the temperature at the centre? If two opposite edges are maintained at T_1 and the remaining edges are at T_0, what will be the temperature at the centre?

Note that no detailed understanding of the physics of heat flow is necessary to determine the steady state temperature at the center of the plate; symmetry and linearity ideas are sufficient. These arguments do not, however, determine the temperature at *any* other point in the plate and tell us nothing about the physics of heat flow except that it's a linear process.

(c) Can the temperature at the centre of a rectangular plate, with one edge maintained at a temperature T_0 and the others at T_1, be determined using just linearity and symmetry arguments ? Using the heat equation write down a dimensional statement concerning the temperature behaviour at the centre of this plate.

Exercise 6.2 *Rod Heating*

⋆ A rod of cross-sectional area A and length L is initially at temperature T_0 and is supplied with heat at a uniform and constant rate q_0 per unit rod length. Both ends of the rod are maintained at temperature T_0. The objective is to determine how the thermal and geometric parameters influence the temperature variation $T(L/2, t)$ at the midpoint of the rod.

(a) It's useful to think about what's likely to happen. Very briefly describe the way you would expect the temperature profile to develop. Scale the describing equations and thereby determine the temperature change and time scales that are appropriate for the problem. Show that the scaled heat equation reduces to

$$U_{t'} = U_{x'x'} + 1.$$

Hint: Take care with the dimensions of q_0.

(b) Using the above scaling results write down a statement (of the dimensional type) displaying the dependence of $T(L/2, t)$ on the parameters of the problem, and describe a minimal set of experiments that could be performed to determine the required midpoint temperature vs. time variation for different size rods under different (but uniform) heat supply conditions.

(c) Obtain the steady state temperature distribution for the scaled equations and thus obtain an explicit expression for the maximum temperature reached at the rod's centre. Interpret your result.

(d) Under steady conditions the rate of heat supply to the rod must balance the heat loss rate due to conduction from the rod ends. Check to see if this balance is realized by your solution.

Exercise 6.3 *Rod Heating 2*

If the ends of the rod considered in Exercise 6.2 are not maintained at T_0 but are in contact with a well stirred fluid at temperature T_0, then one might expect the maximum rod temperature reached in the rod to be dependent on q_0 and the relative effectiveness of the heat conduction process within the rod and heat removal process by the fluid, as expressed in an appropriate dimensionless group. Under the circumstances specified, the Newtonian Law cooling description, see (6.9)

$$-k\frac{\partial T(L,t)}{\partial x} = \beta(T(L,t) - T_0),$$

describes the heat loss from the end at $x = L$, where the transfer coefficient β depends on the properties of the fluid and the stirring rate.

(a) Scale the equations and thus show that the combination $\mu = \beta L/k$ is the appropriate dimensionless group described above.

(b) Solve the steady state equations and thus explicitly determine the maximum temperature reached in the rod. Examine and comment on the dependence of the maximum rod temperature on μ. Especially comment on the small and large μ solution limits.

Exercise 6.4 *Phosphate Problems*

As a result of run off from the surrounding land following a rain storm, a large lake of uniform depth h receives a sudden influx of a soluble phosphate compound in its upper layers. The phosphate then diffuses downward into the rest of the lake. Assuming that the phosphate concentration c is a function of depth z (measured positively downwards from the lake's surface), and time t only, and that the rate at which phosphate diffuses across any horizontal cross-section is proportional to $\partial c/\partial z$, show that c satisfies the Diffusion Equation. Explain why reasonable additional conditions to be satisfied by c are $\partial c/\partial z = 0$ at $z = 0$ and $z = h$. Determine the steady state solution. Comment.

Hint: Either consider the conservation of phosphate in a layer lying between z and $z + dz$, or use the general approach employed in Section 6.3 to determine the 3D phosphate diffusion equation, and then specialize to the one-dimensional case.

Exercise 6.5 *Insulation Effectiveness*

An air conditioner has the capacity to remove heat from a room at a rate P and is to operate when the external temperature is T_a. After the conditioner is turned on the temperature of the room reduces until a balance is reached between the removal rate of heat by the conditioner and the rate at which heat is being conducted into the room through its walls. Determine the steady uniform temperature T_r which the conditioner could maintain within the room if the surface area the room presents to the environment is S for the various wall structures listed below, and comment on the results obtained. Assume $T = T_a, T_r$ represent reasonable approximations to the boundary conditions on the exterior and interior walls of the room respectively.

1. The surface consists of a single layer of material of thermal conductivity k and thickness d.

2. The surface consists of an external metal cladding of thickness d_m and thermal conductivity k_m with an inner insulating layer of thickness d_i and thermal conductivity k_i. If the ratio of the thermal conductivities k_m/k_i is large in comparison with the ratio of the thicknesses, which layer essentially determines the difference in temperature which can be maintained? What would be the situation if there were more layers involved?

3. ⋆ The surface is made up of walls and windows and so consists of two distinct wall structures; a single layer of glass (which is a moderate insulator), and a good insulator inside a cladding. Extend the earlier calculations to cover this case by introducing relevant parameters to define the problem. What dimensionless group determines the effect of glass areas on the room's insulation? Using Table 6.1 and your experience make some quantitative assessments.

Confined layers of air, when thin enough to prevent the development of convective motions, have a low thermal conductivity. Explain the use of double glazing.

Exercise 6.6 *Bush Fires*

The following (purely algebraic) simple model underlies much of the predictive work done on wild fires. Using such models, predictions are made for the speed of travel of fires under varying fuel, moisture level and atmospheric conditions, with different wind speeds, and in varying terrain. With such predictions, fire fighters can determine the locations where they should attempt to control a fire. Forrest workers collect data on the (dry) mass of fuel w per unit ground surface area (mainly litter) and the moisture content m (measured as a fraction of w), and record wind speeds etc. and topography.

Before a fuel element will burn the moisture must be driven from it, and its temperature must be raised to the ignition point T_{ign} for that fuel. The advancing fire front provides the heat input, so that the speed of travel must be such that the total heat received by a fuel element is just sufficient for its temperature to be raised to ignition before it's engulfed by the fire.

(a) If Q (joules/m^2) is the total heat input from the fire being absorbed by fuel elements per unit ground surface area as they approach and are engulfed by a line flame front, see Fig. 6.4, show that a reasonable mathematical expression of the above description is given by

$$Q = w[c_f(T_{ign} - T_0) + mc_w(T_{bp} - T_0)],$$

where T_0 is the initial temperature of the fuel and T_{bp} is the boiling point of water; the c's are specific heats.

Straightforward (but tedious) geometric calculations lead to the result that the total heat input (joules/m^2) per unit ground surface area received from a flame front is given by

$$Q = \gamma \frac{EL}{2V}(1 + \cos\alpha),$$

where V is the flame front speed, E is the radiation intensity of the flame (i.e. the heat output per unit time per unit area of the flame surface), L is the "length " of the flame (see Fig. 6.4), and α is the angle the flame front makes with the ground in the direction of propagation. The constant γ is included to account for the many sources of heat loss. The radiation intensity is assumed to be constant.

The flame length L will be dependent on the rate of fuel burning Vw and experiments (together with dimensional arguments) suggest the result

$$L = \nu(wV)^{2/3}$$

where ν is a dimensional constant.

(b) Combine the above results and thus determine an expression for the speed of travel V as a function of the fuel characteristics and the angle α of the flame front.

(i) According to the model does fire speed increase with the quantity of fuel? In the light of your experience does this result make sense?

(ii) Why do fires travel more rapidly in the wind direction?

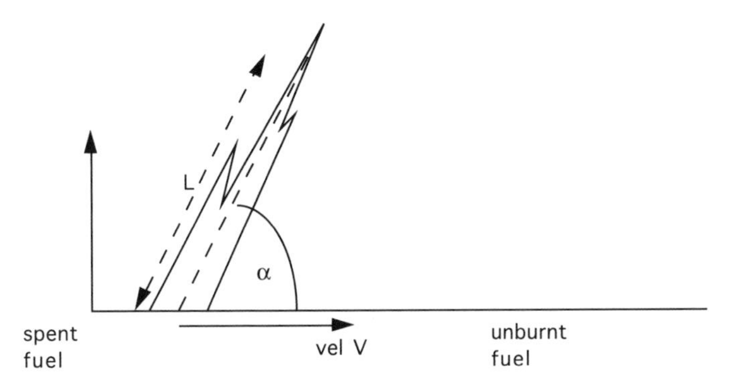

Figure 6.4: Bush fires.

(iii) Should fires travel more rapidly up hill or down hill?

(iv) Would you expect fires to travel more rapidly in variable terrain? In variable wind conditions?

Exercise 6.7 *Flame Propagation*

⋆⋆ Under many circumstances a flame front is seen to propagate with constant speed through a medium. Why should this happen and what determines the propagation speed?

The parameters usually used to characterize a chemical reaction are the reactant concentrations, the ignition temperature T_{ign}, the heat of reaction $\mathcal{R}$, and the time span for the reaction to proceed to completion τ. At temperatures below the ignition temperature the reaction proceeds at a negligible rate, while above this temperature the reaction proceeds at the maximum rate allowed by the concentrations of reactants. Thus the heat from the chemical reaction raises the temperature ahead of the flame front to the ignition temperature and the reaction then proceeds to completion. The heat transfer mechanism from the flame to the fuel will vary with circumstances; we'll assume a diffusive process operates. The above description provides us with a very useful simple model for describing flame propagation. An identical model is used to describe ablation processes.

A one-dimensional steady propagating front will only be possible if the heat release rate per unit area of the flame front q is constant, and the temperature profile remains fixed relative to the moving front.

(a) We first check to see if the one-dimensional heat equation admits a steady solution relative to a co-ordinate system moving at the propagation speed U. Show that a change in variables $(x,t) \rightarrow (\xi,t)$, where $\xi = (x - Ut)$ transforms the homogeneous heat equation to the form

$$\frac{\partial T(\xi,t)}{\partial t} - U\frac{\partial T(\xi,t)}{\partial \xi} = \kappa\frac{\partial^2 T(\xi,t)}{\partial \xi^2}.$$

What does the equation reduce to if the temperature profile is steady relative to this moving frame?

(b) Determine the solution of the resulting steady equation in the region ahead of the flame front $\xi > 0$, and thus determine the heat flux required at the origin (the flame location) to maintain a temperature excess at the origin over that at $x = \infty$ of $T_{ign} - T_{inf}$.

(c) A propagating flame front is usually modelled as a front of the above type advancing ahead of a constant temperature region (the "burnt" region). The reaction zone is thus assumed to be a sheet of infinitesimal thickness (the "flame sheet"), and conductive heat losses into the "burnt" region are ignored. Under these circumstances determine the speed of propagation corresponding to a given flame heat flux q.

To obtain a unique expression for the propagation speed it is necessary to use one's understanding of the chemistry to obtain a second relationship connecting the heat production rate q to chemical composition etc. Our understanding of the chemistry in most circumstances is so limited that empirical results usually need to be used for this purpose.

(d) Obviously treating the reaction zone as a flame sheet only makes sense if the reaction zone is thin compared with the diffusion zone. Under what circumstances would you expect the flame sheet approximation to be adequate?

Exercise 6.8 *Drying plant canopies*

If heat is continuously applied to the surface of a semi-infinite conducting body, then the heat continues to penetrate to greater and greater depths as time goes on, with the result that both temperature levels and the depth of penetration increase with time. If, however, heat absorption occurs within the body (due to some chemical reaction or change in state) then one might expect the depth of penetration to be finite, with a steady state situation being eventually realized when the heat input rate is balanced by the chemical heat absorption rate. We'll determine the steady state depth of penetration which is of importance in many contexts. The situation described here occurs when heat penetrates a wet plant canopy. Also moisture level predictions for bush fire fuel are made using such drying out models. The oxidization of metals is another related problem.

A semi-infinite body, whose material has specific heat c, density ρ and thermal conductivity k, has the further property that, when its temperature T is above a level which may be taken as zero, it absorbs heat at a rate αT per unit volume as a result of a change of state occurring. The body is initially at the zero level temperature and thereafter heat is supplied at the surface at a rate q per unit area.

(a) Write down the partial differential equation, the boundary conditions, and the initial conditions for this problem.

(b) Using these equations determine the temperature rise scale, the penetration depth scale, and the time scale over which temperature increases may be expected.

(c) Determine the scaled equations.

(d) Find a steady state solution of these equations.

(e) Write down the mathematical conditions to be satisfied by a corrective function $U(x', t')$ in order that the sum of the steady state solution and $U(x', t')$ will satisfy all the required conditions. Do not attempt to solve the equations for U. Techniques to obtain this solution are developed in the next chapter. In practical circumstances this transient behaviour is important eg. in agricultural circumstances it's important to know how many days it will take for a canopy to dry out.

Exercise 6.9 *Separated Heat Solutions*

Using Maple plot out temporal changes in a temperature profile whose initial shape contains two or more modes of the wavy form obtained in Section 3.2 (make your own choice of modes). Make any observations that you feel are significant.

Exercise 6.10 *Separated Wave Solutions*

If appropriate scales (l, $\sqrt{m/Tl}$) are employed, the *wave equation*

$$\eta_{tt} + \gamma\eta_t - \eta_{xx} = 0 \;\text{ in } 0 < x < 1, \;\text{ with }\; \eta(0) = \eta(1) = 0,$$

describes the damped small amplitude lateral vibrations of an elastic string, see Exercise (2.7) in Chapter 2.

(a) (i) Determine separated solutions of the undamped $\gamma = 0$ wave equation that are oscillatory in the x direction, and by choosing the separation constant so that the boundary conditions are satisfied, show that *standing* wave displacement patterns of the form $\sin n\pi x \cos \omega t$ can be generated. Determine the frequency of vibration ω associated with different wave vibration modes (as specified by n). How might one such mode be generated? Do the results match common knowledge about the behaviour of stringed instruments? Explain.

(ii) Examine the behaviour of superimposed waves. How will the shape of the vibration pattern change as time goes on? Use Maple to graph evolving spatial patterns.

(b) **(i)** Use the separation of variables procedure to produce separated solutions for the damped string equation ($\gamma \neq 0$) satisfying the boundary conditions. In particular examine modes whose shape at $t = 0$ are given by $\sin n\pi x$, and thus make comparisons with the undamped cases examined above. How does a relatively small damping affect the evolving shape of a pure vibration mode, or a combination of modes?

(ii) Use Maple to graph evolving spatial patterns.

(c) $\star$ Separation of variables doesn't always work. Try the procedure on the equation

$$Y_{tt} + \alpha Y_{xt} - Y_{xx} = 0.$$

It's much more difficult to deal with such cases. It's however often possible to guess special solutions in such cases. Obtain a solution of this equation of the form $e^{at}X(x)$.

Bibliography

Carslaw, H.S. and Jaeger,J.C. (1959). *Conduction of Heat in Solids*, Oxford at the Clarendon Press.

Hill, James M. and DeWynne, Jeffrey N. (1987). *Heat Conduction*, Blackwell Scientific Publications, Oxford.

Kreysig, Erwin (1983). *Engineering Mathematics*, John Wiley & Sons.New York.

Chapter 7

Surface heating

Similarity solutions
Fundamental solutions
The method of images
Boundary integral methods
Generalised functions

7.1 Introduction

Because the heat equation is linear, superposition of solutions is possible and thus from a stock of simple solutions it's possible to build up solutions to complex problems. Thus the temperature distribution in a body can be thought of as being due to the additive influence of the various external and boundary agents affecting the heat flow. There are in fact special solutions of the heat equation that are sufficiently fundamental that the solutions of very broad categories of heat conduction problems can be written down immediately in terms of these *fundamental solutions*. Even in cases when this is not possible it's still the case that these solutions usually play an essential role in determining the solution. These *fundamental solutions* correspond to the temperature distribution due to an external "pulse" (i.e. an instantaneous concentrated source) of heat. One such fundamental solution associated with 1D heat flow problems is that due to an instantaneous plane heat source. This solution will be determined in Section 7.3 and the use of this solution for building other solutions, that are also broadly applicable to particular subclasses of 1D problems of special interest to us, will be discussed in Section 7.4.

The major modelling applications of interest to us involve in one way or another the surface heating of bodies. In such circumstances mathematical complications arise because the amount of heat being exchanged between the body and its environment is not prescribed a priori and needs to be determined as part of the solution process; indirect solution methods are thus required. Thus, although the fundamental solutions developed in Section 7.4 cannot be used directly to produce exact solutions, effective techniques for handling such situations can be produced using such solutions. The method employed to do this is referred to as the **boundary integral method** and is a *very* effective procedure for handling broad classes of linear partial differential equation problems arising out of modelling. Basically this method is favoured because it serves to reduce the computational size of problems enormously; thus the method converts many problems that are too large for present computer facilities to ones that are manageable.

To describe the sources generating fundamental solutions it's necessary to extend our ideas of what we mean by a function to include *generalised functions.* This fascinating and important topic will be discussed briefly in Section 7.7 but will not, however, be required for further work in this book and thus is not required reading. A number of interesting applications of techniques developed in this chapter (nuclear fall-out, fire, insulation design etc.) will be presented in the exercise set. First of all, however, some mathematical preliminaries will be presented.

7.2 Mathematical Background

1. **Order notation** It is extremely useful to have a mathematical notation which serves to indicate the size of terms. There is a convenient notation in mathematics which orders terms according to their size in a mathematical limit. We describe the results here for a small parameter t. The corresponding results for large values of T are obtained by applying those for small t, for the small parameter $1/T$.

   (a) $O(t)$. A function $f(t)$ is said to be $O(t)$ i.e. of large order t, if there is a constant C such that $|f/t| < C$ for t sufficiently small. It's not necessary for the ratio to yield a well defined limit as $t \to 0$. The notation merely indicates that f is about the same size as t for small t. Thus $t\sin(1/t)$ is of order t.

   (b) $o(t)$. A function $f(t)$ is said to be $o(t)$ i.e. of small order t, if $\lim f/t$ is zero in the limit as $t \to 0$. The notation indicates that the function is smaller than t for sufficiently small t. Thus $t\sin(t)$ is $o(t)$ or $O(t^2)$.

   Of course in modelling it's the relative *numerical* size of terms that matters rather than the size in some limit; it's as well to keep this in mind when making approximations. Maple will determine small and large parameter and variable limits and expansions for functions (use `series` and `asympt`); although, at the time of writing, with limited success.

2. **The error functions** A standard mathematical function, which arises often in heat transfer problems, is the error function $\mathrm{erf}(x)$ defined by

$$\mathrm{erf}(x) = \frac{2}{\sqrt{\pi}} \int_0^x e^{-u^2}\, du. \tag{7.1}$$

The multiplicative factor has been chosen so that $\mathrm{erf}(\infty)$ is unity. Most readers would have encountered this function in a statistics context. Note that

$$\mathrm{erf}(x) = -\mathrm{erf}(-x).$$

There is a useful closely related function denoted by erfc defined by

$$\mathrm{erfc}(x) = 1 - \mathrm{erf}(x) = \frac{2}{\sqrt{\pi}} \int_x^\infty e^{-u^2}\, du. \tag{7.2}$$

This function is referred to as the complementary error function. These functions are known to Maple.

3. **The Beta and Gamma functions** The Gamma function $\Gamma(x)$ defined by

$$\Gamma(x) = \int_0^\infty t^{x-1} e^{-t} dt \tag{7.3}$$

is a generalization to real and complex numbers of the common factorial function $n! = n(n-1)(n-2)\ldots 1 \equiv \Gamma(n+1)$ with the important properties

$$\Gamma(x+1) = x\Gamma(x) \tag{7.4}$$

$$\Gamma(1/2) = \sqrt{\pi} \quad \text{and} \quad \Gamma(1) = 1. \tag{7.5}$$

The Beta function $B(x, w)$ is defined by

$$B(x, w) = \int_0^1 t^{x-1}(1-t)^{w-1} dt, \tag{7.6}$$

and is related to $\Gamma(x)$ by

$$B(x, w) = B(w, x) = \frac{\Gamma(x)\Gamma(w)}{\Gamma(x+w)}$$

Both functions arise frequently in conduction problems and elsewhere, and are known to Maple.

7.3 The Fundamental Plane Source Solution

The idea is simple! If we can determine the temperature distribution due to the instantaneous input of heat at a point in space and time then it should be possible to use this as a building block to generate the temperature distribution due to *any* heat input in space and time by simply adding up the contributions due to such sources distributed in space and time. For the purpose envisaged it is necessary to have a heat input that is instantaneous in time and infinitesimal in space because (it seems likely that) any finite space and/or finite time span source could not be used to generate sources and solutions corresponding to a more refined heat input structure than itself—thus continuous sources generally could not be described by any building block source that's finite. Now there are technical mathematical difficulties (the normal functions of analysis can't be used to describe such a source), and real physics difficulties (inevitably infinite temperatures will be generated). However, we will forget about these difficulties for the time being and proceed on the assumption that it all makes sense.

We'll work with the 1D plane source case in which a finite amount of heat per unit area is *instantaneously* transferred into a *plane surface* located in an infinite conducting medium, see Fig. 7.1.

Since there are no length or time scales associated with such an instantaneous plane source it's completely characterized by:

- the location $x = 0$ (say),
- the input time $t = 0$ (say), and
- the magnitude of the heat input Q per unit surface area.

If $T(x,t)$ is the temperature change brought about by the source then the equations governing the temperature distribution after the heat input from the source are:

$$T_t = \kappa T_{xx}, \quad -\infty < x < +\infty, \quad t > 0, \tag{7.7}$$

$$T(x,0) = 0, \quad \text{for} \quad x \neq 0, \tag{7.8}$$

$$\lim_{x \to \pm\infty} T(x,t) = 0, \tag{7.9}$$

with

$$\int_{-\infty}^{\infty} T(x,t)\, dx = Q', \quad \text{where} \quad Q' = Q/\rho c \tag{7.10}$$

is the heat input per unit area scaled relative to the thermal mass ρc per unit volume. The initial condition (7.8) specifies that all the heat has been input on the plane $x = 0$ at time $t = 0$, and the last equation (7.10) serves

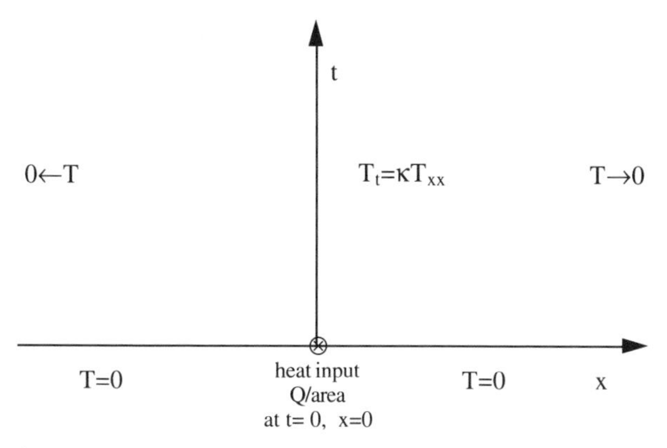

Figure 7.1: The instantaneous plane source problem.

to quantify the external heat input to this surface; our source as prescribed by the above list is thus completely specified in the equation set.

Comment: Note that we've carefully avoided specifying what happens *at* $t = 0, x = 0$.

Obviously since there's no mechanism for heat gain or loss, the total heat content in the material remains fixed as time goes on, as the last equation indicates.

The only parameters occurring in the above equation set are x, t, κ, and Q' so the solution for T must be of the form

$$T = T(x, t, \kappa, Q'), \tag{7.11}$$

and we now use dimensional arguments to determine viable solution forms. It's noteworthy that for any *finite* source additional parameters specifying the size and shape in time and space of the source would be necessary. We'll see this is a crucial matter.

7.3.1 Dimensional arguments

Now, of the bracketed parameters in the expression for T, *only* Q' contains the dimensions of temperature, so that any possible dimensionless groups cannot contain the parameter Q'. Thus the dimensionless parameters must consist of possible combinations of the remaining parameters and variables (κ, x, t), and we know from the work of the last chapter that $x/\sqrt{\kappa t}$ provides an appropriate dimensionless combination. Furthermore (ignoring equivalent combinations) it's the only dimensionless combination that can be formed from the stated parameters and variables. For a finite source other groups would arise. It's also clear from the heat content equation (7.10) that Q' has the dimensions $[\Theta][L]$ so that Q'/x or equivalently $Q'/\sqrt{\kappa t}$ provides the required scaling term for temperature. Thus dimensional arguments tell us that the solution that we seek must be of the form

$$T = \frac{Q'}{\sqrt{\kappa t}} U(\xi), \quad \text{where} \quad \xi = x/\sqrt{\kappa t}. \tag{7.12}$$

The implication of the above arguments is that a change in variables

$$(T, x, t) \rightarrow (U, \xi, t)$$

must reduce the original partial differential equation for T in (x, t) to an ordinary differential equation for U in ξ—*an immense simplification!* Of course if the resulting equation with appropriate boundary conditions has a solution then our assumption (based on instantaneous plane source ideas) that there exists a solution with the simple structure envisaged is confirmed. If, on the other hand, this equation with boundary conditions has no solution, then we can only conclude that no such simplified solution is available. In either case the exercise is worthwhile.

The variable $\xi = x/\sqrt{\kappa t}$ is referred to as a *similarity variable* and the corresponding solution as a *similarity solution* because of the similarity of the solution at different times. Thus, for example, if we knew the temperature profile at any time t_0 (say) we could infer from (7.12) its value at all later times. For example, see Fig. 7.2, if T_0 is the temperature at location x_0 at time t_0 (so that $\xi_0 = x_0/\sqrt{\kappa t_0}$); then at time $4t_0$ the location $2x_0$ also has $\xi = \xi_0$, so from (7.12), $T(2x_0, 4t_0) = T_0/2$. Apart from the mathematical simplification (*P.D.E.* $\rightarrow$ *O.D.E.*) and information compression implied by this result, there are obvious experimental implications. Thus to experimentally determine the complete temperature pattern $T(x, t)$ it's sufficient to measure the temperature at one location for all time; the complete ξ range will be covered in this way. (It's easier to do this than to measure T at a given t for all x.)

7.3.2 Invariant transformations

Before proceeding to determine the similarity solution we'll examine an alternative approach for obtaining such similarity solutions. This purely mathematical approach is equivalent to the dimensional analysis approach in our particular case but in general is *much* more powerful. The approach seeks to determine the variable transformations that leave the partial differential equation and subsidiary conditions unchanged in mathematical form. The idea is that if such transformations can be determined then one can learn something about the solution structure by exploiting them; for example in our case we can identify the similarity variable. Lie Group Theory is the name used to describe this work which has been of major importance in many areas of science but has been of particular importance

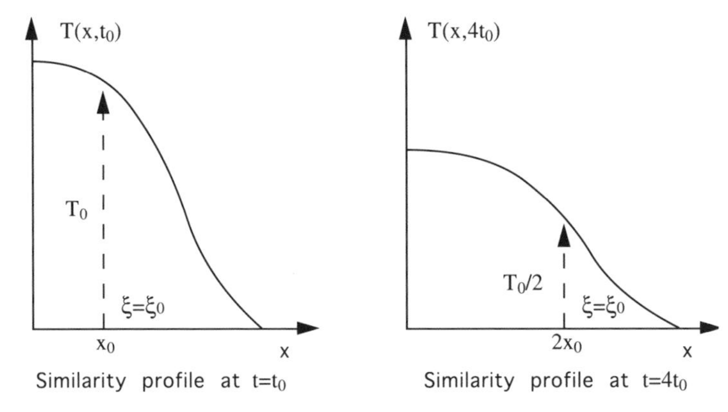

Figure 7.2: Similarity solution profiles.

in understanding the physics of phase change (liquids to solids etc.). The transformations examined in Lie Group theory are quite general and the theory enables one to simplify the process of determining such transformations. To illustrate the procedure we examine the effect of the restricted class of *stretching* transformations

$$T^*(x^*,t^*) = \gamma T(x,t), \quad x^* = \alpha x, \quad t^* = \beta t$$

on the heat equations (7.7,7.8,7.9,7.10) of present interest. In the new stretched variables the equations become:

$$T^*_{t^*} = \kappa(\frac{\alpha^2}{\beta})T^*_{x^*x^*},$$

$$T^*(x^*,0) = 0, \quad \text{for } x^* \neq 0,$$

$$\lim_{x^* \to \pm\infty} T^*(x^*,t^*) = 0,$$

$$\int_{-\infty}^{\infty} T^*(x^*,t^*)\,dx^* = \alpha\gamma Q'.$$

Now if $\alpha = \beta = 1$, we of course recover the original equations, but are there other possibilities? The heat equation remains unchanged in form (i.e. invariant) if $\alpha^2 = \beta$ and the integral condition remains invariant if $\alpha\gamma = 1$. Thus if $\alpha = \sqrt{\beta}$ and $\gamma = 1/\sqrt{\beta}$ all the defining equations are unchanged by the transformation. The possible invariant stretching transformations are thus given by

$$T^* = \frac{1}{\sqrt{\beta}}T, \quad x^* = \sqrt{\beta}x, \quad t^* = \beta t,$$

for arbitrary β. Thus, manipulating these equations, we obtain

$$\beta = (\frac{x^*}{x})^2 = \frac{t^*}{t} = (\frac{T^*}{T})^{-2},$$

which implies a solution of the form $T(x,t) = 1/\sqrt{t}F(x/\sqrt{t})$. This is of course equivalent to the result (7.12) obtained using dimensional arguments.

Comment: It is interesting to note that the more general Lie group theory (which looks for more general invariant transformations than the stretching transformations examined above) uncovers another 5 non-trivial invariant transformations of the heat equation which lead to solutions that do not satisfy the subsidiary conditions (7.8, 7.9, 7.10) and so are not of interest in the present context. Maple has a Lie Group package with limited capabilities. You might like to check it out.

7.3.3 The similarity solution

We now return to solving for the similarity solution. Recall that our dimensional arguments suggested the substitution $(T, x, t) \to (U, \xi)$ where

$$T(x,t) = \frac{Q'}{\sqrt{\kappa t}}U(\xi), \quad \text{and} \quad \xi = x/\sqrt{\kappa t}. \tag{7.13}$$

If our arguments are correct then, after performing the indicated change of variables and simplifying, we should be left with equations that *only* involve the U's and ξ's i.e. there should be no x's or t's left over. This provides us with an excellent check for both the scaling arguments and our algebra. Now using operator notation we have

$$T_t(x,t) = [\frac{\partial}{\partial t} + \frac{\partial \xi(x,t)}{\partial t}\frac{\partial}{\partial \xi}][\frac{Q'}{\sqrt{\kappa t}}U(\xi)],$$

and, after performing the indicated differentiations and eliminating (x,t) in favour of (ξ, t), we end up with

$$T_t(x,t) = -\frac{Q'}{2t\sqrt{\kappa t}}[U(\xi) + \xi U'(\xi)].$$

A similar calculation gives

$$T_{xx}(x,t) = \frac{Q'}{(\kappa t)^{3/2}}U''(\xi).$$

Notice that both T_t and T_{xx} contain the factor $1/t^{3/2}$ so that, after substituting these expressions into the heat equation (7.7), t disappears from the expression leaving

$$U'' + \xi U'/2 + U/2 = 0. \tag{7.14}$$

The integral condition (7.10) becomes

$$\int_{-\infty}^{\infty} U \, d\xi = 1.$$

Notice that as $t \to 0$ or as $x \to \pm\infty$, $\xi \to \pm\infty$ so that the three conditions (7.8,7.9) on $T(x,t)$ collapse onto the two conditions

$$U(\pm\infty) = 0,$$

in the ξ variable. Alternatively, noting the symmetry that's evident from the physics or the differential equation, we can work in the $\xi > 0$ domain with boundary conditions

$$U(\infty) = 0, \quad U'(0) = 0, \tag{7.15}$$

and integral condition

$$\int_0^\infty U \, d\xi = \frac{1}{2}. \tag{7.16}$$

It remains only to solve the second order ordinary differential equation (7.14) subject to the three conditions (7.15, 7.16). All our expectations and hopes concerning the availability of a simple plane source solution hinge *crucially* on the existence of a unique solution to this problem. Here is a situation in which theoretical issues need to be resolved to answer a very practical question. Thus a numerical evaluation displaying a solution with the correct behaviour would be unconvincing, given the delicacy of the mathematics and importance of its consequences. Before running to Maple or using a differential equation database such as Kamke (1948) to search for a solution, we should think about what's involved.

The equation is linear, *second* order, non-singular, and *three* conditions need to be satisfied—doesn't look good! If the conditions were initial conditions there would be no hope (the theory in this case is clear-cut), but the conditions are non-standard (2-point boundary conditions, infinite domain, integral condition) so the outcome is not predicted by standard theory and a solution is possible. Since all points of the finite domain are *ordinary* points for the differential equation we know it will have two well behaved independent solutions in the finite domain, only one of which (with its attached arbitrary constant) will have zero slope at the origin. This is we hope the symmetric solution needed. Thus the problem will have a unique solution if and only if the solution with zero slope at the origin is also integrable i.e. this solution should decrease rapidly enough at infinity for the integral to be finite. Note that the arbitrary constant multiplying the integrable solution can always be adjusted so that the integral condition (7.16) can be satisfied.

Maple's response to the command

```
dsolve(diff(y(x),x,x)+x*diff(y(x),x)/2+y(x)/2,y(x));
```

gives, with some minor modifications for the way it represents arbitrary constants and the standard mathematical numbers,

$$C_1 e^{-x^2/4} - \imath C_2 [\sqrt{\pi}/2] \mathrm{erf}(\imath x/2) e^{-x^2/4}.$$

This is a little disappointing; we would have preferred real solutions to our real equation! The first solution is real but an understanding of complex analysis is necessary to make sense out of the second term. Note, however, that the first solution is integrable and has a zero slope at the origin so it's precisely the one we want, and we *do* have the unique solution required.

Aside: It would be nice to identify the second real solution, even if only to confirm the above theoretical conclusions. To do this we can take advantage of the first solution to generate another real solution using the *reduction of order* technique. You'll recall that one looks for a solution of the form $Y(x)e^{-x^2/4}$. Using Maple this leads to a first order ordinary differential equation in Y', which can be solved to give the second real solution

$$e^{-\xi^2/4}\int_0^\xi e^{y^2/4}\,dy. \tag{7.17}$$

This clearly has a non-zero slope at the origin (look at its Taylor expansion), and furthermore it can be shown that it behaves like $2/x + O(x^{-2})$ for large values of x, and so vanishes at ∞ but not sufficiently rapidly to be integrable. Although Maple has facilities for determining this asymptotic behaviour (use `asympt`), it fails to provide the answer in this case. Such asymptotic results can be obtained by integrating by parts, see Exercise 7.2.

Returning to the problem in hand, the solution satisfying the required equation and boundary conditions is

$$U = Ce^{-\xi^2/4},$$

and the constant C needs to be chosen so that the integral condition (7.16) is satisfied. Maple is up to this task and yields the result $C = 1/2\sqrt{\pi}$. Thus there is just one solution

$$U(\xi) = \frac{e^{-\xi^2/4}}{2\sqrt{\pi}}, \tag{7.18}$$

see Fig. 7.3, that satisfies all the conditions of the problem.

In terms of the original physical variables, see (7.13), the solution is

$$T(x,t) = \frac{Q'}{2\sqrt{\pi\kappa t}}e^{-x^2/(4\kappa t)}. \tag{7.19}$$

This is the fundamental instantaneous plane source solution we've sought.

The solution is well behaved except possibly at $t = 0$. Note that for $x \neq 0$ as $t \to 0$ the exponentially small term on the top line dominates the algebraically small term on the bottom line so that $\lim_{t\to 0} T(x,t) = 0$, as required by the initial conditions. If however $x = 0$ then

$$T(0,t) = \frac{Q'}{2\sqrt{\pi\kappa t}}, \tag{7.20}$$

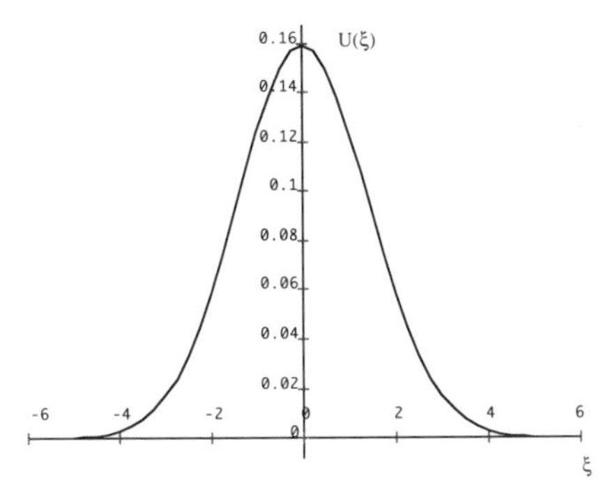

Figure 7.3: The similarity solution.

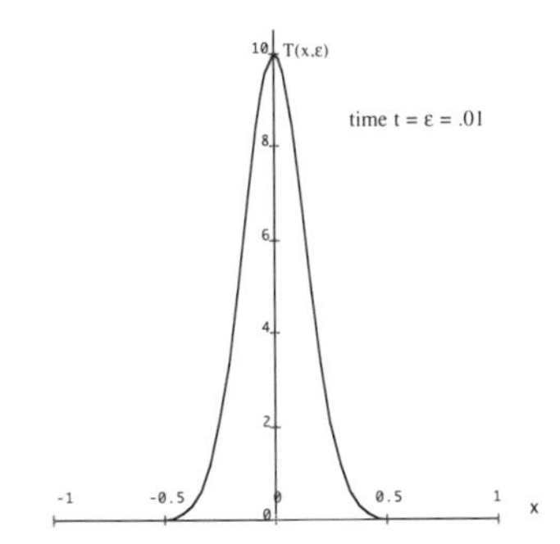

Figure 7.4: Fundamental plane source solution: small time behaviour.

so the limit is infinite as $t \to 0$—*disturbing,* but perhaps not surprising given the nature of the source. To get a better understanding of what's happening here we look at the solution for small t. For t small (ϵ say) the temperature profile $T(x, \epsilon)$ is as shown in Figure (7.4). Thus the temperature at $x = 0$ is large (of order $1/\sqrt{\epsilon}$) at $x = 0$ and the width of the profile is small (of order $\sqrt{\epsilon}$); in this way the area under the T profile (and thus the heat content) remains fixed as $\epsilon \to 0$, as required by the conditions of the problem. One can in fact think of the similarity solution $T(x, t)$ for $t > \epsilon$ as being the temperature distribution corresponding to an "initial" temperature profile of $T(x, \epsilon)$ at $t = \epsilon$—physically sensible, except in the limit as $\epsilon \to 0$. Thus, apart from the isolated non-regularity at $(x, t) = (0, 0)$ which is forced on us if we're going to talk about instantaneous plane sources, the solution makes sense physically and mathematically, and it's the *only* solution that has the desired characteristics. In spite of misgivings and *real* hazards associated with such *weak solutions,* mathematicians have learned to live with them because of the mathematical simplifications that result. We'll return later in Section 7.7 to discuss this matter further.

Interpretation

The similarity variable definition (7.13), together with the similarity solution (7.18), graphed in Fig. 7.3, encapsulate the solution behaviour. The ratio

$$r = \frac{T(x,t)}{T(0,t)} = \frac{U(\xi)}{U(0)} = e^{-\xi^2/4}$$

provides us with a measure for the change in shape of the temperature profile in space and time. Maple gives $r = 0.11$, 0.019 for $\xi = 3$, 4, with an even more rapid decay rate for larger ξ; when $\xi = 8$ the ratio is down to 1×10^{-7}! The temperature profile has thus effectively reached zero for $\xi = 4$ i.e. in terms of physical variables, for values of x given by $x = 4\sqrt{\kappa t}$, which we might as well think of as being the width of the profile at time t. Thus the highly peaked initial temperature profile decays and spreads out with the maximum temperature decaying like $1/\sqrt{t}$ and the width of the profile increasing like $\sqrt{\kappa t}$. In short, the spreading rate of the temperature profile is initially very rapid and then slows down considerably as time goes on. It's also worth noting that, although all locations (however distant) are affected by the source, the time delay for experiencing a significant effect increases rapidly with distance from the source (like x^2/κ), and also the magnitude of the effect decreases rapidly with this distance. An intuitive understanding of these results can be of great assistance when trying to assess what will happen in other related circumstances. Thus, for example, when objects at different temperatures are brought into contact one might expect an initial rapid heat exchange to occur, followed by a very slow approach to thermal equilibrium. Also one might expect the temperature of an irradiated surface at a given time to be *much* more sensitive to recently received radiation than that received at earlier times. These speculations will be confirmed by later analytic work, and of course such observations can be of great value for modelling purposes.

Aside: relativity It is interesting to note that the solution is inconsistent with the theory of relativity in that it predicts a non-zero temperature at all locations *however distant* from the source *immediately* after the heat input, see (7.19). For this to happen some energy from the source would have to travel at a speed greater than the speed of light; a situation inconsistent with relativity theory. This is a general property of the heat equation and not of this particular solution. The difficulty arises because the empirical heat flux vs. temperature gradient relation (6.4) underlying this work is based on an extrapolation from steady state situations to unsteady conditions. There are physical reasons why this extrapolation can't work. In metals, for instance, heat transfer takes place mainly as transport by the electrons and they certainly cannot travel faster than the speed of light. Of course nobody worries about this difficulty since the amount of energy predicted by the heat equation to be transferred at such speeds is so incomprehensibly small that it could not possibly be observed; even if it were present. In all ways the heat equation agrees with observation under dynamic conditions, and may be regarded as totally reliable.

Comment: similarity solutions The above similarity solution technique has been used in many areas of science and is of major importance. We'll in fact encounter the technique again when considering traffic flow problems in Chapter 14. The underlying idea is simple: if one considers a situation in which all externally imposed scales are removed, then the only scales that can arise must come out of the underlying physics (in our case as expressed in the heat equation), and must imply the existence of similarity variables that can be identified by using dimensional arguments or examining invariant transformations. Thus one examines infinite or semi-infinite time and space domain problems (any finite domain has an associated length) with the physical set-up being one which avoids externally imposed scales. It's especially important to note that *these ideas do not depend on linearity.* In fact the similarity technique is one of the very few exact techniques that can be employed to solve nonlinear problems—herein lies its real importance.

7.4 Building Solutions

Although the above similarity solution is of interest in its own right its real importance derives from its use as a building block for the solutions of a wide range of 1D heat conduction problems, some of which we'll now examine. For convenience we'll denote the temperature distribution due to the release of a unit of heat per unit area at $x = 0, t = 0$ in an infinite region by $G(x,t)$. Thus, see (7.10,7.19),

$$G(x,t) = \begin{cases} 0 & \text{for} \quad t < 0 \\ \mu \dfrac{e^{-x^2/(4\kappa t)}}{\sqrt{t}} & \text{for} \quad t > 0 \end{cases}, \tag{7.21}$$

where

$$\mu = \frac{1}{2\rho c\sqrt{\pi\kappa}} \equiv \frac{1}{2\sqrt{\pi\rho c k}}. \tag{7.22}$$

7.4.1 Infinite space problems

- Since the heat equation is invariant under a change of time origin, the temperature rise at $x = 0$ due to the release of an amount of heat $q(t')\,dt'$ per unit area in the time interval, t' to $t' + dt'$ at $x = 0$ is given by

$$dT(0,t) = [q(t')dt']G(0,t-t').$$

More explicitly using equation (7.21)

$$dT(0,t) = \begin{cases} 0 & \text{for} \quad t < t' \\ \mu \frac{q(t')\,dt'}{\sqrt{t-t'}} & \text{for} \quad t > t'. \end{cases}$$

Here $q(t')$ can be thought of as the heat supply rate per unit area at time t' and, using linearity, we can sum the contributions due to successive heat additions associated with the supply rate $q(t)$ for $t > 0$ to give in the limit (so the sum becomes an integral)

$$T(0,t) = \mu \int_0^t \frac{q(t')}{\sqrt{t-t'}}\, dt'. \tag{7.23}$$

Thus we've determined the temperature at $x = 0$ due to an arbitrary heat supply rate $q(t)$ at $x = 0$.

The evaluation of the temperature at any other location x can similarly be obtained merely by including the additional factor

$$e^{-x^2/[4\kappa(t-t')]}$$

in the integrand, as can be seen from (7.21). Thus we have the solution of all problems where the heat supply rate per unit area to a surface within the conducting body is prescribed. To understand the physics better it's as well to carry out a simple exact calculation. Certain mathematical insights will also be gained.

Example 7.4.1 *Consider the situation in which heat is supplied at a constant rate q per unit surface area to the plane $x = 0$ in an infinite region for a time interval Δ, after which the heat supply is turned off.*

Using the above result the temperature at the $x = 0$ surface is given by

$$T_s(t) = \begin{cases} \mu q \int_0^t \frac{dt'}{\sqrt{t-t'}} & = 2\mu q\sqrt{t} \quad \text{for} \quad t < \Delta, \\ \mu q \int_0^\Delta \frac{dt'}{\sqrt{t-t'}} & = 2\mu q(\sqrt{t} - \sqrt{t-\Delta}) \quad \text{for} \quad t > \Delta. \end{cases}$$

Thus the temperature rise at the $x = 0$ surface during the heating stage behaves like $\sqrt{t}$ i.e. it's rapid! Furthermore the solution is continuous but not differentiable at $t = 0$. You'll assuredly recall the difficulty we had explaining away the $1/\sqrt{t}$ singularity in the temperature profile for the instantaneous plane source solution. The time integration required to determine our present solution removes the singular behaviour. It's thus reassuring that physically unacceptable infinite temperatures don't arise when we use our fundamental solution to generate the solution corresponding to physically realizable finite sources.

Although the solution is finite at $t = 0$, the lack of differentiability at $t = 0$ can cause numerical difficulties. We'll see that numerical solution procedures are often necessary when dealing with real modelling problems and the lack of regularity of the solution at the onset of heating means that the usual numerical solution techniques for partial differential equations (being based on the assumption that solutions are smooth functions) work very poorly indeed. It's clear from the above that this lack of regularity is a basic property associated with the heat equation and the difficulty will arise whenever there's an abrupt change in heating rate. Even if the heat is gradually turned on, for example if $q(t) = rt$ for $t > 0$, the solution will behave like $rt^{3/2}$ for small t and this, while more regular, is again not well represented by approximations involving smooth functions. Where such effects are involved, accurate answers call for the use of the analytic results derived in the manner illustrated above. Algebraic packages, by using such results, can assist in producing codes that overcome such difficulties. We'll return to this question later.

Of course we can obtain the temperature distribution due to the application of a succession of heat inputs of the above type by addition. Given the immediate and rapid local temperature response to heating, it's clear that the local temperature is relatively speaking much more influenced by the recent (as opposed to the past) history of heating, and local (as opposed to remote) heating; this confirms our earlier speculations. A little experimentation using Maple can confirm these ideas, see Exercise 7.1.

For large time (use **asympt**) note that temperature levels decay like $1/\sqrt{t}$ i.e. slowly!

- The heat equation is invariant under spatial translations so that $G(x - \zeta, t)$ is the temperature distribution due to a unit of heat per unit area released at $x = \zeta$ and $t = 0$. Now the release of the quantity of heat $q(\zeta)d\zeta$ per unit surface area per unit length between locations ζ and $\zeta + d\zeta$ at $t = 0$ will instantaneously cause a rise in temperature given by $\rho c T(\zeta, 0+)d\zeta = q(\zeta)d\zeta$. By thus identifying the spatial distribution of heat input required to raise the temperature of all locations in an infinite body to a given initial temperature distribution $T(x, 0+) = T_0(x)$ we determine the subsequent temperature development to be

$$\rho c \int_{-\infty}^{\infty} T_0(\zeta) G(x - \zeta, t) d\zeta. \qquad (7.24)$$

- If $q(x',t')$ is the external supply rate per unit volume so that $q(x',t')\,dx'dt'$ is the heat input per unit surface area into the space/time element defined by (x',t',dx',dt') then the resulting temperature response is given by

$$T(x,t) = \int_0^t [\int_{-\infty}^{\infty} q(x',t')G(x-x',t-t')\,dx']\,dt'. \qquad (7.25)$$

  Heat inputs into surfaces within the material are also covered by this expression providing a *generalised function* interpretation of q is adopted. This will be discussed further in Section 7.7.

- By adding the solutions (7.24,7.25) we can thus obtain the solution to the general 1D heat conduction problem

$$T_t = \kappa T_{xx} + q/\rho c \text{ in } -\infty < x < +\infty,$$

  with

$$T(x,0) = T_0(x),$$

  and

$$T(x,\pm\infty) = 0.$$

- Note that if $U(x,t)$ is any solution of the heat equation then so is $\partial U(x,t)/\partial t$—to see this simply apply the operator $\partial/\partial t$ to the heat equation and interchange orders of differentiation. Using the same linear operator property we can see that $\partial U(x,t)/\partial x$, $\partial^2 U(x,t)/\partial x^2$ etc., and $\int_a^x U(x',t)dx'$ etc. are also all solutions. In particular we can generate many useful solutions using the similarity solution. Note especially that

$$\int_0^x t^{-\frac{1}{2}} e^{-u^2/4\kappa t}\,du \quad = \quad 2\sqrt{\kappa}\int_0^{x/2\sqrt{\kappa t}} e^{-\xi^2}\,d\xi,$$

  is a solution, i.e. using the error function definition (7.1)

$$T_0 \text{erf}(\frac{x}{\sqrt{4\kappa t}}), \qquad (7.26)$$

  is a solution. This solution describes the temperature in a semi-infinite body initially at temperature T_0 when the surface at $x = 0$ is maintained at temperature 0 for $t > 0$ (check out the initial and boundary conditions); a solution of obvious physical importance.

The importance of the 1D fundamental solution of the heat equation is perhaps now evident. The solutions listed above are of particular interest for our applications. For a more extensive listing of solutions obtained using the plane source solution the reader is referred to the heat conduction classic of Carslaw and Jaeger (1959).

7.5 Other Fundamental Solutions

The plane source solution is said to be **fundamental** because the solution to almost all 1D infinite region plane problems can be written down immediately in terms of this solution. Dimensional arguments can also be used to determine the similarity solution due to release of a quantity Q of heat at time $t = 0$ at a point in 3D space. This *point source* solution is given by

$$T(x, y, z, t) = \frac{Q}{8\rho c(\pi \kappa t)^{\frac{3}{2}}} e^{-r^2/(4\kappa t)}, \tag{7.27}$$

where

$$r^2 = x^2 + y^2 + z^2,$$

and it can be used to generate the temperature distribution due to an arbitrary heat supply rate in infinite 3D space with arbitrary initial temperature conditions, and so it is also *fundamental* for such problems. In particular, by summing constant point source solutions over the plane $x = 0$, the plane source solution can be constructed; thus the point source solution is more fundamental than the plane source solution. The point source solution is determined in Exercise 7.7, and some applications are considered. For an extensive listing the reader is again referred to Carslaw and Jaeger (1959).

The fundamental plane and point solutions can be thought of as being *particular solutions* of the inhomogeneous heat equation corresponding to pulsed external heat inputs. Thus, in a *generalised function* notation the plane source solution satisfies the inhomogeneous heat equation

$$T_t - \kappa T_{xx} = Q'\delta(t)\delta(x), \quad -\infty < x < \infty, \quad t \geq 0,$$

subject to the boundary and initial conditions

$$T(x, 0) = 0,$$

$$\lim_{x \to \pm\infty} T(x, t) = 0.$$

Aside: The *Dirac delta function* $\delta(t)\delta(x)$ notation is used to indicate that the heat is supplied *at* $x = 0$, $t = 0$. This notation was also used in the beam problem in Section 3.1 Chapter 5, and will be discussed in detail later in this chapter.

In the work that follows we will append to the above *particular solutions*, solutions of the homogeneous heat equation (i.e. *complementary functions*), and thus obtain other solutions of the above inhomogeneous heat equation with the same source but satisfying different subsidiary conditions. Such an approach enables one to produce solutions that are *fundamental* for certain semi-infinite and finite region problems that are of special interest for our models.

7.5.1 The method of images

A very simple and neat intuitive idea referred to as **the method of images** enables one to generate fundamental solutions that are appropriate when dealing with semi-infinite and finite regions bounded by planes, spheres or cylinders.

The semi-infinite solid

Firstly note that the 1D fundamental temperature distribution $G(x,t)$, see (7.21), is symmetrical about $x = 0$, so that this solution automatically satisfies a zero flux condition at $x = 0$, a result one might have anticipated from the physics. Noting also that heat released at $x = 0$ in an infinite conductor can flow either into the positive x region or into the negative x region, we can see that $2G(x,t)$ represents the temperature distribution in an insulated *half space* $x > 0$ due to a unit heat input per unit area at $t = 0$ into the exposed surface. The fact that we obtain the solution to a semi-infinite plane problem by considering an equivalent infinite plane problem may be disturbing. Note, however, that the expression we've found certainly satisfies the heat equation in the region $x > 0$ with the required heat input at $x = 0$, and so by definition it's a solution. Furthermore the uniqueness result tells us there is just one such solution so it must be *the* solution. Note especially for later purposes that the surface temperature is given by

$$T(0,t) = \frac{F_0(t)}{\sqrt{\pi\rho c k}} \text{ where } F_0(t) = \frac{1}{\sqrt{t}}. \tag{7.28}$$

The same argument tells us that the surface temperature of a semi-infinite body due to a surface heating rate $q(t)$ per unit area is, see (7.23),

$$T(0,t) = 2\mu \int_0^t \frac{q(t')}{\sqrt{t-t'}}\, dt' \equiv \frac{1}{\sqrt{\pi\rho c k}} \int_0^t \frac{q(t')}{\sqrt{t-t'}}\, dt', \tag{7.29}$$

see (7.22), a result that we'll find most useful in later work.

The source solution (7.28) obtained above can be thought of as being due to two instantaneous sources in an infinite region each with unit heat output per unit surface area; one just above the plane (at $x = 0+$) and the other (its image in the x plane) just below the plane (at $x = 0-$). This trick of introducing image sources in an infinite region containing material of the same conductivity to ensure no heat flux across a plane surface can be used to generate other fundamental solutions to the insulated semi-infinite solid problem. Thus

$$Q[G(x - x', t - t') + G(x + x', t - t')]$$

represents the temperature distribution in an insulated solid $x > 0$ due to the release of a quantity of heat Q per unit surface area at location x' at

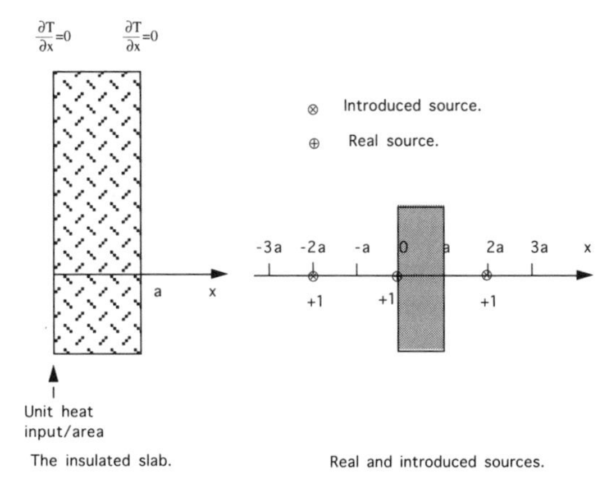

Figure 7.5: The insulated slab fundamental solution.

time t', and this can be used to build up the solution in an insulated half space due to an arbitrary heat input. We will see later how useful these results are for real modelling situations.

By using image sinks anti-symmetric solutions of the heat equation can be produced, so that a zero temperature condition on the plane $x = 0$ can be ensured. Thus

$$\mathcal{Q}[G(x - x', t - t') - G(x + x', t - t')]$$

represents the temperature distribution in a semi-infinite solid $x > 0$ whose surface is maintained at temperature zero, due to the instantaneous release of a quantity of heat $\mathcal{Q}$ per unit area at location x' at time t'.

The finite slab problem

Consider now the problem of determining the temperature distribution in a slab of material ($0 < x < a$) whose faces are both insulated due to a unit heat input at time zero at a boundary $x = 0$, see Fig. 7.5. Let us start with the solution $2G(x, t)$ for the semi-infinite region $x > 0$, see (7.21). This solution satisfies all the necessary conditions save that the heat flux at $x = a$ is given by $-2kG_x(a, t)$; it should be zero. To ensure zero flux at this boundary we need to add to the above G solution another solution of the heat equation which will cancel this flux. From our earlier work we know what's needed; an identical source at $(x, t) = (2a, 0)$ conducting through the same material. Again there is no need for material of the same conductivity to be actually present in the region $x > a$, and some other material might actually be there. We merely recognize that if we imagine it to be there then the result applied in $0 < x < a$ will satisfy both the heat equation and the zero flux boundary condition at $x = a$. The solution obtained is

$$T(x,t) = 2[G(x,t) + G(x-2a,t)],$$

which, since G is an even function of x, obviously satisfies the zero flux condition at $x = a$. This solution is not symmetric about the plane $x = 0$ (the additional source destroys the symmetry), so that there will now be a net heat flux across this plane which may be corrected for by including a source at $x = -2a$, and so on ad infinitum. Thus we are led to examine the function

$$2[G(x,t) + \sum_{n=1}^{\infty}\{G(x-2na,t) + G(x+2na,t)\}],$$

or more explicitly, see (7.21),

$$\frac{2\mu}{\sqrt{t}}\left[e^{-x^2/(4\kappa t)} + \sum_{n=1}^{\infty}\left(e^{-(x-2na)^2/(4\kappa t)} + e^{-(x+2na)^2/(4\kappa t)}\right)\right],$$

requiring us to bring to bear on the problem our knowledge of infinite series. Note particularly that the introduced "imaginary" sources are all located *outside* the solution domain $0 < x < a$ so that the terms corresponding to these sources are solutions of the homogeneous heat equation inside this domain.

Because the terms of the series behave like $\exp -(n^2)$ for large n, this series, and the series obtained by differentiating any number of times with respect to the variables, converge for all finite $t > 0$. The implication is that these differentiation operations are valid, and more particularly the process of applying the heat equation operator to the series is a valid procedure. Now, apart from the first term which corresponds to the required source, all terms satisfy the homogeneous heat equation and so, by applying the heat equation operator to the series, we see that it satisfies the heat equation with the required source. Also the required initial and boundary conditions are satisfied; this series is *the* desired unique solution. The surface temperature is of particular interest and is given by

$$T(0,t) = [\frac{1}{\sqrt{\pi\rho c k}}]F(t), \tag{7.30}$$

where

$$F(t) = \frac{1}{\sqrt{t}}\left[1 + 2\sum_{n=1}^{\infty} e^{-n^2a^2/(\kappa t)}\right], \tag{7.31}$$

after substituting for μ, see (7.22). Note that as $a \to \infty$ $F(t) \to F_0(t)$, see (7.28), i.e. we recover the semi-infinite region solution.

Important Comment: usefulness and convergence There are very significant differences between the objectives and approach adopted by pure mathematicians and those adopted by modellers that are illustrated well by this example. The series obtained above converges for any finite value of t and so in a *real* sense it is *the* solution to our problem—end of story as far as a pure mathematician is concerned. For a modeller what's needed is an expression that's *useful* for evaluation purposes. In fact, although a modeller welcomes convergence (because of the security convergence implies), it's not necessary—many solution expansions used for practical purposes are not convergent (normally they're just asymptotic). More explicitly, the above expansion will be acceptable if few terms are necessary to obtain $T(x,t)$ to the required accuracy. To obtain the desired accuracy we need to continue to add terms until the remainder, typified in size by the last added term, is small. Now the terms decrease very rapidly when $a^2/(\kappa t)$ is not small, so the expansion is excellent for small to medium times relative to the scale a^2/κ. When t is relatively large, more than $\sqrt{\kappa t}/a$ terms, see (7.31), are required to obtain acceptable accuracy. In physical terms, more and more of the sources contribute to the solution's behaviour for large times. Thus the image method is powerful when dealing with small to moderate times (and often provides the *only* effective means of describing the non-regular solution behaviour for small times, witness Example 7.4.1); but is not really of practical use for large times. In the next chapter we will examine a useful technique for extracting the long term solution behaviour.

Aside: The behaviour for large times can often be deduced from other considerations. In our particular case, for example, the heat introduced at time $t = 0$ cannot escape the slab because its faces are insulated. Furthermore, any unevenness in the temperature profile will eventually be evened out by diffusion, so the eventual steady state will be one of uniform temperature. In order that this steady state temperature be consistent with the known heat input (in this case unity), the temperature rise must be $1/[\rho c a]$. It is actually possible to extract the behaviour of the solution for large times from the above image solution, but this requires a high level of advanced mathematical technique. We will see in the next chapter, however, that there are much more effective mathematical means for dealing with the large time problem.

Comment: extensions Fundamental solution methods and the image methods represent an imaginative use of simple intuitive physical ideas and observations made long ago which have been seen to be so useful that they have become part of the standard inventory of modelling tricks. These techniques are not restricted in application to the heat equation. They play a fundamental role in both the theory and practical solution of *any* linear partial differential equation.

7.6 Boundary Integral Methods

The method of images works well for "standard" zero flux or temperature boundary conditions and the fundamental solutions corresponding to point, line, plane; steady, instantaneous, periodic and travelling sources in simple geometries are classical results that can be looked up in Carlslaw and Jaeger (1959). Many solutions built up from these are also listed; it's really an indispensable source for a modeller! Our interest lies in providing useful solutions or procedures for dealing with "non-standard" boundary condition problems for which no such simple image ideas work and exact solutions are rarely possible, so it's necessary to look for effective approximate (often numerical) schemes. We'll examine an important non-standard case of interest for our later models; the case of the Newtonian surface heating or cooling of a semi-infinite body.

The heat gain or loss rate per unit area from a surface is usually dependent mainly on the temperature difference between the surface and its environment, and the linear Newtonian condition

$$q(t) = -kT_x(0,t) = \beta(T_0 - T(0,t)), \tag{7.32}$$

provides a useful model of this situation. Here $q(t)$ is the heat flux into the surface, T_0 is the temperature of the environment and $T(0,t)$ is the temperature of surface of the solid occupying $x > 0$. Often the Newtonian condition is written in the form

$$T_x(0,t) = -\alpha(T_0 - T(0,t)), \quad \text{where } \alpha = \beta/k. \tag{7.33}$$

From this equation it follows that $1/\alpha$ has the dimensions of length and represents the depth scale of heat penetration over a unit time interval (the time interval being defined by β and the flux scale using (7.32)).

We'll consider the case of a semi-infinite body initially at temperature 0 gaining (or losing) heat from (to) an environment at temperature T_0 according to the above Newtonian condition, see Fig. 7.6. Clearly there will be heat flow into the conducting material until the surface temperature of the body reaches T_0. The interest lies in determining how the time span for this process depends on the physical parameters.

Now, from our earlier work, we know that the surface temperature of the semi-infinite conduction body $T_s(t) \equiv T(0,t)$, due to a heating rate $q(t)$ per unit area, is given by (see (7.29))

$$T_s(t) = 2\mu \int_0^t \frac{q(t')}{\sqrt{t-t'}} dt'.$$

Although the actual value of the heat flux is not prescribed by the problem, the above Newtonian boundary condition (7.32) imposes the consistency

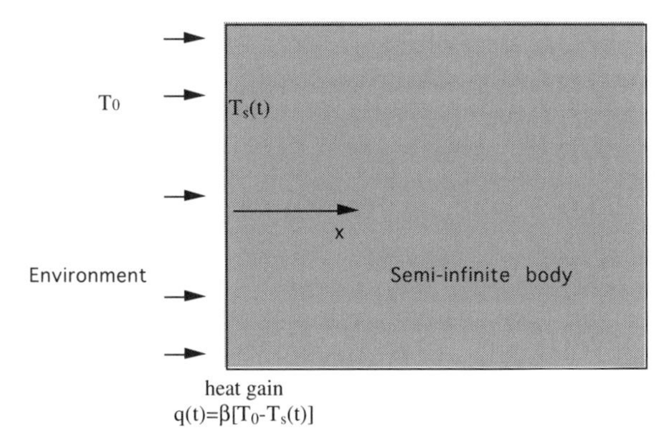

Figure 7.6: Newtonian heating of a semi-infinite body.

condition

$$T_s(t) = 2\mu\beta \int_0^t \frac{T_0 - T_s(t')}{\sqrt{t-t'}}\,dt', \tag{7.34}$$

which must be satisfied by the surface temperature $T_s(t)$. This equation encapsulates all the physics of the problem in that if a $T_s(t)$ can be found that satisfies this equation, then this T_s will be consistent with both the heat equation in the conductor $x > 0$ and the Newtonian condition on its boundary. Of course, if required, the temperature distribution $T(x,t)$ in the body can then be determined from the known surface heat input by direct integration, although often this information is not of interest. Because the unknown function $T_s(t)$ occurs under the integral sign this equation is described as an *integral equation*, and this particular form of integral equation is known as a Volterra equation. One might hope to reduce this integral equation to a differential equation by simply differentiating it with respect to t. Sometimes this works, but it doesn't here. It is in fact much more difficult to find exact closed form solutions to integral equations than to differential equations, but generally integral equations have a significant advantage over differential equations for numerical approximation work; basically because numerical integration is a much more satisfactory procedure than numerical differentiation.

The fact that our work has lead to an unfamiliar mathematical problem should not be unduly concerning; familiar approximation techniques have universal applicability so it's a matter of adapting these techniques. To get a feel for the problem it is often useful to think about how one could numerically compute the answer. For example in our problem a sensible numerical approach would be to replace the integral by an equivalent sum. Thus one can see that the equivalent discrete problem involves determining the solution of a set of linear algebraic equations for the unknown T values.

In this way one can convince oneself that the problem makes sense and has a unique solution, see Exercise 7.17. The proof of this result is another matter. The existence/uniqueness theory for Volterra integral equations of the above type is in fact known, so we're assured the equation serves to determine a unique function T_s. The contraction mapping theorem can be used to establish this result from first principles.

Comment: alternative techniques Before proceeding it's appropriate to compare the integral equation formulation of the problem with other formulations. In our present situation a conventional numerical integration of the defining partial differential equation with boundary conditions and initial condition would involve the evaluation of T at each location in a 2D grid covering the (truncated) (x,t) domain of interest. To numerically solve the integral equation, however, it would be only necessary to determine $T_s(t)$ at a 1D grid of t values; a considerable reduction in computational effort! The reduction in effort is in fact *much* greater than one might naïvely expect—for technical (numerical stability) reasons the grid spacing needs to be very fine in the t direction to ensure the numerical partial differential scheme works, let alone achieve the required degree of accuracy. Perhaps a $10^2 \times 10^4$ array of points would be necessary to achieve a 1% accuracy using a *finite difference* scheme. (The finite difference approach replaces derivatives by approximate quotients, and solves the resulting set of linear algebraic equations for the unknown values of T on the (x,t) grid.) A numerical integration of the integral equation would on the other hand require only about 10 points to achieve a 1% accuracy even if a crude numerical integration scheme is employed, see Exercise 7.17. There is a variety of other standard numerical partial differential equation procedures that handle the problem more efficiently than the finite difference scheme, but none of these is nearly as efficient as the integral equation procedure, and furthermore none is able to cope well with the non-regularity problems that arise for small time because of the abrupt heat input (as observed above in Example 7.4.1). Not only does the computational effort go up markedly with dimension but also the accuracy and believability decrease rapidly—so much so that it's a major achievement if the dimension of the problem is reduced and an integrated form of the defining equation is employed. For larger problems (eg. 3D problems) the gain is *crucial*—so much so that many large problems can only be attacked using methods akin to that developed here. The technique we've used here to convert the problem of determining $T(x,t)$ in (x,t) space to the problem of determining T on the boundary of the (x,t) domain is called **the boundary integral method**, and for the above reasons this technique is of major importance. Of course, if one can produce analytic results rather than straight numeric results much more understanding is possible.

Whether or not one uses analysis or numerics the first thing to do is to scale the problem. The appropriate temperature scale is obviously T_0

and if t_0 is the time scale then the multiplicative factor arising on the right hand side of the integral equation (7.34) after scaling is $2\mu\beta\sqrt{t_0}$. Equating this to unity removes all parameters from the problem giving the time scale $1/(\mu\beta)^2$, or equivalently $\pi/\alpha^2\kappa$, see (7.22,7.33). Evidently this provides us with a crude estimate for how long it takes for the body to thermally respond to its environment. Thus with the variable change

$$\mathcal{T} = \frac{T_s}{T_0} \quad \text{and} \quad t = \frac{\pi}{\alpha^2\kappa}\tau, \tag{7.35}$$

the integral equation becomes

$$\mathcal{T}(\tau) = \int_0^\tau \frac{1-\mathcal{T}(\tau')}{\sqrt{\tau-\tau'}}d\tau'. \tag{7.36}$$

For later approximation purposes it's useful to reduce the integration domain on the right hand side to unity by substituting $\tau' = \xi\tau$ in the integral to give

$$\mathcal{T}(\tau) = \sqrt{\tau}\int_0^1 \frac{1-\mathcal{T}(\xi\tau)}{\sqrt{1-\xi}}\,d\xi. \tag{7.37}$$

7.6.1 Moderate time behaviour

Now for small scaled times the surface temperature will have not changed much so $\mathcal{T}(\tau)$ will be small compared with 1. Thus, at least for early times, see (7.37)

$$\mathcal{T}_1(\tau) = \gamma_1\sqrt{\tau} \quad \text{where} \quad \gamma_1 = \int_0^1 \frac{1}{\sqrt{1-\xi}}\,d\xi,$$

should provide a reasonable approximation to the integral equation, and therefore for $\mathcal{T}(\tau)$. This result can be improved using an iterative scheme. The scheme

$$\mathcal{T}_{n+1}(\tau) = -\sqrt{t}\int_0^1 \frac{\mathcal{T}_n(\xi\tau)}{\sqrt{1-\xi}}\,d\xi, \quad n \geq 1, \tag{7.38}$$

where

$$\mathcal{T} = \sum_1^\infty \mathcal{T}_n(\tau),$$

suggests itself, at least for small times. Each iteration introduces an additional $\sqrt{\tau}$ so that the solution can be written in the form

$$\mathcal{T} = \sum_1^\infty \gamma_n\tau^{\frac{n}{2}}, \tag{7.39}$$

where

$$\gamma_{n+1} = -\gamma_n\int_0^1 \xi^{\frac{n}{2}}/\sqrt{1-\xi}\,d\xi. \tag{7.40}$$

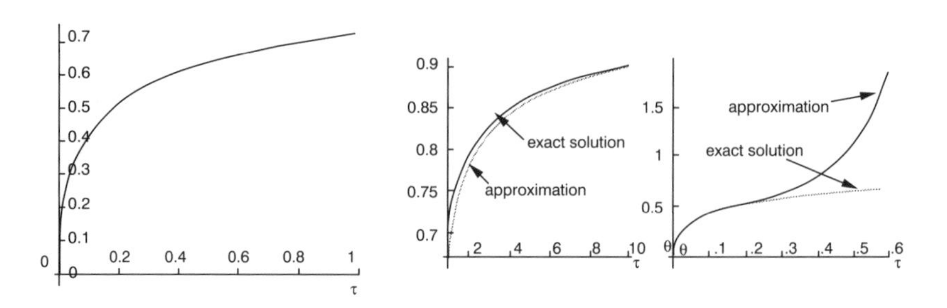

Figure 7.7: Newtonian heating of a semi-infinite body: exact solution and approximations.

Algebraic packages are of course tailor-made for such iterative calculations and Maple is *almost* up to the task. Thus in response to

```
int((sqrt(xi)/sqrt(1-xi)),xi=0..1);    % xi ≡ ξ
```

it gives $-\frac{\pi}{2}$; correct in magnitude but wrong in sign. Clearly it's necessary to be cautious when using packages especially when multivalued functions are involved! After correcting deficiencies of this type one can extract results to any prescribed order. To be useful the successive terms of this series should decrease reasonably rapidly; this is clearly the case for small τ, as expected. Good results with few terms are obtained for scaled times up to about 0.3, see Fig. 7.7, plotted using Maple. For larger scaled times the number of terms that have to be retained to yield reasonable results increases dramatically and the calculation time becomes excessive. The iteration scheme is thus not useful for values of $\tau > 0.3$ and thus one is really unable to say much about the "eventual state" and the time taken to reach this state. What is needed is a better description of the behaviour of the solution for large values of time.

Aside Although the above $T(\tau)$ expansion (7.39) is useless in its present form for large τ evaluations, the potential for extracting information about the large time behaviour does exist. With luck it's just possible that we can identify the function whose series expansion is the one extracted above. To understand what's required note that the expansion $\sum_{n=1}^{\infty}(-1)^n x^{2n+1}/(2n+1)!$ is a convergent but completely useless representation of a function for anything other than small x, but any student of mathematics will recognize that the function corresponding to this series is $\sin x - x$, which represents a theoretically equivalent but *much* more useful representation for the function over the complete x domain. In order to recognize and thus *sum* such a series in this way it's necessary to know its complete expansion. In modelling we rarely can determine more than a few terms of a solution expansion, so clearly in most cases any attempt at such a summation is foiled at first base. In our present situation we do in effect (because of the recurrence relation) have a complete expansion, so a summation is possible. Such summations, although limited in applicability, have led to major breakthroughs in many branches of science. The following observations enable one to sum the solution expansion:

Experienced mathematicians will recognize that the integrals (7.40) are all known in terms of standard functions that occur often in applications; the Beta or Gamma functions, see (7.6). The known recurrence relations for these functions enable us to obtain the following relationships connecting the successive even and odd coefficients γ_n of our present expansion:

$$\gamma_{2n}/\gamma_{2(n-1)} = \pi/n \quad \text{and} \quad \gamma_{2n+1}/\gamma_{2n-1} = \pi/(n+1/2).$$

These recurrence relations have for their initial values $\gamma_1 = 2$ and $\gamma_2 = -\pi$. Thus the sum of the even terms can be recognized to be related to the Taylor series of the exponential function and so it follows that

$$\sum_1^\infty \gamma_{2n}\tau^n = -\sum_1^\infty (\pi\tau)^n/n! = 1 - e^{\pi\tau}.$$

One has to work a little harder to sum the odd γ's. The odd terms define a function

$$H(\tau) = \sum_{n=0}^\infty \gamma_{2n+1}\tau^{n+1/2},$$

whose value at $t = 0$ is zero and whose derivative is given by

$$\begin{aligned} H'(\tau) &= \gamma_1\tau^{-1/2}/2 + \sum_{n=1}^\infty (n+1/2)\gamma_{2n+1}\tau^{n-1/2} \\ &= \tau^{-1/2} + \pi\sum_{n=1}^\infty \gamma_{2n-1}\tau^{n-1/2} \\ &= \tau^{-1/2} + \pi\sum_{n=0}^\infty \gamma_{2n+1}\tau^{n+1/2} \\ &= \tau^{-1/2} + \pi H(\tau). \end{aligned}$$

This observation provides a differential equation for H, which when solved subject to the initial condition $H(0) = 0$ (which is required by the definition of H), yields the result

$$H(\tau) = e^{\pi\tau}\int_0^\tau s^{-1/2}e^{-\pi s}\,ds = \frac{2}{\sqrt{\pi}}\int_0^{\sqrt{\pi\tau}} e^{-u^2}e^{\pi\tau}\,du = e^{\pi\tau}\operatorname{erf}(\sqrt{\pi\tau}),$$

where use has been made of the transformation $\pi s = u^2$. When the even and odd summations are combined the solution for $\mathcal{T}$ is found to be

$$\mathcal{T} = 1 - e^{\pi\tau}\operatorname{erfc}(\sqrt{\pi\tau}), \tag{7.41}$$

an exact closed form of solution which can be checked by direct substitution into the integral equation.

Reiterating, the above *summation* procedure uses special results not generally available and so is not of general applicability.

7.6.2 The asymptotic behaviour

The series expansion (7.39) for $\mathcal{T}$ is only useful for moderate τ values, so we need a reliable scheme of calculation to determine the asymptotic time behaviour.

There are strong physical reasons to believe that the surface temperature of the body increases monotonically to its eventual equilibrium state $\mathcal{T} = 1$. Furthermore, since we know that heat will penetrate to greater and greater depths within the body at slower and slower rates, one might expect the approach to equilibrium to be slow. How slow? Since we are interested in determining the small difference of the scaled temperature $\mathcal{T}$ from its final value of unity it makes sense to work in terms of the difference

$$U(\tau) = 1 - \mathcal{T}(\tau),$$

where U is presumed to be small for large times. In terms of U the integral equation (7.37) becomes

$$1 - U(\tau) = \sqrt{\tau} \int_0^1 \frac{U(\xi\tau)}{\sqrt{1-\xi}}\, d\xi. \tag{7.42}$$

Since U is presumed small

$$1 = \sqrt{\tau} \int_0^1 \frac{U_1(\xi\tau)}{\sqrt{1-\xi}}\, d\xi, \tag{7.43}$$

represents a sensible first approximation to this equation from which we may hope to extract a useful first approximation $U_1(\tau)$ for $U(\tau)$. Now the balance represented by this equation can only be realized for large τ if $\int_0^1 U_1(\xi\tau)/\sqrt{1-\xi}\, d\xi$ is of order $1/\sqrt{\tau}$. This can only happen if the $U_1(\xi\tau)$ within the integral is of order $1/\sqrt{\tau}$, since $1/\sqrt{1-\xi}$ is integrable with integral of size 1 and the domain of integration is of size 1. Thus for large τ the function U_1 must be approximated by

$$U_1(\tau) = \delta\tau^{-1/2} + u(\tau)$$

where δ is some constant to be determined and the unknown left over piece u is relatively small for large τ, i.e. of $o(1/\sqrt{\tau})$. The constant δ must be determined so that the integral equation is satisfied to unit order so that, substituting this expression for U_1 into the integral equation, we obtain

$$1 = \delta \int_0^1 \frac{d\xi}{\sqrt{\xi(1-\xi)}}.$$

Maple evaluates the integral exactly to give π. Thus $\delta = 1/\pi$ and the behaviour of the surface temperature for large times is given by

$$\mathcal{T}_s = 1 - 1/[\pi\sqrt{\tau}] + o(\tau^{-1/2}). \tag{7.44}$$

Can we improve on this estimate? To do this, as usual, we can use iteration. Note however that the additional term to be included on the left hand side of the integral equation (7.42) is of order $\tau^{-1/2}$, and we know from Example 7.4.1 that the early stages of a heating process also have an influence on the temperature that dies off like $\tau^{-1/2}$ for large τ, so that we cannot expect to extract further information about the large time behaviour without examining the details of the solution over the *whole* time range. To do this global solution estimates would be required. A numerical scheme using the above analytic results for the small and large τ behaviour could be used, see Exercise 7.17).

7.6.3 Interpretation

The exact solution together with the series and asymptotic results are plotted in Fig. 7.7 using Maple. It is apparent from the figure that after a very rapid initial temperature rise displaying the expected $\tau^{\frac{1}{2}}$ behaviour associated with an abrupt change of heat input, there is a dramatic slowing in the rate of increase in temperature. Thus although the surface temperature reaches 0.8 of its eventual value for $\tau = 2.24$, and 0.9 by time $\tau = 9.82$; it does not achieve the value 0.99 until $\tau = 1013$. Very slow! Thus for this problem the time scale extracted using scaling arguments provides a poor indication of the actual time required to "reach" equilibrium.

7.6.4 General comments

It is worth reviewing the steps taken to solve this problem because they are very typical of what's often necessary. We found:

- The equations expressing the basic physics (the heat equation with subsidiary conditions) were not in a useful form for direct evaluation so some pre-processing was required (leading to the integral equation).
- Having determined the scale for the process (especially the time scale to reach equilibrium), interest centred mainly on particular independent variable ranges (in our case the small time and large time ranges).
- To extract information about these ranges physical/mathematical arguments were employed to identify the dominant terms of the defining equation, leading to simpler approximating equations. Iterative corrections were then employed to both improve the results and to provide a check on the range of applicability of these local expansions. Much of the tedium was left to Maple.

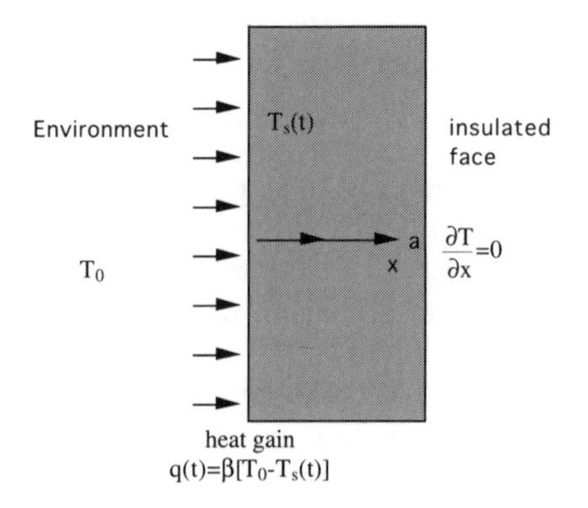

Figure 7.8: Newtonian heating of a slab.

- To complete the calculations usually some straightforward numerical work is required. In our case exact results were obtained. An appropriate numerical scheme is described in Exercise 7.17

7.6.5 The slab problem

The above solution method carries across readily to more complex problems; one simply needs to employ the appropriate fundamental solution. Consider, for example, the Newtonian heating (or cooling) of a finite slab of material of thickness a with one face ($x = a$) thermally insulated, while the other face ($x = 0$) is exchanging heat with the environment, see Fig. 7.8. Again the slab is initially at temperature 0 and is gaining heat from the environment at temperature T_0. One might expect equilibrium to be reached more rapidly for thinner slabs; we'd like to know the relevant dimensionless parameter, and the dependence of the heating time on this parameter. The relevant length scales are a and the length scale $1/\alpha$ that arises out of the Newtonian condition (7.33). Clearly if the dimensionless parameter $a\alpha$ is large the slab can be thought of as being "thermally thick"; if it's small then the slab is "thermally thin".

Earlier we found the fundamental solution for such an insulated slab to be given by

$$T_s(t) = \frac{2\mu Q}{\sqrt{t}}[1 + 2\sum_{n=1}^{\infty} e^{-n^2a^2/(\kappa t)}],$$

see (7.30,7.22), where Q is heat input per unit area into the exposed surface $x = 0$. Thus in the same way as for the semi-infinite problem we can determine by integration the surface temperature due to a known supply rate of heat $q(t)$ per unit area, and by ensuring that this supply rate matches that required by the Newtonian condition, we determine the integral equation consistency condition

$$T_s(t) = 2\mu\beta \int_0^t \frac{T_0 - T_s(t')}{\sqrt{t-t'}}[1 + 2\sum_{n=1}^{\infty} e^{-n^2a^2/\{\kappa(t-t')\}}]\, dt',$$

for the surface temperature. If we choose the same scales $(\pi/\alpha^2\kappa, T_0)$ for the time and temperature as we did for the semi-infinite body, then this simplifies to

$$\mathcal{T}(\tau) = \int_0^\tau \frac{1 - \mathcal{T}(\tau')}{\sqrt{\tau - \tau'}}[1 + 2\sum_{n=1}^{\infty} e^{-\nu n^2/(\tau-\tau')}]\, d\tau', \qquad (7.45)$$

where

$$\nu = a^2\alpha^2/\pi.$$

It's useful to compare this with the corresponding semi-infinite body equation (7.36), with solution (7.41). Clearly the dimensionless combination ν usefully characterizes the effect of slab thickness on heat transfer. A little thought about the likely effect of the additional finite slab terms on the above equation may convince the reader that the equation predicts that the surface temperature will adjust to the environment more quickly for thinner slabs—makes sense! Also note that the additional terms (because of their exponential form) increase rapidly in size as $\tau \to \nu$; clearly the thickness of the slab begins to have a significant effect for scaled times τ of order ν i.e. for unscaled time t of order a^2/κ (a result one might have anticipated on simple dimensional grounds). How do we proceed to extract accurate solutions?

If ν is large then the slab is thermally thick, and for scaled times of unit order it will heat up as if it were a semi-infinite body. Under such circumstances the integral equation is well approximated by the $\nu = \infty$ half plane solution (7.41), and an iterative improvement will provide a useful solution for scaled times of order ν. For larger times the corrective terms will grow unacceptably large as can be seen from the integral equation, and the scheme is then not useful.

When ν is small we face a real problem, since for scaled times of *unit order* we will need to evaluate many terms to obtain an acceptable result, as can be seen from the integral equation. Thus our solution is *only* useful for small scaled times.

In both cases the solution ceases to be useful for times somewhat greater than the time taken for heat to penetrate the thickness of the slab, so the solution is not particularly useful for determining finite slab heat transfer effects. The problem arises basically because the fundamental image source solution we're using is not useful if many sources contribute to the solution, and for thin plates after a short time many sources contribute. An alternative fundamental solution that's useful for large time is required. We'll determine such a solution in the next chapter.

In the above work we've analysed the thermal adjustment of a finite slab to its environment when the unexposed face is insulated. The reader should note that (because of symmetry) the above solution also describes the thermal adjustment of a slab of thickness $2a$ to the environment when *both* faces are exposed to the same environment. Note also that it's a simple extension to cover cases in which (in addition to the above) heat is being supplied to one or both surfaces from some external source eg. the Sun. Such cases, of importance for thermal insulation problems, will be examined in the exercises.

7.7 Generalised Functions

In our instantaneous plane source work (by using scaling arguments) we managed to avoid the need to specify the source distribution that generates the similarity solution. The source distribution is not a function in the normal sense. The notation $\delta(t)\delta(x)$ is conventionally used to represent the *generalised function* associated with the (unit) source where the *Dirac delta function* $\delta(x)$ is defined by the following rules:

$$\delta(x) = 0 \quad \text{for} \quad x \neq 0, \tag{7.46}$$

$$\int_{-0}^{0+} \delta(x)dx = 1, \tag{7.47}$$

where the notation indicates that the integral across the origin is unity. Very singular! Using this notation the equations describing the heat pulse problem can be written

$$T_t = \kappa T_{xx} + Q'\delta(t)\delta(x), \qquad -\infty < x < \infty, \quad t \geq 0,$$

$$T(x,0) = 0,$$

$$\lim_{x \to \pm\infty} T(x,t) = 0,$$

where $Q' = Q/\rho c$. In formal terms it can be seen that, with the prescribed rules for determining the delta functions, the heat supply rate term $Q\delta(t)\delta(x)$ introduced by the above formalism does satisfy the heat input requirements (no input except at $t = 0, x = 0$ and with total input Q per unit area). Furthermore, no time and length scales are present in the description—an essential requirement for the similarity arguments used earlier. One should, however, be uneasy at dealing with such formal rules. In order to manipulate such *generalised functions* a description that relates to conventional functions is necessary. Notice that the function

$$g(x,\epsilon) = \frac{1}{\epsilon\sqrt{\pi}} e^{-(x/\epsilon)^2}$$

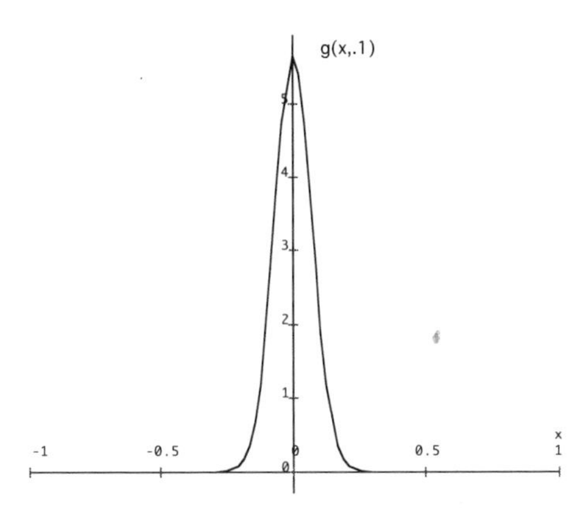

Figure 7.9: The delta function: a particular representation.

has an integral from $-\infty$ to ∞ equal to 1 (use Maple) for *all* values of ϵ. Also note, see Fig. 7.9, that this function is centred on $x = 0$, has a width of order ϵ, and height of order $1/\epsilon$; so that as ϵ becomes smaller its profile becomes more and more concentrated about $x = 0$. Thus in the limit as $\epsilon \to 0$ the function vanishes except at $x = 0$, and although infinite at $x = 0$ the integral across 0 remains unity; so the limit of this function represents precisely what we're after and can be thought of as being *the* delta function. In the context of our heat source problem all the above makes physical sense. The heat supply rate term $Qg(x, \epsilon_1)g(t, \epsilon_2)$ represents a heat source that transfers the amount of heat Q per unit area into the infinite conductor over the distance interval defined by $(x - \epsilon_1, x + \epsilon_1)$ over the time interval $(0 - \epsilon_2, 0 + \epsilon_2)$. As ϵ_1 and ϵ_2 approach zero the supply rate must compensatingly increase to transfer the required amount of heat. Thus the mathematical simplicity achieved in putting in all the heat at one point can only be realized at the expense of accepting an infinite heat flux, and an associated infinite temperature will also result. However, at least the *total* heat input per unit area due to the source is finite and, in spite of the evident physical problems, the price is well worth paying; given the mathematical simplicity that results from its acceptance.

Comment: Note that the similarity arguments used in Section 7.3 work only if additional parameters do not enter the problem through the initial and boundary conditions. Now, only in the limit as the ϵ's $\to 0$, do the scales disappear from the source specification, so that the similarity solution arguments can be used; it's *essential* to use generalised functions!

Of course there are many equivalent prescriptions for the delta function which achieve the same ends, and other prescriptions may be more convenient in particular circumstances, eg.

$$\lim_{\beta \to 0} \frac{\beta}{\pi(x^2 + \beta^2)},$$

or the $\lim_{\epsilon \to 0} f(x, \epsilon)$ where

$$f(x, \epsilon) = 1/\epsilon \quad \text{for} \quad |x| < \epsilon,$$

$$f(x, \epsilon) = 0, \quad \text{for} \quad |x| > \epsilon.$$

When trying to determine the effect of various operations on the delta function the trick is to work with any of the above well defined ordinary functions and defer taking the limit until all operations of interest have been carried out; in this way the operations of integration, differentiation (producing, in conventional notation, $\delta'(x)$) can be defined so as to fit in with normal calculus, and in fact there's no need to distinguish between ordinary and generalised functions *providing appropriate care is exercized.* Thus it's evident from the deferred limit definition that $\int_a^b \delta(x - \xi)dx = 1$ if the interval (a, b) encloses $x = \xi$, and is zero if it does not. If $\xi = a$ or b, then $0, 1, or/1/2$ *all* look like reasonable answers. To get the correct answer it's necessary to examine the underlying physics. Thus in our heat pulse problem in order that the required amount of heat be transferred to the solid we need $\int_0^\infty \delta(t)dt = 1$. Thus we should think of the delta function as being centred on $t = 0+$, not $t = 0$. Barring end point difficulties of this type, one can build up simple rules associated with delta functions that enable one to avoid the deferred limit procedure. Thus, for example, it's clear from the above that $\int_a^b g(x)\delta(x)dx = g(0)$ if the interval encloses the origin; so that the integral "isolates" the value of the function $g(x)$ at $x = 0$—it's often useful to think of $\delta(t)$ in operator terms.

Other generalised functions arise in a natural way. Thus the *Heaviside step function* $H(x)$ defined by

$$H(x) = 0 \quad \text{for} \quad x < 0,$$

$$H(x) = 1 \quad \text{for} \quad x > 0,$$

which is often used to model abrupt changes, can be thought of as the limit of a sequence of well behaved functions eg. $lim_{\epsilon \to 0}\text{erf}(x/\epsilon)$. It's perhaps clear from the definition of the delta function that $H'(x) = \delta(x)$ and any doubt can be removed if one uses the deferred limit procedure.

Also since integration by parts yields

$$\int_a^b g'(x, \epsilon)f(x)dx = g(x, \epsilon)f(x)|_{x=a}^{x=b} - \int_a^b f'(x)g(x, \epsilon)dx,$$

for each of the sequence functions $g(x, \epsilon)$ defining the delta function, and any reasonable $f(x)$, it follows that

$$\int_a^b f(x)\delta'(x)dx = -f'(0),$$

if (a, b) encloses 0; which provides a useful (formal) definition for $\delta'(x)$. In this way one can produce a hierarchy of generalised functions all related to the delta function, and also produce useful rules for working with such functions that can be employed to avoid always returning to the deferred limit definitions. When in doubt the deferred limit description can always be employed to check the validity of any mathematical manipulations, but basically the normal mathematical procedures work providing the physics of the situation makes sense. The importance (and pitfalls) of such generalysed functions (or *distributions*) is such that a formal theory rigorously justifying the use of the notation and the use of relevant operations has been produced and is referred to as **Distribution Theory** or **Generalised Function Theory**, see Lighthill (1959). Although modellers are aware of the results of such theories, mostly they simply use the notation and if necessary use the physically based intuitive deferred limit ideas to overcome difficulties. The distribution ideas are not unfamiliar to most, although the notation may be—the point masses and charges used in elementary physics are simple examples. The exercises will familiarize the reader with the use of these ideas.

7.8 Summary

The solution of the heat equation corresponding to an instantaneous point source is fundamental in that in either direct ways (straight summation or integration) or indirect ways (leading to integral equations), solutions to virtually all heat conduction problems can be generated using this solution. The solutions obtained using an infinite set of (image) source solutions are generally useful for moderate times but are not appropriate for determining the long term behaviour in most cases, even though the solutions are very often exact in the mathematical sense that the resulting series converges.

The shape of the boundaries and the nature of the boundary conditions play an important role in determining the appropriate fundamental solutions to be used for a particular application. The application of source solution ideas to non-standard problems leads to boundary integral formulations that are particularly useful for computation.

7.9 Exercises

Exercise 7.1 *Similarity Solution Behaviour*

Using Maple it is a simple matter to examine various features of the plane source solution which can be written in the form

$$\frac{1}{\sqrt{t}}e^{-x^2/t},$$

when appropriate scales are adopted.

(a) Plot the solution as a function of x at various times. Especially examine the behaviour for small times. Comment.

(b) Investigate the temperature variations with time at locations at various distances from the source. Plot appropriate curves and comment on the temperature rise and decay rates. Determine analytic expressions for the maximum temperature realized at various locations from the source, and the time taken to reach these maximum values. Comment.

(c) Although the relative effectiveness of sources separated in space and time as far as the temperature levels at a particular surface is concerned is evident from the solution form, it's useful (and simple using Maple) to examine a few situations to get a feel for the results. Plot out the spatial temperature distribution at different times due to two heat sources of equal strength separated in space or time. Choose representative cases.

Exercise 7.2 *Asymptotics for Integrals*

Integration by parts can often be used to obtain an asymptotic estimate for integrals. In Section 7.3 of the text a reduction of order technique was used to determine a second real solution of the similarity equation. We obtained the result, see (7.17),

$$U_2(\xi) = e^{-\xi^2/4}\mathcal{I},$$

where

$$\mathcal{I} = \int_0^\xi e^{y^2/4}\,dy,$$

and it was of crucial importance to determine whether the solution U_2 is integrable. Clearly the integral $\mathcal{I}$ is exponentially large, but the factor multiplying the integral in U_2 is exponentially small, so it's not clear how the combination will behave.

The trick is to integrate $\mathcal{I}$ by parts in such a way that the new integral that arises is smaller for large ξ than the one we presently have. Firstly note that for values of y of unit order the integrand is of unit order so that the integral over a domain of unit length will also be of unit order, and thus will not contribute significantly for large values of ξ.

(a) Noting that

$$\int^{\xi} e^{y^2/4}\,dy = \int^{\xi} \frac{d}{dy}[e^{y^2/4}]\,\frac{2}{y}\,dy,$$

perform the integration by parts suggested by this expression and show that the result obtained is

$$[\frac{2}{y}e^{y^2/4}]^{y=\xi} + \int^{\xi} e^{y^2/4}\,\frac{2}{y^2}\,dy.$$

(b) Noting the difficulty at $y = 0$, split $\mathcal{I}$ into a finite integral piece and an infinite integral piece and thus show that

$$\mathcal{I} = e^{+\xi^2/4}[\frac{2}{\xi} + o(\frac{1}{\xi})],$$

as $\xi \to \infty$.

(c) $\star$ Perform a second integration by parts and thus show that the solution U_2 behaves like

$$\frac{2}{\xi} + (\frac{2}{\xi})^2 + o(\frac{1}{\xi})^2,$$

as $\xi \to \infty$. Thus the solution is not integrable at ∞.

(d) Note that the original integral $\mathcal{I}$ provides a perfectly acceptable expression for numeric evaluation purposes for values of ξ of unit order and Maple will produce accurate results in this range. For large values of ξ the integral cannot be used directly for numeric evaluation and the above asymptotic result fills this gap. Using the integral, its asymptotic approximate, and Maple, plot out the solution U_2 over the range $0 < \xi < 10$.

Comment: The technique described above is often used to obtain useful asymptotic expressions for functions defined by integrals, and since such integral expressions for solutions often arise out of the solution process, this elementary technique is important. The asymptotic approximations found in handbooks of *Special Functions* have been obtained using variants of the above technique and *Complex Function Theory.* It is hoped that eventually algebraic packages will become sufficiently sophisticated to deal with most asymptotic evaluations.

Exercise 7.3 *A Similarity Problem*

By using linear operator ideas we were able to determine the temperature distribution in a semi-infinite body initially at a given temperature T_0 when its surface is raised to (and maintained at) a different temperature 0 for

$t > 0$, see (7.26). It's instructive to use similarity arguments to obtain this solution.

Firstly note that the problem is set up for the similarity procedure: the domain is semi-infinite and the initial and boundary conditions introduce no scales, apart from the temperature change at $x = 0$ for $t > 0$.

(a) Write down the equations that completely specify the problem, confirming that they are in line with the uniqueness theorem.

(b) Determine scales for the relevant quantities that reduce the number of parameters to a minimum.

(c) Show that the temperature field may be described in terms of a function $f(\eta)$ which satisfies the ordinary differential equation

$$f''(\eta) = -\eta f'(\eta)/2,$$

where η is a suitably defined similarity variable.

(d) Show that the boundary conditions and initial conditions become

$$f(0) = 0, \quad f(\infty) = 1.$$

(e) Hence, using Maple, find the temperature distribution in the body and plot solution curves. Determine the heat flux at $x = 0$ and comment on the result obtained.

Exercise 7.4 *A Nonlinear Diffusion Problem*

⋆⋆ As pointed out in the text the similarity procedure can be used for extracting special solutions of nonlinear partial differential equation problems, and important advances in physical understanding have been made using the technique for such problems. A notable case in the diffusion area is the flow of water through soils where nonlinear conduction effects strongly influence flow behaviour. Here we examine the related heat conduction problem. For nonlinear conduction problems it's mathematically more convenient to work with the enthalpy, see Chapter 6 (6.2,6.14).

A semi-infinite body, of a material whose thermal diffusivity varies considerably with the temperature, is initially at a temperature T_0 and its surface temperature is suddenly raised to a temperature T_1. The enthalpy, $\mathcal{H}$, which is a strictly increasing function of the temperature T, satisfies the partial differential equation

$$\frac{\partial \mathcal{H}}{\partial t} = \frac{\partial}{\partial x}\left[\kappa(\mathcal{H})\frac{\partial \mathcal{H}}{\partial x}\right].$$

Note that physical set up is such that a similarity solution is to be expected.

(a) Show that the choice of scales indicated by the similarity variable

$$\zeta = x/\sqrt{t},$$

leads to a nonlinear ordinary differential equation for the enthalpy.

(b) Determine the appropriate boundary conditions for this ordinary differential equation.

(c) As $x \to \infty$, $T \to T_0$. Using this determine a first estimate for the solution for large ζ, and set up an iterative scheme for determining the asymptotic solution behaviour. Note that one arbitrary constant arises that needs to be determined by matching the solution onto the numerically obtained medium ζ solution.

(d) ⋆⋆ Working with a particular $\mathcal{H}(T)$ of your choice (any function with a significant variation over the temperature range will suffice) use a numerical package to explore the effect of variable conductivity on heat dispersal.

Exercise 7.5 *How Hot Is It?*

Some bodies feel hotter to the touch than others, even though they are in the same environment and might be expected to be at the same temperature. This may be observed with objects exposed to the hot sun for some time. Why should this be so? Assuming the cause is physical (it may be physiological or psychological) the difference could arise either because the objects are at different temperature, or because the thermal contact produces an incorrect impression of the temperature of the object. In a later example (7.9) we examine the first question. Here we investigate the contact aspect of this problem. Both problems are of interest in many contexts. In the usual way we consider the simplest relevant problem.

Two semi-infinite bodies with different thermal properties (k_1, ρ_1, c_1), and (k_2, ρ_2, c_2) are initially at temperatures T_1 and T_2 and are brought into contact at $t = 0$; we wish to determine how the parameters determine the contact temperature and the subsequent heat exchange.

(a) Write down the defining equations for the temperature $T(x,t)$ in the regions $x < 0$ (occupied by material 1), and in $x > 0$ (occupied by material 2).

(b) Heat conservation requires that the heat flux out of body 1 must equal the flux entering 2 at any time. Write down a mathematical expression for this result.

(c) It's not clear how many (if any) additional conditions are required at $x = 0$. Note however that the field equations are second order in x

for both $x > 0$ and $x < 0$, so we'd expect 2 independent solutions in $x > 0$ and also in $x < 0$; a total of 4 arbitrary constants will arise if we solve the equations in the separate regions. So far we have 2 imposed boundary conditions at $\pm\infty$ and one at $x = 0$; thus we need one additional condition. A sensible additional condition would seem to be that T is continuous across $x = 0$. Here's a case where a uniqueness result would be handy! (And is available.)

(d) Use dimensional arguments to show that the common contact temperature T_c does not vary in time—a surprising result!

(e) Using the error function solution (7.26) show that

$$T_c = \frac{T_1\zeta_1 + T_2\zeta_2}{\zeta_1 + \zeta_2}$$

where

$$\zeta_1 = \sqrt{k_1\rho_1 c_1} \text{ and } \zeta_2 = \sqrt{k_2\rho_2 c_2}.$$

Determine the heat transfer rate. Examine relevant cases.

(f) Given that the thermal conductivity varies with materials to a much greater extent than either density or specific heat, explain why metals tend to feel much hotter or colder to the touch than non-metals. One might expect the conductivity of a hand to be about the same as that of water.

(g) Why is it reasonable to model the hand as a semi-infinite body in this context?

Exercise 7.6 *Starting a Fire?*

What thermal and chemical characteristics of a material determine its combustibility, and what heating pattern is most likely to cause ignition? Ignition occurs if the temperature exceeds the *ignition temperature* of the material, so the problem reduces to one of determining the maximum temperature realized with different heating patterns.

A given amount of heat per unit area H is applied to a surface of a body over a time interval Δ.

(a) Assuming the heating rate is uniform over the interval, use (7.29) to show that the maximum temperature is reached at the surface of the material at the end of the heating interval. Determine this maximum value as a function of Δ, H, and the material properties.

(b) Using Maple determine the maximum temperature reached as a function of the index n if the rate of heating is a suitable multiple of t^n for $n = -1/2, 1, 2, 3$. Compare the results with those obtained if the heating rate is uniform or if all the heat is supplied instantaneously. Interpret your results.

(c) If it is known that there is a reasonable spread of the heat supply over the time interval, what temperature rise would you predict? Estimate the amount of heat necessary to start the reaction as a function of the ignition temperature T_{ign}, the material properties, and Δ. Comment on the material properties that are desirable for combustibility and incombustibility.

(d) What do the above results tell us is the best way to start a fire in a combustible material?

Exercise 7.7 *The Fundamental Point Source Solution*

For the 3D heat equation a fundamental solution can be found by considering the instantaneous release of a quantity of heat H at a point in an infinite body.

(a) The solution to this problem must be of the form $T = T(r, t, H, \rho, c, k)$ where $r = \sqrt{x^2 + y^2 + z^2}$, the distance from the release point. Why?

(b) The spherically symmetric form of the Laplace operator is

$$\nabla^2 T = T_{rr} + \frac{2}{r} T_r.$$

Write down

(i) the heat equation,

(ii) the initial conditions, and

(iii) an integral condition for the total heat content.

(c) Identify the similarity solution form.

(d) Using Maple verify that the function

$$T(r,t) = He^{-r^2/(4\kappa t)}/[8(\pi\kappa t)^{3/2}], \qquad (7.48)$$

satisfies all the required conditions. (One can also obtain this solution by direct means by first determining the similarity equation and then using Maple to identify its solution.)

Exercise 7.8 *Steady State Problems*

As pointed out in the text the instantaneous 3D point solution can be used to generate the solution to virtually all heat conduction problems. Steady state heat conduction problems are especially important because of the relevance of the solutions obtained in the entirely different physical context of inverse square law problems (gravitation, electrostatics). Here we determine the steady state temperature due to the release of heat at a constant rate q at a point in 3D space, and use this to solve *Poisson's Equation*.

(a) If heat is supplied to a point at a fixed rate q over an infinite period of time the steady state solution will be generated. Thus by using the instantaneous point source solution (7.48) obtain an integral representation for the temperature distribution due to the supply rate q from $t = -\infty$ to 0, and thus show that the steady state solution is

$$T(r) = \frac{q}{k}\frac{1}{4\pi r}.$$

Hint: Note that although Maple cannot cope directly with the required integral it can cope with the related integral

$$\int_0^{\infty} e^{-a^2/\sigma}\, d\sigma/\sigma^{3/2}.$$

Show that under steady conditions the total heat flow per unit time out of a sphere of radius R balances the heat flow per unit time from the source. Why should this be so?

(b) Using the above show that a particular solution of the steady state heat conduction equation

$$k\nabla^2 T = q(\mathbf{x}) \quad \text{in} \quad \mathcal{D}$$

is

$$\frac{1}{4\pi k}\int_{\mathcal{D}} \frac{q(\mathbf{x}')}{|\mathbf{x}-\mathbf{x}'|}\, dV_{\mathbf{x}'}.$$

The above equation is referred to as *Poisson's equation*. Physics students may recognize this result (when appropriately scaled) to be the gravitational potential due to a distribution of matter. Such connections occur often and are important.

Exercise 7.9 *Irradiated Bodies 1*

During the day radiation from the sun is partially reflected and partially absorbed within the material of objects on the Earth, often in thin "surface" layers. Conduction redistributes the heat within the objects and some of the heat is lost to the environment (mainly due to convection) from the surfaces. We'll assume a Newtonian cooling description, see (7.32). After sufficient time losses will balance gains so that steady conditions will be realized. Roughly speaking one *might* expect the time scale to reach equilibrium to be of order L^2/κ where L is the appropriate length scale (estimate this for lumps of iron, wood, leaves etc. using Table 6.1), but surface heat transfer parameters may play an important role. In order to determine the temperature of an object placed in the sun, one needs to quantify the above. In this exercise and in Exercise 7.11 we will examine this situation which is also relevant for determining the time taken to cook a chicken in the oven, and for determining ignition criteria for combustible materials.

The intensity E of radiation is the energy flow rate across a unit area at right angles to the flow. Experiments show that the intensity of a parallel beam of radiation in a material which absorbs it varies according to the law

$$E'(x) = -\mu E(x),$$

where x is the distance along the beam and μ is referred to as the absorption coefficient of the material the beam is passing through.

(a) Assuming that μ is constant, calculate the variation of the intensity with depth of a beam whose intensity at $x = 0$ (the surface of the body) is E_0. (For non-transparent materials the length scale μ^{-1} is often sufficiently small that radiation can be thought of as being absorbed *at* the surface.)

(b) Assuming that the radiation absorbed at any depth within the body is converted to heat energy at that same depth, calculate the rate of heat generation per unit volume per unit time as a function of x due to the absorption of radiation from the beam, and thus show that the equation governing the temperature distribution within a semi-infinite body is

$$T_t = \kappa T_{xx} + [\mu E_0/(\rho c)]e^{-\mu x} \quad \text{in} \quad x > 0. \tag{7.49}$$

(c) $\star$ If the beam of radiation is incident on the object *for sufficient time*, then the loss rate from the body's surface (due mainly to convection) will increase until eventually an equilibrium state will be reached in which the temperature is dependent only upon the distance below the surface. Under such steady state conditions, assuming there is no heat flux as $x \to \infty$, what flux condition needs to be imposed at the surface $x = 0+$ of the conducting body ? Using this condition and also the condition that there's no heat input at $x = \infty$, show that the temperature difference that exists between the surface and material "deep" within the body under such steady conditions is given by $E_0/k\mu$.

Hint: Integrate the heat equation across the solution domain.

Check that the sign of your result makes sense in terms of the direction of heat flow. Interestingly enough this temperature difference does not depend on knowing the heat loss mechanism at the surface. Why should this be so?

(d) The above result suggests that low conductivity materials reach higher temperatures than higher conductivity materials when placed, for example, in the sun. Does the result match experience? Explain.

(e) Solar radiation consists of various wavelengths with differing intensities. If the incident intensity in the wavelength range $[\lambda, \lambda + \delta\lambda]$ is $E_0(\lambda)\delta\lambda$ and the absorptivity is $\mu(\lambda)$, show that the temperature difference between the surface and deep material, under steady conditions is given by

$$\int_0^\infty \frac{E_0(\lambda)}{k\mu(\lambda)} d\lambda.$$

Hint: Use linearity.

Exercise 7.10 *Lake Pollution*

We return to the phosphate problem of Chapter 6, see Exercise (6.4). Phosphate from agricultural land is flushed into the upper layers of a lake of depth h at $t = 0$, and then diffuses downwards. No further phosphate is flushed into the lake. The mass flux per unit area transferred downwards by turbulent diffusion is described by $m = -\kappa\partial c(z,t)/\partial z$ where $c(z,t)$ is the concentration at depth z at time t; so the diffusion equation governs the process.

(a) Explain why, in the absence of deposition on the lake's bottom the constraint

$$\int_0^h c(z,t)dz = M$$

is reasonable, where M is the mass of phosphate compound per unit lake surface area flushed into the lake at $t = 0$.

(b) For small time it is to be expected that the fundamental one dimensional source solution will accurately describe dispersal. Write down this solution and indicate the equations this solution satisfies.

(c) Over what time scale would you expect the similarity solution to accurately describe $c(z,t)$?

(d) By adding to the above solution a contribution due to a second source, obtain a description that's useful over a larger time range, and estimate the time scale for which this solution is useful. Assume no deposition.

(e) ⋆ Make further improvements and plot successive approximations.

(f) For large time what would you expect the concentration to be, assuming no deposition?

(g) ⋆⋆ Assuming deposition to the lake's bottom occurs at a rate that's proportional to the concentration there, obtain an integral equation for the deposition rate, and suggest a useful first approximation for early time.

Hint: Introduce an additional sink of unknown strength at the lake's bottom, and ensure there's no additional flux introduced at $z = 0$.

Exercise 7.11 *Irradiated Bodies 2*

⋆⋆ Estimates for the time scale required for heat to conduct through common natural objects such as lumps of wood etc. (with small diffusivities) will convince the reader that steady state conditions are unlikely to be realized for such objects if left in the sun. The results of Exercise (7.9) are thus somewhat misleading; the temperature rise calculated there represents an upper bound that's unlikely to be reached in practice, either in the sun or in the oven. We therefore need to examine the unsteady problem. We'll stick to the semi-infinite body situation.

The temperature distribution of interest within the body is due to the additive influence of the heat supply of known strength due to radiation, and a heat loss of unknown strength from the surface due to Newtonian cooling (we'll assume). One can either work as in the previous exercise with the semi-infinite domain and ensure that the conductive heat transferred to the surface matches that required by the cooling condition, or work in the extended infinite domain, using appropriate image sources; the latter is neater. If the solution domain is extended to $-\infty$ the equation governing the temperature distribution can be written

$$T_t - \kappa T_{xx} = 2(G - L)/\rho c \quad \text{in} \quad -\infty < x < \infty,$$

where G is the heat gain rate per unit volume by the body due to the absorption of radiation, and L is the heat loss rate by Newtonian cooling from the surface; recall the heat inputs need to be doubled when the domain is extended. Using generalised function notation, see Section 7.7, $L = q(t)\delta(x)$, where $q(t)$ is the unknown loss rate from the surface.

(a) Identify $G(x)$ explicitly, using the results of Exercise 7.9. Identify $q(t)$ implicitly in terms of the surface temperature $T_s(t)$, using the Newtonian cooling condition (7.32).

(b) ⋆⋆ Using either the infinite or the semi-infinite domain approach write down an integral representation for the temperature distribution in the body due to $G(x)$ and $q(t)$ and obtain from this the surface temperature $T_s(t)$. By thus identifying $q(t)$, obtain an integral equation for $T_s(t)$. Introduce appropriate temperature, time and length scales.

Hint: Follow the pattern of calculation used in Section 7.6.

(c) Which terms of the integral equation would you expect to be dominant for small to moderate scaled times? Obtain a first estimate for $T_s(t)$ for small t. Examine the solution behaviour as $t \to 0$. Is this what you'd expect?

(d) Set up an iterative scheme for improving your first estimate.

(e) ⋆ For large time, which terms of the integral equation would you expect to dominate? Write down a first estimate of the large time behaviour.

(f) Plot the solution estimates obtained above.

Comment: Because of Maple's inability to handle in a simple way the integrals arising in this example it is a tedious matter to produce higher order estimates. The procedure is however straightforward; in essence the problem has been solved. Note that nonlinear surface cooling conditions could also be handled by the same procedure, but it's unlikely that analytic first estimates would be available for the resulting nonlinear integral equation, so a numerical integration of the integral equation would be necessary, see Exercise 7.17. Such nonlinear models are often necessary if large temperature changes occur (for example in ignition problems), or if the effective conductivity changes as a result of state change (for example in cooking problems).

Exercise 7.12 *Cooking a Chicken?*

⋆ Typical cooking instructions:

To roast a chicken place it in a 180°C oven for a period of 30 minutes per kilogram. Do remember to baste it! For the last 10 minutes raise the temperature to high levels to brown the surface.

To boil a chicken place it in slowly boiling water (necessarily at 100°C) for about 15 minutes per kilogram.

Explain the paradox: Why should it take longer in the oven when the oven's temperature is greater than the temperature of boiling water?

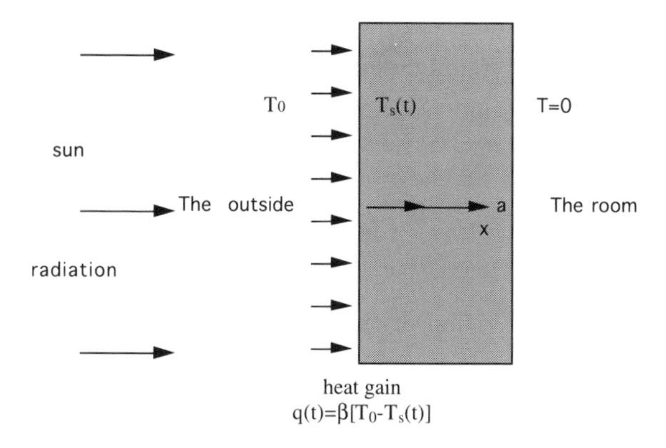

Figure 7.10: Thermal insulation.

Hint: What does one mean by the oven's temperature? What determines the heat transfer rate to the chicken?

For a fascinating account of the science of food and cooking the reader is referred to McGee (1984).

Exercise 7.13 *Thermal Insulation*

⋆ We return to the thermal insulator problem, see Exercise (6.5) of Chapter 6. In that problem we determined the heat extraction rate required of an air conditioner to maintain a given room temperature for diferent outside temperatures and for a variety of wall structures (with different thermal properties). A simple steady state model was used. For large air conditioned buildings considerable savings can be made by choosing the wall material and designing the heat exchange units to minimize the expense over the lifetime of the building. An efficient procedure for determining the transient response of the walls to varied heat inputs is a core element for such calculations. The same problem of course arises in freezer design. With this in mind we'll examine the thermal response of a sheet of thickness a with one face ($x = a$) maintained at a fixed temperature 0 (by the air conditioner) and the other face ($x = 0$) gaining heat from external sources (the sun), and losing heat to the environment, see Fig. 7.10.

Our first objective is to determine the fundamental solution associated with the prescribed boundary conditions. Using this we can build up the solution due to any input. Thus we need to determine the temperature distribution in a sheet of thickness a due to an instantaneous heat input $\mathcal{Q}$ per unit surface area at the insulated surface $x = 0$, when the surface $x = a$ is maintained at temperature 0.

(a) We apply the method of images.

(i) First write down the solution due to a source with heat input $2Q$ at $x = 0$ in the extended infinite domain. Why the factor 2?

(ii) Noting that this solution fails to satisfy the zero temperature condition at $x = a$, introduce an appropriate sink to cancel the effects of the first source. Show that the temperature field due to the two sources is

$$T(x,t) = 2Q[G(x,t) - G(x-2a,t)],$$

where the notation of the text is used.

(iii) This solution fails to satisfy the zero flux condition at $x = 0$. Introduce an appropriate source etc. etc. and thus show that the required fundamental solution is given by

$$2Q[G(x,t) + \sum_{n=1}^{\infty}(-1)^n\{G(x-2na,t) - G(x+2na,t)\}].$$

(b) ⋆⋆ The surface $x = 0$ receives heat at a known rate r per unit surface area (from the Sun) and loses heat at a Newtonian rate $\beta(T_s(t) - T_0)$ per unit area to its surroundings, where $T_s(t)$ is the surface temperature and T_0 is the temperature of its environment. Assuming the radiation from the sun is absorbed *at* the surface, $x = 0$ write down an integral expression for the temperature distribution, and thus determine an integral equation for surface temperature $T_s(t)$. This equation represents an effective mathematical formulation of the problem. We'll now undertake a preliminary examination of this equation.

(i) Scale the equation and determine the dimensionless group defining the relative thickness of the sheet. Under what conditions could the sheet be regarded as being "thermally thick"? Specify what you mean by this term. For such circumstances write down a sensible first estimate for the solution, and set up a sensible iteration scheme to investigate the effect of the finite thickness of the sheet on the heat transfer.

(ii) Even if the "thermal thickness" is large the surface temperature (and insulation effectiveness) must eventually be affected by the effects of the cooling at the surface $x = a$. After what time scale would you expect the finite thickness of the sheet to play a significant role?

Comment: Economics demands that the sheet be *just* thick enough to insulate the room; so it's clear that the above solution will not cover the complete time range of interest. We'll return to this problem in the next chapter, where we'll develop more appropriate solution techniques for examining the large time solution behaviour.

Exercise 7.14 *A Nuclear Accident*

This problem and the problem that follows trace some of the steps involved in attempting to model the fall-out from a nuclear accident. The primary objective is to determine atmospheric concentration levels and fall-out levels at various locations. Usually the amount of material released will be unknown, so it would be necessary to infer this from measured concentration levels. A crude dispersal model which ignores fall-out is examined here, and a more realistic model is examined in Exercise 7.15, where fallout will be taken into account.

The transport of fine nuclear material expelled into the atmosphere is dominated by air movement which may be thought of as having a steady velocity component (with non-zero mean) with turbulent fluctuations (with zero mean) superimposed. Basically the mean flow carries the nuclear material with it, while the fluctuations cause mixing and thus dispersal of the material. It may be shown theoretically and displayed experimentally that the rate and direction of dispersal of a material due to turbulent fluctuations depends on the concentration gradient, so the situation is analogous to heat dispersal with the mass flux per unit area given by $m = -\kappa \nabla c$ where c is the concentration of material and the ***dispersion coefficient***, κ, will depend on the size of the turbulent fluctuations; experimental data is available.

We'll assume the mean velocity and the dispersion coefficient remain fixed, and we'll use a co-ordinate system moving with the mean flow. We'll examine the situation in which the accident releases a mass M of material at $t = 0$ at $(x, y, z) = (0, 0, 0)$, with $0 < z < h$, where h is the effective height of the atmosphere. After this release the leak is plugged.

In the initial stages the dispersal process will be very complicated and dependent on the details of the accident. However, after a time scale of order h^2/κ, one might expect the concentration profile in the vertical direction to settle down; for simplicity let's assume for this preliminary model that the concentration is independent of z, so $c(x, y, z, t) \equiv c(x, y, t)$; and that the fall-out is negligible.

Under the prescribed circumstances $c(x, y, z) \equiv c(r, t)$ where $r = \sqrt{x^2 + y^2}$ (why?), so that it's appropriate to use the cylindrical form of the Laplacian and the dispersal is governed by

$$\frac{\partial c}{\partial t} = \kappa(\frac{\partial^2 c}{\partial r^2} + \frac{1}{r}\frac{\partial c}{\partial r}).$$

(a) Use scaling arguments to show that the similarity solution form is given by

$$c(r, t) = \frac{M}{\kappa h t} \mathcal{C}(\xi)$$

where $\xi = r^2/\kappa t$, and determine the equation for $\mathcal{C}$.

(b) Use Maple to obtain the similarity solution explicitly.

(c) Plot $\mathcal{C}(\xi)$ and comment on the solution behaviour.

(d) Plot $c(r_0, t)$ for various distances r_0 measured from the origin in the moving frame (ie. the effective source location). Comment.

(e) Plot the maximum (scaled) concentration levels expected at various distances r_0 from the effective source location, and determine the expected time for concentration peaks. A typical figure for κ is $5 \times 10^4 \text{cm}^2/\text{s}$ and for mean wind speed, 15m/sec.

(f) Although from the scientist's point of view the above plots are of most interest, civil authorities would like to know about the changes in concentration levels at *fixed* locations on the earth. Plot scaled concentration levels at various scaled locations downstream from the accident as a function of scaled time.

Exercise 7.15 *Nuclear Fall-out*

⋆⋆ Here we examine some of the features ignored in Example 7.14.

In the atmospheric context complications arise in describing the dispersion of material because vertical air movements in the atmosphere are much less strong than movements in the horizontal, and also the variability in wind direction has a different quantitative dispersal effect to that due to variability in wind speed. Such directional complications may be accounted for by simply using different dispersion coefficients in the three directions leading to a simple model of the transport process given by

$$c_t = \kappa(\alpha_1 c_{xx} + \alpha_2 c_{yy} + \alpha_3 c_{zz}),$$

relative to the mean flow, where κ and the α's can be inferred from atmospheric data. Another serious limitation arises because we ignored fall-out in our first model. We'll examine these questions.

(a) Again, after a time of order $h^2/(\alpha_3\kappa)$ one might expect the concentration profile in the vertical direction to settle down, but one wouldn't expect a uniform profile. One might expect the profile to settle down to a particular shape $Z(z)$, where the function Z allows for such effects as the variation of the density of the atmosphere with height, and where Z is scaled so that $\int_0^h Z(z)\,dz = 1$. With this in mind it seems sensible to look for a solution of the form $c = C(x, y, t)Z(z)$. By substituting this form into the equation for c and integrating across the atmosphere, show that the equation governing C is

$$C_t - \kappa(\alpha_1 C_{xx} + \alpha_2 C_{yy}) = -F,$$

where we've assumed that there is no loss of material out the top of the atmosphere, and F is the mass flux of material deposited to the ground.

(b) Show by appropriately scaling the spatial variables (x, y) that the equation reduces to the normal $2D$ diffusion equation

$$C_t - \kappa(C_{x'x'} + C_{y'y'}) = -F.$$

Thus, mathematically speaking there are no significant complications resulting from the non-isotropy of the flux vs. concentration gradient relation, and in fact the results obtained in the $F = 0$ case in Exercise 7.14 carry across immediately.

(c) ⋆⋆⋆ Suppose that the rate at which mass is deposited on the Earth's surface is proportional to the local concentration $C(x, y, t)Z(0)$. Derive an integral equation for the concentration $C(x, y, t)$. If the deposition rate is relatively small an iteration based on the $F = 0$ similarity solution already obtained in Exercise 7.14 would be appropriate. The interested reader might like to proceed with the calculations.

Comment: The biggest difficulty with this model is the fact that rain brings much more of the material back to Earth than slow settling does. If one knew the rainfall history how might one attempt to allow for this? Also the "mean" wind speed varies significantly over the time span and distances of interest, and varies with height. The dispersion can be greatly increased by such effects—essentially the material is transported in the high speed layers and then diffuses into neighbouring zones. This makes the outcome hard to predict, and the dispersal much greater than might be naïvely expected.

Exercise 7.16 *A Generalised Function Application*

As an example of the use of generalised functions we determine the steady state temperature distribution in a rod $0 < x < L$ of unit sectional area due to a steady point source located at $x = a$, with $0 < a < L$, emitting heat at a rate q. We then use this result to calculate the steady state temperature due to arbitrary heat input $f(x)$ per unit length per unit time along the rod. We'll examine the case in which both ends of the rod are maintained at temperature 0.

(a) Under steady conditions show that in generalised function notation the describing equations are

$$k\frac{d^2T(x)}{dx^2} + q\delta(x - a) = 0 \quad \text{for} \quad 0 < x < L,$$

with

$$T(0) = T(L) = 0.$$

In particular, using the definition of the delta function, check that the heat input is correct.

(b) Noting that $\delta(x-a) = 0$ for $x \neq a$ obtain the solution in the domain $0 < x < a$ satisfying the boundary condition at $x = 0$. Also obtain the solution in the domain $a < x < L$. Note that two arbitrary constants remain to be determined.

(c) The source introduces heat at $x = a$ so one would expect a slope change in the temperature profile across $x = a$. Integrate the heat equation across the source to give the result

$$k[\frac{dT(a+)}{dx} - \frac{dT(a-)}{dx}] + q = 0.$$

Does this result make sense physically? Explain. Show that a second integration across $x = a$ leads to the result that T is continuous at $x = a$.

(d) Use these conditions to show that the point source solution is given by $qG(x,a)$ where

$$G(x,a) = \begin{cases} \frac{1}{kL}[(L-a)x] & \text{for } x < a, \\ \frac{1}{kL}[a(L-x)] & \text{for } a < x < L, \end{cases}$$

is referred to as the *Green's function.*

(e) Add up the contributions due to sources distributed along the rod (as in the building solutions Section 7.4), and thus show that the temperature distribution due to a heat input rate $f(x)$ per unit length is

$$\int_0^L f(a)G(x,a)da.$$

Comment: It's easy to show that this result is identical to the result obtained by the variation of parameters technique normally used to solve such ordinary differential equations. The Green's function procedure extends to partial differential equations whereas the other doesn't; so the present approach is much more powerful.

(f) Determine the steady state temperature distribution in the rod due to a steady point source at $x = a$ if the end $x = 0$ is insulated and the other end is maintained at zero temperature.

Comments: This problem can be solved directly by matching solutions in the regions $0 \leq x < a$ and $a < x \leq L$ across $x = a$, or by using an image technique. To use the image technique extend the solution domain to $-L$ and introduce an image source at $-a$. The results of part (d) can then be used to generate the required answer.

Very often when dealing with images in complex situations it's not clear what strength sources need to be introduced. For example in semi-infinite region problem with heat input at the surface examined in Section 5.1 it's perhaps not clear if a factor of 1 or 2 is required when extending to the infinite domain. In such cases it's useful to consider simple examples of the present type to check to see if one's ideas are correct.

Exercise 7.17 *Integral Equations: A Numerical Scheme*

Standard numerical packages can be used to numerically process the integral Equation 7.36 that arose in the boundary integral methods Section 7.6, once the difficulties associated with large and small times discussed in the text have been overcome. Let's see what's involved.

(a) As with the numerical solution of ordinary differential equations the aim is to step forward in small time intervals Δ, determining the solution at each stage using an appropriate recurrence relation. Using the defining integral equation (7.36), obtain an exact expression for the temperature change over the small time interval $(t, t+\Delta)$ of the form

$$T(t+\Delta) - T(t) = I_1 + I_2,$$

where I_1 is an integral over the domain $(0, t)$ and I_2 is an integral over the small time interval $(t, t+\Delta)$.

(b) By discretising the time interval (write $t = n\Delta$), replacing I_1 by an approximate sum (use the trapezoidal rule, or Simpson's rule), and approximating I_2 (use Taylor's expansion), obtain an expression for $T_n \equiv T(n\Delta)$ of the form

$$T_{n+1} = T_n + \Delta \sum_{j=0}^{j=n} \gamma_j T_j + O(\Delta^2) \quad \text{for} \quad n \geq 0.$$

This scheme enables one to step forward in time from any initial t state *except* $t = 0$. The γ's become infinite when $t = 0$ which is a singular point of the scheme. Note that real physical difficulties arise when $t = 0$ because the solution is not regular there as we saw in Example 7.4.1. The way out of this difficulty is to use the analytic results obtained in the text. We simply use the series expansion to step away from the problem point $t = 0$, and then use the above scheme.

Another numerical difficulty arises in that one can't integrate forward in time indefinitely; the numerical error build up eventually swamp the solution. It's sensible therefore to use the the asymptotic scheme of the text for large time evaluations.

Comment: This situation is typical in that usually a mix of analytic and numeric work is required to extract answers, and the choice is imposed by the situation rather than by the preference of the modeller.

Bibliography

Carslaw, H.S. and Jaeger, J. C. (1959). *Conduction of Heat in Solids*, Oxford at the Clarendon Press.

Kamke, E. (1948). *Differential Gleichungen Lösungsmethoden und Lösungen*, Chelsea Publishing Co.

Lighthill, M.J. (1959). *Introduction to Fourier Analysis and Generalised Functions*, Cambridge University Press.

McGee, Harold (1984). *On Food and Cooking*, Charles Scribner & Sons New York.

Chapter 8

Fourier methods

Convergence problems
Newtonian slab heating
Composite fundamental solutions

8.1 Introduction

The methods developed in the last chapter for boundary value problems, while working well for moderate times, failed to be useful for longer times. In this chapter we show how to obtain useful large time solutions using a method first developed by Fourier. Fourier's initial development was directed at the solution of initial value problems and at the time was extremely controversial. Many leading mathematicians at that time failed to accept it as a valid technique. Over the years it has been developed to the stage where it is now accepted as a valid mathematical tool which has served to motivate many new mathematical ideas. *Fourier or Harmonic Analysis* is still an active theoretical research area, although the present research work holds no great interest for modellers. While it does generate a mathematical solution to a large range of initial value problems of the heat equation, the solutions it normally generates are not usually satisfactory for the purposes of the mathematical modeller. Here we will show that it does, however, provide a very useful way of generating fundamental solutions, which, when used with boundary integral techniques, can be very effective for handling such modelling situations. To illustrate this we'll re-examine the Newtonian slab heating problem of Section 6.5 Chapter 7.

Apart from its use for handling partial differential equations, Fourier theory finds a wide range of uses in applications ranging from its use in the digital recording industry, to all forms of image enhancement, to the analysis of trends from data etc. These applications will not be discussed here.

8.2 Preliminaries

You'll recall in Section (6.3.2) that we used a method referred to as *separation of variables* to obtain solutions of the heat equation, and by judicious choices of the *separation constant* we were able to generate a large number of solutions (dropping the primes on all variables)

$$\sin(n\pi x)/e^{-(n\pi)^2 t} \quad 0 < x < 1 \quad \text{for } n = 0 \cdots$$

to the heat equation satisfying zero temperature boundary conditions at $x = 0, 1$. These solutions were of interest for the understanding of the transient heat flow in a rod. A linear combination of such solutions would provide the complete solution to the problem if the coefficients A_n could be chosen in such a way that the initial condition

$$\theta(x, 0) = f(x) = \sum A_n \sin n\pi x \tag{8.1}$$

could be satisfied, see (6.23, 6.24). (The this earlier work $f(x) = -x$.) For the particular $f(x)$'s that can be generated by adding together a finite number of such $\sin n\pi x$ terms this is a trivial matter, and it's evident that the resulting expression for $\theta(x, t)$ solves the problem. For the other $f(x)$'s no *finite* sum will work and two major questions arise:

- How should one best choose the coefficients ?
- Is it possible to generate $f(x)$ exactly and thus determine exact solutions to the heat equation for arbitrary initial conditions?

The first question is an approximation question and obviously the decision about the usefulness of various approximations lies not in the mathematics but in the application; what is the approximation to be used for? More explicitly, given that the $\sin n\pi x$ functions seem appropriate in context, and given that a small number of terms is *essential* for practical evaluation, one could choose to minimize the maximum error, the modulus of the error averaged over the domain etc.—many possibilities arise, all *seem* sensible *mathematical* things to do, and (for a given $f(x)$) the mathematics required to do any of these things is straightforward *in theory*. From the point of view of the application the relevant question is more likely in a physical context to be: of the conserved quantities (eg. energy, mass, momentum) which should one try hardest to conserve?—a very different consideration! Having identified the appropriate error then, if there are N coefficients that one is willing to evaluate one simply minimizes the expression for the appropriately defined error as a function of these N coefficients, ending up with N (nonlinear in general) equations for the N coefficients. Now, anyone who has tried to solve just two nonlinear equations in two unknowns will recognize that, although simple in theory, this is a formidable computational task. Furthermore, if an improvement in accuracy is required, then one would have to redo the calculation with the required additional terms. Fourier recognized that if the error to be minimized is

$$\mathcal{E} = \int_{\mathcal{D}} (Error(x))^2 dx, \tag{8.2}$$

where $Error(x)$ is the local error and $\mathcal{D}$ is the solution domain, then the coefficient evaluation is trivial. $\mathcal{E}$ is referred to as the *mean square error*. Thus for our particular example the *local error* is given by

$$Error(x) = [f(x) - \sum_{n=1}^{N} A_n \sin n\pi x]$$

if N terms are used, and to minimize the mean square error $\mathcal{E}_N$ all we have to do is differentiate the above expression with respect to each of the coefficients $A_j, 1 \leq j \leq N$ and equate the result to zero. This gives

$$\int_0^1 [f(x) - \sum_{n=1}^{N} A_n \sin n\pi x] \cdot \sin j\pi x \, dx \; = 0 \;\; \text{for } j = 1, 2, ...N$$

a set of N linear equations for the N unknowns A_j. However

$$\int_0^1 \sin n\pi x \sin j\pi x dx = \begin{cases} 0 & \text{if} \quad j \neq n \\ \frac{1}{2} & \text{if} \quad j = n, \end{cases}$$

so, as a result of this *orthogonality* result, nearly all terms in the equations disappear leaving

$$A_j = 2 \int_0^1 f(x) \sin j\pi x dx \;\; \text{for } j = 1, 2, ...N;$$

a remarkable result!—piece of luck perhaps? That is, we have an *explicit* expression for all the coefficients for *any* $f(x)$ that's *trivial* to evaluate. Furthermore coefficients of any order can be determined without reference to lower order coefficients, and without reference to the number of approximating terms N. Note the implications:

- Without qualifications (like the order of the approximation) one can talk about *the* Fourier components of a function.

- The re-evaluation of all earlier coefficients that we'd normally expect to have to make in order to improve an approximation is unnecessary; one simply adds terms until the desired accuracy is achieved.

- As well as being able to easily reconstruct a function knowing its Fourier components (basically a list of numbers that's very easy to computationally store), we can also "filter" out unwanted components of a function (eg. often one wants to remove "noise"from a function) or even isolate our attention on a particular frequency or range of frequencies (useful when looking for possible resonances, see Chapter 12).

In addition to computational procedures there are mechanical and electrical devices that can do all the above processing for us, because of the special relevance of oscillatory functions to linear systems; which is why Fourier analysis is central to image enhancement, sound recording, and data examination processes.

One should always remember, however, that *the basic reason the Fourier approach is used so broadly is that the computational advantages are so great that they outweigh any consideration of the physics of the situation.* Thus, although there are physical situations in which the Fourier error corresponds to the unaccounted for energy (so that by minimizing this one feels that one is getting close to the most important physical criterion as far as approximation is concerned), this is not often the case, so one should not imagine that there is some special *physical magic* about the Fourier approach. Given the computational power now available, more sensible choices from the application point of view may now be moe appropriate. We will see, however, in Chapter 12, that in the important areas of vibrations and wave propagation Fourier techniques are especially suitable.

8.2.1 Sturm-Liouville theory

The orthogonality result underpinning the Fourier series work is not just a piece of luck associated with the $\sin n\pi x$ functions. It's a characteristic of the *Sturm-Liouville Problems* that results whenever one uses separation of variables to solve linear partial differential equations. A *Sturm-Liouville system* consists of a homogeneous linear second order differential equation of the form (referred to as *adjoint*)

$$(p(x)y'(x))' + (q(x) + \lambda r(x))y(x) = 0 \quad \text{in} \quad a < x < b, \tag{8.3}$$

with homogeneous two-point boundary conditions of the *separated* form

$$c_1 y(a) + c_2 y'(a) = 0, \quad c_3 y(b) + c_4 y'(b) = 0, \tag{8.4}$$

where the functions $p(x), q(x), r(x)$ are suitable functions of x. Such systems only have non-trivial solutions for certain values of λ, λ_n (say) called the *eigenvalues* with corresponding *eigensolutions* $\phi_n(x)$. As indicated earlier such systems arise naturally out of the separation of variables procedure;

thus the $\sin n\pi x$ arose because we needed to determine solutions of the harmonic equation satisfying zero solution boundary conditions. The p's, q's and r's in the general Sturm-Liouville system need to be such that the differential equation exhibits the oscillatory type behaviour needed to satisfy the homogeneous boundary conditions. ($r(x)$ single signed, no equation singularities, are sufficient conditions) For such systems there are a denumerable set of eigenvalues λ_n with associated eigenfunctions ϕ_n and the orthogonality condition goes through thus:

Theorem 8.1 *Sturm-Liouville Theorem. The eigenfunctions arising out of a Sturm-Liouville system are orthogonal in the sense that*

$$\int_a^b r(x)\phi_n(x)\phi_m(x)dx = 0 \quad \text{for} \quad n \neq m. \tag{8.5}$$

Note especially that the *weighting* function that needs to occur in this result is the coefficient $r(x)$ multiplying the separation constant λ in the defining equation (8.3). Using this orthogonality condition it is a simple matter to extract the coefficients of the eigenfunction expansion of a function

$$f(x) = \sum A_n\phi_n(x);$$

simply multiply this expression by $r(x)\phi_j(x)$ and integrate across the domain to determine A_j. The resulting expansion is best in the sense that the mean square error defined by

$$\mathcal{E} = \int_a^b r(x)[Error(x)]^2 dx = 0$$

is minimized. A reasonably complete account can be found in Ince (1956).

In simple linear algebra terms separation of variables leads to eigenfunctions that are orthogonal in an appropriate sense and so can be used as a basis for representing functions of x in general, and solutions of partial differential equations in particular.

Comment: There is an important unwritten (because it's unprovable) "theorem" of applied mathematics that all scientists work with that states that, if a result makes sense then it's correct. For example in the above context the orthogonality result will go through in cases in which the p's, q's and r's are such that the differential equation is singular (and so can have unbounded solutions), providing the equation has solutions that can satisfy the boundary conditions and are such that the orthogonality integral makes sense. We have, of course, used this theorem often in this book (for example in the delta function work). Euler (complex numbers), Archimedes and Newton (calculus) all used this theorem effectively. Of course one is obliged to check the validity of the outcomes resulting from the use of such "theorems".

8.2.2 Basic Fourier series

It's clear from the above that a finite Fourier sum does as good job as it can in a mean square sense when representing a function over a domain. From the theoretical point of view we'd like to know if the *full* series converges locally (and so represents the function), and from the practical point of view we'd like to know about the rapidity of convergence, so we can gauge its usefulness. These issues were controversial in Newton's time and in fact the proofs are technically difficult and not very enlightening. The following represents a review of the properties of Fourier series which underlie most of the applications that follow. An understanding of these results can be best gained through the exercises. The results are presented for the "normal" Fourier series ($\sin 2n\pi x, \cos 2n\pi x$), but extend to the general orthogonal function expansions arising out of Sturm-Liouville systems. For a detailed (but difficult) account see Titchmarsh (1939).

1. Consider a function f defined on the interval $[0, 1]$. Associated with this function is a set of Fourier coefficients $\{A_n,\ \ n = 0 \ldots \infty\}$ and $\{B_n,\ \ n = 1 \ldots \infty\}$, given by

$$A_n = \epsilon_n \int_0^1 f(x) \cos(2n\pi x)\, dx, \tag{8.6}$$

$$B_n = \epsilon_n \int_0^1 f(x) \sin(2n\pi x)\, dx, \tag{8.7}$$

where

$$\epsilon_n = \begin{cases} 1 & \text{for } n = 0, \\ 2 & \text{for all other } n. \end{cases} \tag{8.8}$$

These coefficients are obtained using the orthogonality condition. From these sets of Fourier coefficients one can generate a function F defined *for all* x by

$$F(x) = \sum_{n=0}^{\infty} A_n \cos(2n\pi x) \ + \sum_{n=1}^{\infty} B_n \sin(2n\pi x), \tag{8.9}$$

provided some meaning can be given to this series. Usually one requires that the series converges. Of course one hopes that $F(x) \equiv f(x)$ for $0 \leq x \leq 1$. All the oscillatory functions have periodicity 1 so $F(x+1) = F(x)$ and so for x outside the range $0 \leq x \leq 1$, the function simply "repeats". It's often useful in fact to think of $F(x)$ as being a Fourier representation over the infinite domain of the function defined by $f(x)$ for $0 \leq x < 1$, together with its *periodic extension* defined by $f(x+1) = f(x)$ outside this range.

2. **Theorem 8.2** *If the series (8.9) is uniformly convergent then* $F = f$.

   The theorem can be proved under much less restrictive conditions, but the result is *much better* than is indicated by any such theorem. Thus, even in cases in which the series (8.9) *diverges*, so that the expression $F(x)$ is meaningless in normal terms, it's still often possible to reconstruct $f(x)$ from the Fourier coefficients. Generalised function theory is a natural setting for these results. The crucial feature is that the $f(x) \rightleftharpoons (A_n, B_n)$ association is unique.

3. **Theorem 8.3 (The Riemann-Lebesgue Lemma:)** *If the function* f *is absolutely integrable then*

$$\lim_{n\to\infty} A_n = \lim_{n\to\infty} B_n = 0. \tag{8.10}$$

   This result enables one to assess the behaviour of the Fourier coefficients for large n and thus enables one to establish the convergence results that are essential for using theorem (8.2), and also are essential for assessing the usefulness of a Fourier expansion.

   To determine the convergence properties it's convenient to combine the discussion of the sin and cos terms by working with the complex Fourier coefficients defined by

$$C_n = \epsilon_n \int_0^1 f(x) e^{2\imath n\pi x}\, dx, \tag{8.11}$$

   so that

$$A_n = \Re C_n \quad \text{and} \quad B_n = \Im C_n,$$

   provides the required connection back to our real coefficients. Let's assume for the present that f is at least twice differentiable on the interval. If we integrate by parts we get

$$\begin{aligned} C_n/\epsilon_n &= \textstyle\int_0^1 f(x) \frac{d}{dx}[e^{2\imath n\pi x}/(2\imath n\pi)]\, dx \\ &= [f(x)e^{2\imath n\pi x}/(2\imath n\pi)]_0^1 - \textstyle\int_0^1 f'(x)e^{2\imath n\pi x}\, dx/(2\imath n\pi) \\ &= [f(1) - f(0)]/(2\imath n\pi) - \textstyle\int_0^1 f'(x)e^{2\imath n\pi x}\, dx/(2\imath n\pi), \end{aligned}$$

   and a second integration by parts gives

$$C_n/\epsilon_n = [f(1) - f(0)]/(2\imath n\pi) - [f'(1) - f'(0)]/(2\imath n\pi)^2 + R_2, \tag{8.12}$$

   where

$$R_2 = \int_0^1 f''(x)e^{2\imath n\pi x}\, dx/(2\imath n\pi)^2.$$

Now since f'' is finite it will be absolutely integrable so the Riemann-Lebesgue Lemma tells us that R_2 is less in order than $(1/n^2)$; and thus smaller for large n than the preceding terms, unless these happen to vanish identically. Thus **if** $f(0) = f(1)$ then the coefficients of the Fourier series are of $O(1/n^2)$ or smaller, so that the series converges absolutely and uniformly, and thus the above Fourier series theorem (8.2) tells us that $f(x) \equiv F(x)$. This result is fine as far as it goes, *but the first condition* $f(0) = f(1)$ *is a disaster!* We certainly do not want to restrict our attention to functions that are equal in value at the ends of the interval! This condition is really a surprise; we'd of course expect the behaviour of the series to be dependent on the smoothness of $f(x)$, with the convergence rate becoming poorer for more poorly behaved $f(x)$, but we wouldn't expect it to depend on the boundary values of f! Note that **if** $f(0) \neq f(1)$ then C_n **is in fact of order** $1/n$ because the first term of (8.12) is of order $1/n$, and the remaining terms are too small to affect the asymptotic behaviour. The resulting expansion in the $f(1) \neq f(0)$ case does in fact converge to the function everywhere except at the end points, but the convergence is not uniform, and in any case is so slow as to make the series completely useless from a computational viewpoint! One would expect to need about 100 terms to get a 1% accuracy!

The above arguments go through if there are discontinuities of f within the domain $0 < x < 1$ so the same convergence problems arise, although such difficulties are not unexpected. The results are in fact better stated and understood in the infinite domain using periodic extension ideas, see Exercise 8.2: thus if f together with its periodic extension has discontinuities then C_n is $O(1/n)$. If f together with its periodic extension is continuous but has slope discontinuities then C_n is $O(1/n^2)$ etc. The real annoyance is that it doesn't matter how well behaved f is, its Fourier series will be poorly behaved *unless* $f(0) = f(1)$! In many such circumstances it's possible to rearrange the mathematics so as to ensure this periodicity condition is satisfied. We'll see how to go about this in Exercise 8.2 and in the material of the next Section.

4. Where the function f with its periodic extension is not continuous (at $x = \xi$ say) then the series for F converges to $[f(\xi+) + f(\xi-)]/2$, where the $+$ and $-$ denote taking the limiting values from the two sides of $x = \xi$.

5. All the above results can be extended to the semi-infinite and infinite domain situations. In this context Fourier series become Fourier integrals and the terminology of *Transform Techniques* is used to describe the procedures.

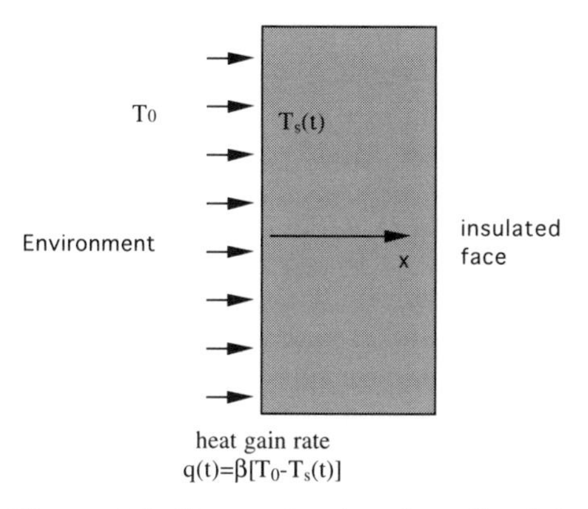

Figure 8.1: Newtonian heating of a slab.

We'll now use the Fourier approach to extract solutions for the slab heating problem.

8.3 Newtonian Slab Heating

You'll recall the set-up, see Fig. 8.1, also Section 7.6.5. We have a slab of material of thickness a with one face $x = a$ insulated and the other face exchanging heat with the environment. The slab is initially at temperature 0 and the environment at T_0. It's clearly sensible to use $a, a^2/\kappa$ as the length and time scales, and T_0 as the temperature scale. Using these scales the problem is defined by the usual heat equation

$$T_t = T_{xx}, \text{ for } 0 < x < 1$$

with boundary conditions

$$T_x(1,t) = 0, \quad \text{and} \quad T_x(0,t) = a\alpha(T(0,t) - 1) \tag{8.13}$$

see (7.33), and the initial condition

$$T(x,0) = 0 \text{ for } 0 < x < 1.$$

The Fourier procedure is longwinded but straightforward. The steps will be listed for future reference:

• **Ensure the boundary conditions are homogeneous**

For reasons related to the convergence difficulties discussed in the last section the Fourier technique can't be employed directly unless the boundary conditions are homogeneous, and in fact the heat equation also needs to be homogeneous, although it's a simple matter to deal with this difficulty, see

Exercise 8.6. In our present problem the Newtonian boundary condition (8.13) at $x = 1$ is not homogeneous. Very often an elementary device can be used to recast the problem in a homogeneous boundary condition form. Thus in our present case if we write

$$T(x,t) = \bar{T}(x) - T'(x,t)$$

and choose $\bar{T}(x)$ in such a way that $T'(x,t)$ satisfies homogeneous boundary conditions, then the difficulty is resolved. Notice that the steady state solution $\bar{T} = 1$ fulfills our requirements; not only does this choice leave us with homogeneous boundary conditions for T' but also the equation for T' is the *homogeneous* heat equation.

Comment: Separating out the steady solution from the *transient* in this way is a very natural thing to do from the physical point of view; as we've seen on earlier occasions a consideration of the physics often leads naturally to a sensible mathematical approach.

The transformed problem for T' becomes:

$$T'_t = T'_{xx},$$

with boundary conditions

$$T'_x(1,t) = 0, \quad \text{and} \quad T'_x(0,t) = a\alpha T'(0,t), \tag{8.14}$$

see (8.13), and the initial condition

$$T'(x,0) = 1 \quad \text{for} \quad 0 < x < 1. \tag{8.15}$$

• Determine separable solutions satisfying the boundary conditions

As seen earlier in Section 6.3.2, separation of variables leads to the coupled ordinary differential equations

$$\mathcal{X}_{xx}(x) - \mu\mathcal{X}(x) = 0; \quad \text{and} \tag{8.16}$$

$$\mathcal{T}_t(t) = \mu\mathcal{T}(t) \tag{8.17}$$

for the separated solutions $\mathcal{X}(x)\mathcal{T}(t)$ of the heat equation. The relevant Sturm-Liouville problem consists of the homogeneous second order ordinary differential equation (8.16) with two point homogeneous boundary conditions

$$\mathcal{X}_x(1) = 0, \quad \text{and} \quad \mathcal{X}_x(0) = a\alpha\mathcal{X}(0), \tag{8.18}$$

see (8.14). It's now a matter of carefully checking out all possible choices of the separation constant μ; it's very easy to miss out particular separation constant cases and end up with an *incomplete* set of eigensolutions, and thus an incorrect answer. For positive values of μ (λ^2 say) the equation

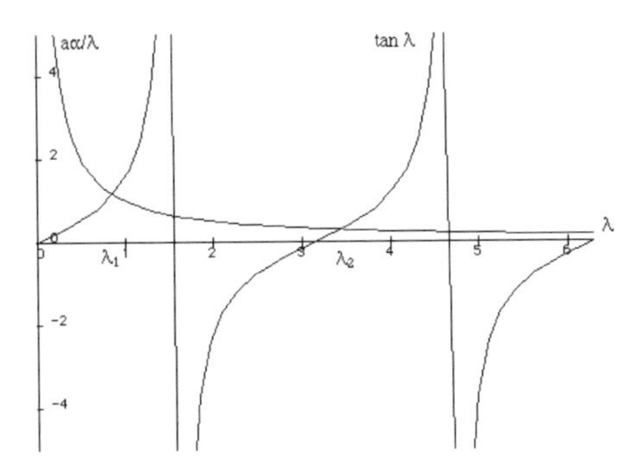

Figure 8.2: Eigenvalues: solutions of $\tan\lambda = a\alpha/\lambda$.

is exponential with solutions $(\sinh\lambda x, \cosh\lambda x)$ (it's more convenient to use the hyperbolics than the exponentials here because of the boundary conditions) and the only solution satisfying the boundary conditions is the trivial solution. Similarly for $\mu = 0$ no non-trivial solutions result. For negative μ ($-\lambda^2$ say), oscillatory functions $(\sin\lambda x, \cos\lambda x)$ result and the solution that satisfies the boundary condition at $x = 1$ is $\cos\lambda(x-1)$. The remaining boundary condition is satisfied for these functions if and only if

$$\lambda\sin\lambda = a\alpha\cos\lambda. \tag{8.19}$$

This equation for the eigenvalues can't be solved exactly which is somewhat inconvenient, but not important. The nature of the solutions of this equation may be seen most easily by considering the intersection of the two curves $Y(\lambda) = \tan\lambda$ and $Y(\lambda) = a\alpha/\lambda$, graphed in in Fig. 8.2. There is an infinite set of such values and the larger ones become increasingly close to $\lambda = N\pi$ where N is a large integer. Maple can be used for the first few roots and an asymptotic approximation is appropriate for the remaining roots. Let λ_n for $n = 1\cdots$ denote these eigenvalues.

The separated solutions of the heat equation satisfying the boundary conditions are thus given by, see (8.17),

$$\cos[\lambda_n(1-x)]\ \exp(-\lambda_n^2 t).$$

• **Determine the Fourier coefficients**

Using the separated solutions we generate a rather general solution of the form

$$T'(x,t) = \sum_{n=1}^{\infty} C_n\cos[\lambda_n(1-x)]\exp[-\lambda_n^2 t],$$

where the constants C_n need to be chosen so that the initial condition (8.15) is satisfied, requiring

$$\sum_{n=1}^{\infty} C_n\cos[\lambda_n(1-x)] = f(x) = 1. \tag{8.20}$$

Sturm-Liouville theory tells us that the eigenfunctions are orthogonal over the domain $(0,1)$, so to determine C_j we simply multiply this expression by the jth eigenfunction and integrate across the domain to give

$$C_j = \int_0^1 f(x)\cos[\lambda_j(1-x)]\,dx \Big/ \int_0^1 \cos^2[\lambda_j(1-x)]\,dx,$$

which yields

$$C_j = \frac{4\sin\lambda_j}{2\lambda_j + \sin(2\lambda_j)}. \tag{8.21}$$

If we now revert to the original unscaled physical variables we have the solution

$$T(x,t) = T_0\left\{1 - \sum_{n=1}^{\infty} C_n \cos[\lambda_n(a-x)]e^{-\lambda_n^2\kappa t/a^2}\right\}. \tag{8.22}$$

• **Examine the convergence properties**

Now for large n, from (8.21), C_n is of order $1/\lambda_n$ i.e. of order $1/n$, so the series (8.22) (although non-uniformly convergent) converges too slowly to be of use for evaluation at $t = 0$. (The form suggests that more than 100 terms would be required to give a result accurate to 1%, and the numerical error introduced by a computation would make the answer unreliable.)

Aside: We could have anticipated the convergence rate from the Riemann-Lebesgue work, see Section (8.2.2) item (3). Noting the rearrangement

$$\cos[\mu_n(1-x)] = \cos(\mu_n)\cos(\mu_n x) + \sin(\mu_n)\sin(\mu_n x),$$

we see that our series is of the form we examined using the Riemann-Lebesgue work so that $O(1/n)$ coefficient behaviour is expected unless the function $f(x)$ being represented is such that $f(0) = f(1)$. Now $f(1) - f(0)$ is the temperature drop across the slab at $t = 0$ which is 1; thus C_n is $O(1/n)$. Alternatively simply note that the periodic extension is discontinuous, see Exercise 8.2.

Interestingly the series (8.22) for the temperature field is absolutely and uniformly convergent for all times bounded away from zero because the convergence is dominated by the exponential terms $e^{-\lambda_n^2\kappa t/a^2}$ for large n as long as $t \neq 0$, i.e. terms go down like e^{-n^2}. Thus the series is *only* poorly convergent at $t = 0$! This is a special feature of the heat equation; normally one would expect mathematical difficulties associated with approximating initial conditions to be carried into the solution domain. The convergence is in fact so rapid for $t > 0$ that one only needs $\lambda_1^2\kappa t/a^2$ to be moderately large for *only* the first term in the infinite series to be significant. This method is thus *far superior* to the image technique used in Section 6.5 Chapter 7 for obtaining the long term behaviour of the slab solution. (Recall that in this earlier work we were unable to determine the time required for the slab to reach equilibrium.) From the above solution we can see that equilibrium "has" been reached (within 1%) when the argument of the exponential is about 5 i.e. when $t = t_{eql} = 5a^2/(\lambda_1^2\kappa)$, and the dependence of this time span on the properties of the slab and the heat transfer coefficient β can be easily explored, see Exercise 8.4.

Comment: a complete solution For practical purposes of evaluation, when $\kappa t/a^2$ is small, the Fourier solution is not useful, but of course the image solution obtained in Chapter 7 is very effective for this t range, so one should switch from one description to the other to obtain a useful description for all t values.

By taking advantage of the steady state solution we were able to overcome technical convergence difficulties associated with Fourier series and solve the prescribed slab problem. We would, however, like to determine the solution for much more general circumstances eg. under circumstances in which the environmental temperature changes in time or where there is an external heat input, as in the building insulation problem. The above Fourier method doesn't extend immediately to cover such cases. What's needed for such general circumstances is a *fundamental source solution* that (unlike the fundamental image solution of Chapter 7) is valid for all time. We'll now seek out this fundamental solution using Fourier techniques and then go on to discuss general modelling situations using this fundamental solution.

8.4 Another Fundamental Slab Solution

You'll recall from Chapter 7 that the fundamental solution is the solution that corresponds to an instantaneous heat input into the slab's exposed surface. More explicitly the fundamental solution for our slab problem has to satisfy:

- the homogeneous heat equation,
- the zero flux condition at $x = a$,
- the zero flux condition at $x = 0$ for $t \geq 0$,
- the initial condition that the temperature is zero everywhere save in a neighbourhood of $x = 0$ and which is such that

  $$\lim_{t \to 0} \int_0^{0+} T(x,t)dx = Q' \equiv Q/\rho c$$

  in the generalised function sense. If we denote this initial temperature distribution by $T(x,0) = f(x)$ then in generalised function notation

  $$f(x) = Q'\delta(x), \tag{8.23}$$

where $\delta(x)$ is the Dirac delta function, see Section 7.7.

All these conditions are in a form suitable for the application of the Fourier solution technique. It is a routine application of the method to show that there are no solutions corresponding to the separation constant being negative. From a separation constant of zero we obtain the solution $T = 1$, and all the remaining solutions are of the form

$$\cos \mu_n x e^{-\mu_n^2 \kappa t}$$

where μ_n must satisfy

$$\mu_n \sin \mu_n a = 0.$$

Thus we have the set of eigenvalues $\mu_n = n\pi/a$ for all integers $n \geq 0$. The coefficients C_n of the series solution must be such that the initial condition is satisfied which requires

$$C_n \int_0^a \cos^2(\mu_n x)\, dx = \int_0^a f(x) \cos(\mu_n x)\, dx \quad \text{for } n = 0, 1, \cdots, \tag{8.24}$$

where you'll notice that the $n = 0$ coefficient is included. Because of the shape of the f function the integrand on the right hand side is concentrated about $x = 0$ where $\cos(\mu_n x) = 1$, so the integral becomes $1 \cdot \int_0^a f(x)dx = Q'$, by the definition of the delta function, see Section 7.7. This leads us to the fundamental solution in the form

$$T(x,t) = \frac{Q'}{a}[1 + 2\sum_{n=1}^{\infty} \cos(n\pi x/a) e^{-n^2\pi^2\kappa t/a^2}]. \tag{8.25}$$

The results of *generalised function theory* can be invoked to justify this intuitive result.

We now have two different series representations of the fundamental solution corresponding to an insulated slab; the image solution (7.30) obtained in the last chapter, and the present solution. Although *both* are *theoretically* equivalent representations (in the sense that both series converge to the same function), they are **not** equivalent for practical purposes! Which one should we use? (For example if we wish to determine the surface temperature for a variable rate of heating combined with a Newton cooling loss to the surroundings?)

Let's compare the two representations at $x = 0$. The form from the last chapter may be arranged as

$$\frac{Q'}{a}[1 + 2\sum_{n=1}^{\infty} e^{-n^2[a^2/(\kappa t)]}]\sqrt{a^2/(\pi\kappa t)}.$$

Thus when $t = a^2/(\kappa\pi)$ the two series are identical. For larger values of t the series derived in the present chapter has terms which decrease more rapidly, while for smaller values of t it is the image series from the last chapter which has this desirable property. Thus for practical reasons we should work with the fundamental solution in the form $Q'\mathcal{F}(\tau)/a$ where $\mathcal{F}$ is defined by

$$\mathcal{F}(\tau) = \begin{cases} [1 + 2\sum_{n=1}^{\infty} e^{-n^2/\tau}]/\sqrt{\pi\tau} & \text{for } \tau < 1/\pi \\ 1 + 2\sum_{n=1}^{\infty} e^{-n^2\pi^2\tau} & \text{for } \tau > 1/\pi, \end{cases} \tag{8.26}$$

where τ is the scaled time $\tau = (\kappa/a^2)t$ based on the slab thickness. Note that for both ranges the least rapid convergence will be indicated by the largest size of the exponential term, namely $\exp[-n^2\pi]$. Now when $n = 1$ this gives 3×10^{-6} which means that unless one wants extreme accuracy (normally not warranted because of experimental uncertainties) *only* the $n = 1$ term needs to be retained! There will in fact be only a 4% error introduced if the $n = 1$ term is also dropped and such an error is often acceptable.

We now apply this result to a typical practical problem in which heat is being supplied to the surface of the slab at a known rate $q_0h(t)$ per unit area (here q_0 is the heat flux scale), and is being lost to the environment (at temperature T_0) at a rate prescribed by the Newton Law of cooling. The net supply of heat per unit area of the surface over the time interval $(t, t + dt)$ is

$$[q_0h(t) - \beta(T_s(t) - T_0)]dt,$$

where $T_s(t) = T(0, t)$, and thus the scaled surface temperature $\mathcal{T} = T_s/T_0$ satisfies the equation

$$\mathcal{T}(\tau) = \frac{q_0 a}{kT_0} \int_0^{\tau} [h(\tau') - \frac{\beta T_0}{q_0}(\mathcal{T}(\tau') - 1)]\mathcal{F}(\tau - \tau')\, d\tau'. \tag{8.27}$$

This integral equation represents an appropriately pre-processed representation of the physics of the slab problem. This equation extends the result (7.45) obtained in Chapter 7, where a slightly different scaling was used. At this stage (and not before) the modeller can feel well satisfied with his work. The problem is basically complete; the rest can be left to a machine.

For $0 < t < a^2/(\pi\kappa)$ only the image form of the function $\mathcal{F}$ is needed. As noted above, in this range $\mathcal{F}$ differs little (about 4%) from $1/\sqrt{\pi\tau}$, which corresponds to the fundamental solution for a semi-infinite region, see (7.28), so that normally this will suffice. If this is not accurate enough, iterative corrections due to the small effects of $(\mathcal{F} - 1/\sqrt{\pi\tau})$ can be trivially determined.

For longer times it is almost certainly most efficient to use direct numerical methods to find the solution, for it is unlikely that any useful form of analytic solution will be found. The technique described for Example 7.17 in Chapter 7 could be used.

8.5 Summary

We've seen that, providing technical difficulties associated with Fourier series can be overcome, the Fourier technique leads to solutions that are particularly effective for determining long term solutions of the heat equation. When combined, the source techniques of Chapter 7 and the Fourier technique lead to useful solution descriptions over the complete time range for many problems with simple (standard) boundary conditions. Fundamental solutions that are useful over the complete t range can also be obtained using a mix of the two techniques, and, by using the boundary integral method, lead to solution procedures that are broadly applicable to modelling situations. The procedure was illustrated in the particular case of the finite slab heated by the environment; a problem of particular interest for the "cooking" applications considered in the next three chapters.

8.6 Exercises

Exercise 8.1 *Sturm-Liouville Theory Example*

The homogeneous ordinary differential equation

$$u_{xx}(x) + \lambda^2 u(x) = 0,$$

with λ a parameter, together with the homogeneous boundary conditions

$$u_x(0) = 0, u(1) = 0,$$

is a Sturm-Liouville system. The features that distinguish such systems and underlie their importance will be established here in this special case. The language of linear algebra will be used.

(a) Non-trivial solutions (the *eigensolutions*) for the above system exist only for special values (the *eigenvalues*). Show that for the values

$$\lambda_n = \frac{(2n+1)\pi}{2}, n = 0, 1, \ldots,$$

the solutions are given by

$$u_n = \cos \frac{(2n+1)\pi x}{2}.$$

(b) Show that these solutions are orthogonal in the sense that their *inner product*

$$\langle u_n.u_m \rangle = \int_0^1 u_n(x) u_m(x) dx = 0, \quad \text{if} \quad n \neq m.$$

These eigensolutions provide a convenient basis set for the representation of functions $f(x)$. Show that if we write

$$f(x) = \sum_{n=0}^{\infty} a_n u_n(x)$$

then, using the orthogonality result, we obtain an explicit expression for the coefficients a_n given by

$$a_n = \frac{\langle f.u_n \rangle}{\langle u_n \cdot u_n \rangle}.$$

The resulting expansion is referred to as the Fourier expansion for f. (Other sensible choices for the coefficients are possible; the resulting expansions are *not* referred to as Fourier expansions.)

(c) We define the error in our representation of f if N terms are used to be (the *mean square error*)

$$\mathcal{E}_N(a_0, a_1, \ldots, a_N) = \langle (f - \sum_{n=0}^{N} a_n u_n) \cdot (f - \sum_{n=0}^{N} a_n u_n) \rangle.$$

Show that this error is minimized by the choice of coefficients a_n given above. Thus the Fourier expansion is best in the sense of minimizing

$$\int_0^1 Error^2(x) dx.$$

It can be shown that the expansion converges to $f(x)$ in the above mean square sense providing $\int_0^1 f^2 dx$ exists, see Titchmarsh (1939). Do not attempt to do this.

Exercise 8.2 *Fourier Series Convergence*

(a) Using the eigenfunctions arising out of the Sturm-Liouville system

$$u_{xx} + \lambda^2 u = 0 \quad \text{for } 0 < x < 1 \quad \text{with } u(0) = u(1) = 0,$$

show that the Fourier series expansion for $f(x) = x$ is

$$2 \sum_{1}^{\infty} \frac{(-1)^{n+1}}{n\pi} \sin n\pi x. \tag{8.28}$$

Comment on the convergence properties of the Fourier series. Roughly how many terms would be needed to obtain $f(x)$ to within 1% over the stated range?

(b) Using Maple (use **sum**) plot $f(x)$ and $S_N(x)$ for $N = 1 \cdots 4$ over the range $0 \leq x \leq 1$, where S_N is the sum of N terms of the above Fourier expansion for f. Comment on any features that you think are significant.

(c) Plot S_4 over the larger range $-1 < x < 3$. The *periodic extension* of $f(x)$ is obtained by extending outside the domain $0 \leq x < 1$ in accordance with the periodicity properties associated with the eigenfunctions, and is important because the convergence properties depend on the continuity properties of the function *together with* its periodic extension, see item 3 in Section 2.2. Noting that the eigenfunctions have periodicity 2 sketch the periodic extension $f_{ext}(x)$ of $f(x)$, and in particular show that it has discontinuities at $x = -3, -1, 1, 3$ etc. Thus explain the poor convergence of the Fourier series.

(d) $\star$ It's possible to improve the convergence of an expression by recognizing the source of the convergence difficulty.

(i) Using the same eigenfunctions as above determine the Fourier series for the function $g(x) = e^x - 1$ over the range $0 < x < 1$, and plot the sum to 4 terms of the series over the range $-1 < x < 3$. Sketch the periodic extension associated with $g(x)$. Show that for large n the coefficients b_n of the expansion behave like

$$b_n = (e-1)[2\frac{(-1)^{n+1}}{n\pi}] + O(1/n^2).$$

Explain the poor convergence.

(ii) The convergence is like $1/n$ because the periodic extension is discontinuous at $x = 1$. By utilizing the result (8.28) show that the expansion for $g(x)$ can be written in the form $g(x) = (e-1)x + h(x)$ where the expansion for $h(x)$ converges like $1/n^2$. Using 4 terms plot the Fourier approximation or $h(x)$. Comment. Explain this result. This procedure for *extracting the discontinuity* is often used to improve the convergence of an expansion.

Exercise 8.3 *Rod Problem: Loose Ends*

(a) Using the results obtained in the text and in Exercise 8.2 complete the rod problem. That is, determine the temperature distribution in a rod of length L initially at temperature T_0 after one end is raised to, and maintained at, temperature T_1, while the other end is maintained at T_0. Plot the solution for various times.

(b) For small scaled times the Fourier solution is unsatisfactory. Using the semi-infinite rod error function solution (see Chapter 7 (7.26)) with introduced image solutions, obtain a useful small time solution. Indicate the time span of usefulness of this solution.

Exercise 8.4 *Newtonian Slab Heating: Loose Ends*

The eigenvalues for the radiating slab problem considered in the text are the roots of the transcendental equation $\tan \lambda_n = a\alpha/\lambda_n$.

(a) The first eigenvalue λ_1 is especially important for determining the time for the slab to reach equilibrium with the environment. Using Maple (or other software) plot out the $\lambda_1(a\alpha)$ relationship. In the text we found that the time to reach equilibrium is approximately given by $t_{eql} = a^2/(\kappa\lambda_1^2)$. Using appropriately scaled ordinates produce a plot displaying the dependence of t_{eql} on the plate parameters and the heat transfer parameters. Comment. Pay particular attention to the thermally thin and thick slab cases.

(b) The larger order roots are located close to $\lambda = N\pi$. Write $\lambda_N = N\pi + \epsilon_N$ where $\epsilon_N \ll 1$ and thus obtain a useful asymptotic description for these roots.

Exercise 8.5 *Steady State Plate Problem*

In an earlier Exercise 6.1 in Chapter 6 we saw that linearity and symmetry arguments alone, determine the steady state temperature at the centre of a square plate with one edge maintained at a different temperature to the remaining edges. We now determine the steady state temperature distribution throughout the plate using Fourier methods. The equations determining the steady state temperature distribution $T(x, y)$ in the plate are given by

$$T_{xx} + T_{yy} = 0, \quad \text{in } 0 < x, y < l, \tag{8.29}$$

$$\text{with } \; T(0, y) = T(l, y) = 0, \tag{8.30}$$

$$\text{and } \; T(x, 0) = 0, T(x, l) = T_0. \tag{8.31}$$

(a) (i) Show that separable solutions $X(x)Y(y)$ of equation (8.29) necessarily satisfy

$$\frac{X_{xx}(x)}{X(x)} = -\frac{Y_{yy}(y)}{Y(y)} = -\mu,$$

and thereby show that the solutions

$$\phi_n = A_n \sin \frac{n\pi x}{l} \sinh \frac{n\pi y}{l}$$

satisfy equations (8.29, 8.30), and $T(x, 0) = 0$. Thus $\sum_{n=1}^{\infty} \phi_n$ is a solution that satisfies the partial differential equation and the homogeneous boundary conditions.

(ii) Determine the coefficients A_n so that $\sum_{n=1}^{\infty} \phi_n$ satisfies the remaining boundary condition (8.31).

(iii) Plot $T(x, l/2), T(l/2, y)$.

Comment: There are solution convergence difficulties at the corners $(0, l), (l, l)$ of the plate and it's not clear if the difficulties are real in a physical sense (the boundary conditions would be impossible to realize in practice) or simply technical in the sense that the solution form doesn't useful exhibit the behaviour. Improvement of convergence procedures show that $T(x, y) = T_0(2\theta/\pi)$ near $(0, l)$ where θ is the angular displacement from the edge at temperature 0; the difficulties are thus technical.

(b) Using linearity determine the solution of the steady state heat equation (8.29) satisfying

$$T(x,0) = T_1, \; T(x,l) = T_3 \;\; \text{for } 0 < x < l,$$

$$T(0,y) = T_4, \; T(l,y) = T_2 \;\; \text{for } 0 < y < l.$$

Exercise 8.6 *Externally Heated Rod*

A rod of cross-sectional area a, length L, initially at temperature 0, is supplied with heat at a rate given by $q = q_0 f(x/L)$ per unit rod length, where f is dimensionless and of unit order. One of the rod ends ($x = 0$) is insulated and the other ($x = L$) is maintained at temperature 0. The standard separation of variables procedure doesn't usually work for the non-homogeneous heat equation but the following elementary extension can be employed in such cases.

(a) Describe briefly how you expect the temperature profile to develop. Scale the equations appropriately and thereby give a rough estimate for the maximum temperature realized in the rod and the time required to reach this temperature.

Seek out an exact solution to this problem in the following way:

(i) Guess the appropriate solution form in the special case in which

$$f(\frac{x}{L}) = \gamma_0 \cos\frac{\pi x}{2L}, \quad 0 \le x \le L,$$

and thereby solve the problem for this form of heat input. Be sure to check that the initial and boundary conditions are satisfied. Note that the particular heating function chosen satisfies the imposed rod boundary conditions. This ensures an easy solution determination.

(ii) Determine solutions for the cases in which

$$f(\frac{x}{L}) = \gamma_n \cos\frac{(2n+1)\pi x}{2L}, \quad 0 \leq x \leq L,$$

with $n = 0, 1, 2\ldots$

(iii) By writing

$$f(\frac{x}{L}) = \sum_{n=0}^{\infty} \gamma_n \cos\frac{(2n+1)\pi x}{2L},$$

identify γ_n and thus obtain the solution for the general case. Explain why the procedure works.

(iv) In the special case in which $f(\frac{x}{L}) = 1$ determine the expansion coefficients γ_n explicitly and indicate roughly how many terms are required to obtain $T(x,t)$ to within1%.

Comment: Another approach to the problem would be to first determine the steady state solution and then determine the transient using the standard approach. Note, however, that this steady state extraction approach could not be used for a time dependent heating rate $g(x,t)$ situation, whereas the above approach extends trivially.

(b) Using the above solution how would you go about solving:

$$\frac{\partial T}{\partial t} = \kappa\frac{\partial^2 T}{\partial x^2} + h(x) \text{ in } 0 < x < L,$$

with

$$\frac{\partial T}{\partial x}(0,t) = T(L,t) = 0,$$

and

$$T(x,0) = g(x) \text{ in } 0 < x < L.$$

Hint: Use linearity to break the problem up into two simpler problems.

Exercise 8.7 *Thermal Insulation 2*

We return to complete the thermal insulation problem examined in Chapter 7, see Exercise 7.13. A sheet of building material of thickness a has the "inside" face $x = a$ maintained at temperature 0 by the air conditioner and the outside face $x = 0$ receives heat from the sun at a given rate r per unit area and loses heat to the environment at the Newtonian rate $\beta(T(a,t)-T_0)$ per unit area, where T_0 is the temperature of the environmental. Initially the plate is at temperature 0. The time span to reach equilibrium is of interest.

(a) Write down the defining equations. Determine the steady state temperature distribution and discuss its dependence on the parameters of the problem.

(b) Use Fourier techniques to determine the transient. Determine approximate estimates for the eigenvalues.

(c) Determine an accurate expression for the time to reach steady state as a function of the relevant parameters.

Bibliography

Ince, E.L.(1956).*Ordinary Differential Equations*, Dover, New York.

Titchmarsh, E. C. (1939). *The Theory of Functions*, Oxford.

Chapter 9

The art of cooking

Inverse problems
Variational techniques
Rayleigh-Ritz approximations
Finite element methods

9.1 Introduction

Cooking is a process of applying heat to a material to induce some desired form of chemical or physical change. The efficient control of the cooking processes calls for an adequate understanding of the chemical processes involved and the ability to produce the desired temperature changes in the body. This is made particularly difficult because the heat is normally applied to (and lost from) the surface. For the situations discussed here the chemical and state change processes are not presently well enough understood for models to be reliable. In this chapter we'll restrict our attention to some of the basic questions involved in determining adequate means for producing the desired temperature changes.

A chef when roasting or braising wishes to initially maintain a high surface temperature so that the browning reactions that produce the rich surface flavours can occur (caramelization and Maillard browning reactions occur at about 154°C), but not so high that carbonization reactions occur. Later, lower temperatures are used so that the interior of the food is cooked. A relevant question to consider is therefore: what heat supply rate will produce a desired variation in the surface temperature? We'll examine this question in the Section 9.2. Later, in Section 9.3, we'll examine the question of determining a heating rate that will produce a specified temperature within the body. We'll assume throughout that the bodies behave like ordinary conductors; a somewhat dubious assumption given that foods contain water, some of which is evaporated during the process of cooking. Apart from the fact that some of the supplied heat needs to be expended to do this, temperatures in the body tend to hover around 100°C (the boiling point of water), because of this state change. Present day models account for such effects by using an empirically determined temperature dependent diffusivity and, in spite of this and other limitations, the models have been found to be useful.

The heating processes described above are significant not merely for the preparation of food but also for producing metals with desirable properties. For example steel as cast is in a course granular condition with little strength and unsuitable for machining or shaping, so heat treatment is employed to break up and refine the coarse structure. If the steel is heated above a certain critical temperature, varying from 725°C to 825°C depending on the composition, there is a change of the crystal structure from pearlite (mainly) to a form known as austenite. If this material is then cooled slowly a soft ductile material suitable for machining and working is produced; a process known as annealing. If, on the other hand the austenite is quenched (i.e. cooled rapidly), a hard but brittle crystalline material known as martensite is produced. This hardened steel is too brittle to be used for practical purposes, however further suitable heat treatment (called tempering) permits the steel to retain most of this hardness but lose its brittleness. The required combination of toughness and hardness can thus be realized by appropriate heat application. For further information see Jastrzebski (1959).

We have already determined the solution of closely related problems in Chapters 7 and 8. In these chapters we examined the question of determining the temperature distribution brought about by a known heat input in a finite slab with Newtonian surface heat exchanges. Here we want to determine the heat input to produce desired temperatures. Such *inverse* (as opposed to *forward*) problems are of current active research interest especially in the mining industry. For obvious reasons mining companies would like to know the size, location and shape of an ore body that produces a given anomalous gravitational, electrical or magnetic response on the surface of the earth. Real mathematical difficulties can arise with inverse problems because they (unlike the forward problems) are often not well posed in a physical and mathematical sense. Thus, although it would be surprising to find a non-unique response of a prescribed physical system to a given stimulus (eg. the electrical response of the earth to an applied electrical potential), it would not be such a surprise to find that many different physical set-ups give rise to almost the same physical response. In such cases any finite mathematical solution process that doesn't restrict its attention to a narrow class of possibilities is likely to be under-determined and thus illconditioned. A numerical scheme is such a finite process. We'll learn more about this as we go along.

9.2 Surface Temperature Control

We start slowly and gradually build up in a sequence of steps to the finite slab problem of interest. Our progression will parallel that of Chapter 7. Thus we initially consider the semi-infinite problem and ignore heat losses from the surface.

If a semi-infinite material is heated on its surface by a heat flux $q(t)$ then we have seen that the resulting surface temperature increase, $T(t)$, is given by, see equation (7.29) Section 7.5.1,

$$T(t) = \frac{1}{\sqrt{\pi \rho c k}} \int_0^t \frac{q(s)\,ds}{\sqrt{t-s}}. \tag{9.1}$$

Our problem is the inverse problem of determining the $q(t)$ corresponding to a prescribed $T(t)$. Thus for our purposes the above equation is an integral equation for $q(t)$. An exact solution can be obtained using advanced complex variable techniques however, we can find the answer quite simply by using an imaginative constructive approach based on our knowledge of the solution of the corresponding forward problem.

1. We start by asking what heat flux $q_\Delta(t)$ must be applied in order to produce and maintain a surface temperature change of Δ. Since there is no distance or time scale imposed by the statement of the problem we may expect a similarity solution for $T(x,t)$ of the form $T = \Delta F(\zeta)$ where $\zeta = x/\sqrt{\kappa t}$. In fact we found the solution for F in Exercise 7.3 to be $F = \text{erfc}(\zeta/2)$, see also (7.26).

The heat flux at the surface $x = 0$ is given by $-kT_x(0,t)$ and this is given by

$$q_\Delta = \sqrt{\rho c k}[\Delta/\sqrt{\pi t}],$$

as is seen from the definition (7.2) of the erfc function. Thus we have our required solution for $q_\Delta(t)$. Obviously because of the $1/\sqrt{t}$ singularity, the required initial heating rate can't be realized in practice, but the total heat input to raise the temperature the required amount is finite, so that the ideal solution can be approached in practice. Given the impracticability of the ideal solution some experimentation is in order, and in fact such experimentation is described in the thermal ignition Exercise 7.6 examined in Chapter 7. There we found that the heat input required to raise the surface by a prescribed amount was strongly affected by the time span of heat application and weakly dependent on other features. Thus if it's really important to quickly raise the surface temperature, a strong initial pulse of heat is required (i.e. the heat should be input in the shortest time span possible). In the cooking context it's unlikely that this is an important matter, whereas it certainly is in the ignition context.

It's also worth observing that the solution tells us that the required supply rate to maintain the surface at Δ decreases (like $1/\sqrt{t}$) as the interior of the body warms up; a result that makes physical sense.

2. To obtain the solution in the general $T(t)$ case simply note that the temperature at any time t may be regarded as the sum of all the temperature increases $dT = (dT(s)/ds)\,ds$ for all time intervals ds from time zero to the time of interest. Thus if an initial temperature jump Δ is required and subsequently $T(t)$ is desired, then the required heat input is given by

$$q(t) = \sqrt{\rho ck}[\Delta/\sqrt{\pi t} + \int_0^t \frac{T'(s)\,ds}{\sqrt{\pi(t-s)}}], \tag{9.2}$$

where the $'$ denotes d/ds.

Aside Although there is no need to isolate the initial temperature jump in this way, in most cooking situations a reasonably rapid initial temperature jump is required, so this separation is numerically more satisfactory. (A generalised function description is a very natural mathematical one in such circumstances.)

The above expression is the required solution of the integral equation (9.1), as may be verified by substituting this expression for q in the right hand side of the integral equation and changing the order of integration.

Comment: The above represents an interesting example of the development of a formal mathematical solution from an argument based on a physical interpretation of the mathematics. As mentioned on earlier occasions, if you ignore the physics you're working with hands tied.

3. Consider now the case of a slab of thickness $2a$, that may be gaining surface heat from some external radiating source and exchanging heat with its environment via its two exposed faces (described by the Newtonian flux condition). In this case the required integral equation is

$$T(t) = \frac{1}{\sqrt{\pi\rho ck}} \int_0^t q(s)F(t-s)\,ds \tag{9.3}$$

where the *kernel* $F(t)$ has been obtained in Chapter 7, see (7.30), and $q(t)$ is the *net* surface heat input per unit area due to radiative input and Newtonian exchange. (For very thin slabs the result (8.26) determined in Chapter 8 may be necessary.) Firstly note that once the solution $q(t)$ of (9.3) is determined, the required radiative input or environmental temperature change to produce the required $T(t)$ can be trivially determined using the Newtonian condition.

Now we know from earlier work that the surface temperature is much more influenced by recent surface heating than by heating in the past, or heating from distant locations; so we expect the semi-infinite case $a = \infty$ with kernel $F_0(t-s) = 1/\sqrt{t-s}$ to provide an excellent starting point for any approximation scheme. Thus it makes sense to write the integral equation for the required heat supply in the algebraically equivalent form

$$T(t) = \frac{1}{\sqrt{\pi \rho c k}} \{ \int_0^t q(s)[F(t-s) - F_0(t-s)]\, ds + \int_0^t q(s) F_0(t-s) ds, \}$$

which, after rearranging and substituting for F_0, gives

$$\frac{1}{\sqrt{\pi \rho c k}} \int_0^t \frac{q(s)\, ds}{\sqrt{t-s}} = T(t) - \frac{1}{\sqrt{\pi \rho c k}} \int_0^t q(s)[F(t-s) - F_0(t-s)]\, ds. \quad (9.4)$$

If we regard the term on the right hand side as known, then this is an integral equation for $q(t)$ of the form (9.1) encountered earlier, with the solution (9.2). Using this gives the result

$$q(t) = \sqrt{\rho c k} \left[\Delta/\sqrt{\pi t} + \int_0^t T'(s) \frac{ds}{\sqrt{\pi(t-s)}} - \int_0^t \mathcal{O}(q)(s) \frac{ds}{\sqrt{\pi(t-s)}} \right]$$

where the operator $\mathcal{O}(q)$ denotes

$$\mathcal{O}(q)(s) = \frac{d}{ds} \int_0^s q(s')[F(s-s') - F_0(s-s')]\, ds'.$$

Now the right hand side of this expression is not known a priori so this is *not* an explicit solution for $q(t)$; it's another integral equation for $q(t)$ that needs to be solved iteratively or numerically. Furthermore it's an horrendous expression, so what's the point! The point is that this arrangement is sensible because $\mathcal{O}(q)(s)$ is small as a result of $F - F_0$ being small. Thus the inversion of this integral equation will be straightforward, whether one uses an iterative scheme to improve the first order estimate (9.2), or one replaces $q(t)$ by a discrete representation and inverts the resulting matrix equation using a package. Elaborating further: The original integral equation (9.3) is not in a suitable form for direct numerical evaluation because the integrand contains a square root singularity, and the effect of the above processing is to "extract the singularity" from the integrand, leaving a non-singular integrand which can be handled by standard numerical schemes. The original integrand is in fact dominated in behaviour by the square root singularity, so that the process of extracting this contribution also leads an excellent first estimate (9.2) for the solution.

Aside: Paradoxically applied mathematicians would prefer to work with singular equations because they recognize that, for the above reasons, good analytical approximations are likely to be available for such equations (the more singular the better!). The above situation is analogous to situation that arises when one seeks to evaluate an ordinary singular integral. For example, the way to numerically evaluate

$$I = \int_0^1 e^x \sin x / x^{3/2} dx$$

is to write it in the form

$$I = \int_0^1 1/\sqrt{x}dx + J = 2 + J,$$

where the left over integral J is non-singular and so can be readily computed.

Comment: The above type of preprocessing is particularly important when real time control is involved. The results of investigations such as the above are often used in practice to *steer* the operations of a plant back to an optimum. In such cases efficient numerical schemes are required to keep pace with the real time operations. The pre-processing described above is not algebraically time consuming. In other cases, however, the required preprocessing (although straightforward in principle) can be *very* time consuming in practice unless an algebraic package is employed. The required processed equations from such a package can be output in Fortran or Pascal code suitable for subsequent numerical work.

This completes our analysis of the surface temperature control problem. We'll now examine the more difficult temperature distribution problem.

9.3 Temperature Distribution Control

The problem of obtaining a prescribed temperature distribution in a body arises much more frequently. A number of related features of this problem need to be noted before proceeding. Firstly it should be noted that two strategies exist for raising the temperature of the interior of a body a prescribed amount; one can either raise the surface temperature well above that of the interior in the hope of achieving the result quickly, or one can maintain a moderate surface temperature and wait a long time for the slow conduction process to take effect. Also note that, since diffusion is a smoothing process, all temperature distributions tend in the long term to the same one with the appropriate amount of heat content or heat flux as determined by the boundary conditions. For example the plane source fundamental solution determined in Section 3.3 Chapter 7 asymptotically describes the temperature distribution corresponding to almost *any* one-dimensional initial temperature distribution with unit heat content. Thus, since all manner of heating procedures tend to produce the same temperature distribution, trying to calculate a unique form of heating which will give the required temperature distribution is essentially an unintelligent approach. It is thus not sensible to try to form an integral equation for the heat flux which will yield a given temperature distribution throughout the material (the resulting equation would almost assuredly be under-determined). A more practical approach is to seek particular heating rates which lead to easily calculable temperature distributions, and then to combine them in the most effective way to approximately produce the desired distribution. This may find a way (but not the only way) of achieving the desired temperature distribution.

Notice that the present situation is *very* different from the surface temperature problem, because the surface temperature is much more strongly affected by the immediate surface heat input than by heat inputs at other locations or times. For this reason the resulting integral equation formulation makes sense and leads to a well posed mathematical problem for the surface temperature problem. It should be emphasized that common sense, plus the physical insight gained through earlier work in Chapter 8, have *suggested* the correct solution approach for these situations. The actual mathematical equations do not provide useful insight in this regard. As observed, in Section 4.3 Chapter 4, it's often very difficult to tell if a particular integral equation has a unique solution, and seemingly similar integral equations can be very different.

Comment: It should be added that while the above intuitive arguments concerning uniqueness seem sensible, they're far from convincing. Given the delicacy of matter and its importance in a practical sense, a uniqueness (or non-uniqueness) proof would be a very worthy undertaking. In the absence of such a proof how can one be sure that one isn't (numerically) searching for a unique solution that doesn't even exist?

Returning to the problem in hand. We need to determine simple solutions of the heat equation that, when combined, will generate a range of practically useful surface heating rate possibilities. We've already obtained a number of simple solutions using similarity ideas. These solutions are similar in form in that they all involve the dimensionless combination $\zeta = x/\sqrt{\kappa t}$, but differ in the temperature scales used. In the hope of increasing the number of simple solutions available to us, it thus makes sense to look for solutions of the form $t^m F(\zeta)$. Such solutions will satisfy the heat equation provided that

$$F'' = mF - \zeta F'/2,$$

obtained by direct substitution in the heat equation. Earlier we have found solutions for the $m = 0$ case (the erf function solution used above), and the $m = -1/2$ case (the fundamental plane source solution see (7.14). Observe that if we differentiate the above differential equation we obtain

$$F''' = (m - 1/2)F' - \zeta F''/2,$$

which is the same differential equation with F replaced by F' and with m reduced by 1/2. Thus if $F_{m-1/2}$ is the solution of the differential equation which vanishes at ∞ then

$$F_m = C_m \int_{\zeta}^{\infty} F_{m-1/2}(u)\, du,$$

where C_m is any constant, is another of the "m solutions". Thus from the $m = 0$ solution we can generate other "m solutions" by simply integrating. In this way we obtain solutions in which the temperature at $x = 0$ increases as t^m for $m = 0, 1/2, 1, 3/2, \ldots$. The heat flux at $x = 0$ corresponding to these solutions varies as $t^{m-1/2}$; this gives some flexibility to fit approximations to a range of surface heat flux variations with time. Maple readily performs the integrations (**infinity** is recognized in this context) and the graphs are displayed in Fig. 9.1, where the solutions are normalized so that $F_m(0) = 1$. The close similarity in shape of these solutions for moderate to large ζ is apparent. Given the considerable difference in the variation of the associated heat flux with time, it's clear that the ζ profile shape is quite insensitive to the heat input rate; so fiddling with the shape of the heat input function will achieve little in terms of the ζ profile. Thus to substantially flatten out the temperature distribution with depth x for a given heat input (the usual aim during certain stages of cooking) one must apply the heat slowly so that the variations are spread as a result of κt being large. *In simple terms, uniform cooking normally calls for a slow rate of heating.* An alternative approach is to apply some heat to the surface and then cease the heating process for a while to allow the heat to spread into the interior, and then one can repeat the process until the desired temperature distribution is reached. This particular procedure has the added advantage that the large surface temperatures associated with continuous heat application are avoided. The mathematics presented above permits the calculation of the temperature distribution at the end of the heating stage; however, in order to be able to compute the effect of the on/off cooking procedure, we need to be able to predict the changes of temperature distribution in the interior when there is no heat applied at the surface. We'll turn our attention to this question (which is of fundamental importance) in the next section.

Aside: Both the above procedures for producing uniform cooking require substantial heating times and so may be unacceptable in many metal production facilities. Other solutions/compromises may be acceptable if one examines the situation in its broader context. Thus for steel production a non-uniform hardness/toughness structure resulting from non-uniform heating (as for example occurs in quenching) may meet the total design requirements. A non-uniform hardness/toughness structure may even be advantageous. Thus a hard/tough skin is desirable in structures that are required to withstand large bending moments, for example the flagpole of Chapter 2.

9.4 Effects of Size and Shape

Up to the present we've restricted our attention to situations in which the heat flows in one dimension. Rarely do we cook plates so this not

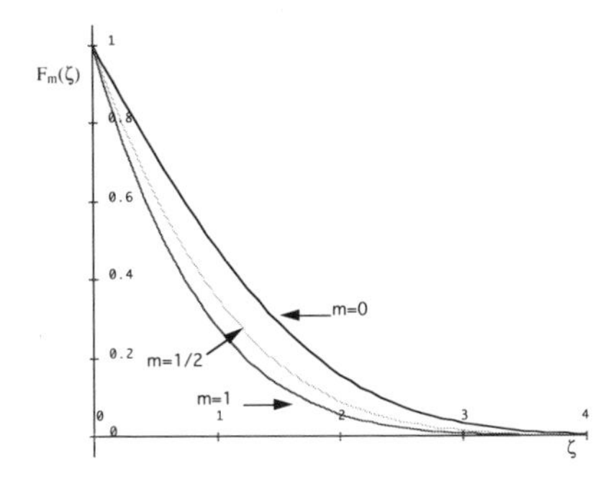

Figure 9.1: 1D heat equation similarity solutions.

good enough. Now for the surface temperature control situations, since local heating basically determines local surface temperatures, the 1D results carry across with either no modification or minor modifications depending on the issue of concern (for example corner regions may require special attention). For the temperature distribution work, however, shape and size effects will influence the results. We'll therefore examine some simple cases where the geometric effects must be considered. For simplicity's sake the problem of cooking objects by placing them in an oven maintained at a fixed temperature will be considered. This is an idealization of a forced convection oven often used for industrial and commercial cooking. In practice it is not possible to maintain the environment of a body receiving heat at a fixed temperature, precisely because the environment is supplying this heat. In a forced convection oven the gas surrounding the body is sucked away to be replaced by other gas which has been re-heated to the initial temperature of the gas. In this way the heat exchange to the body is increased and an almost fixed environment temperature is maintained. Of course it takes time to reheat and circulate the gas, so the model is actually based on the assumption that the time for this recirculation and reheating is small in comparison with the time to transfer significant heat to the body (and so does not influence its rate of heating). Let the temperature of the oven be T_1 and the initial temperature of the roast be T_0. For comparison purposes we'll work with bodies with the same typical dimension d.

Aside: In a conventional household oven, radiation (directly from the element and indirectly from the walls) is the main heat source, but convective motions within the oven can also supply (or remove) heat from the body's surface. Both these processes have been modelled for the slab in earlier chapters and, although it's a relatively straightforward matter to carry across this work in other geometries, we'll avoid the additional complications here.

9.4.1 Special shapes

There are a number of special body shapes for which the one-dimensional treatment using Fourier's ideas can be extended. These shapes are such that there exists a co-ordinate system (α, β, γ) with the following properties:

1. The boundary surface of the body is fixed by one value of one of the co-ordinates,

2. The heat equation admits of the existence of solutions of the form $A(\alpha)B(\beta)C(\gamma)U(t)$ where α, β and γ are the suitable co-ordinates.

Included in this short list are spheres, cylinders, cuboids and ellipsoids. For a listing of these shapes see Morse and Feshbach (1953). The resulting ordinary differential equations for the functions A, B and C arising in this way have been studied and the properties of resulting solutions (referred to as *Special functions*) are recorded in mathematical handbooks, eg. Abramowitz and Stegun (1970), and are known to a limited extent to Maple. Sturm-Liouville theory results also extend to encompass the (often singular) differential equations that arise, so that the orthogonality results and Fourier series expansion work carries through. Although it's tedious and a little daunting at first to work with such special functions, the extension is straightforward (no essentially new mathematics arises). For cases of interest to us we won't need to confront new functions.

A sphere

For a spherical object, with d chosen as the radius, see Fig. 9.2, the temperature distribution will be a function solely of the radial distance rd from the centre of the sphere and the time $\kappa t/d^2$ after the roast is placed in the oven. For such a situation the heat equation for the scaled temperature T, defined so that the actual temperature is $T_0 + (T_1 - T_0)T$, simplifies to (in scaled variables)

$$T_t = T_{rr} + 2T_r/r, \tag{9.5}$$

which has to be solved subject to the the boundary condition

$$T(1,t) = 1, \tag{9.6}$$

and the initial condition

$$T(r,0) = 0.$$

In this form the system is not suitable for the application of the Fourier approach because the boundary condition (9.6) is not homogeneous; however, by writing $T = 1 - \mathcal{T}$ (i.e. by separating out the steady state solution

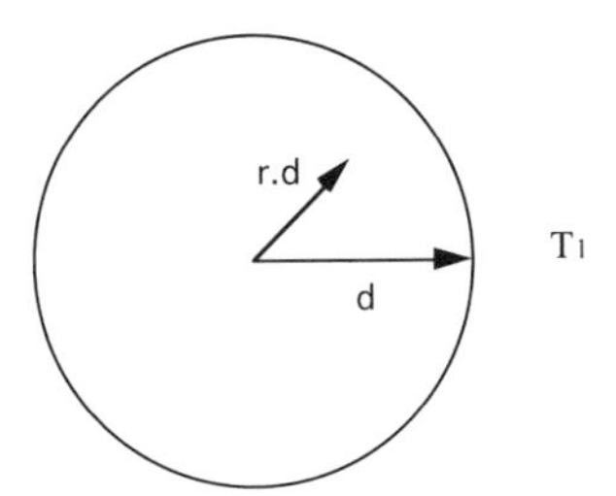

Figure 9.2: The heated sphere.

$T = 1$), we arrive at the homogeneous boundary condition and homogeneous heat equation formulation required for the standard Fourier approach to work. The initial condition for the transient $\mathcal{T}$ becomes

$$\mathcal{T} = 1. \tag{9.7}$$

Because the equation for the $R(r)$ part of the separated solution $\mathcal{T} = R(r)S(t)$ is not the ordinary harmonic type, it's not so easy to see the sign that's needed for the separation constant to ensure the required oscillatory behaviour in the r direction. However, knowing that all solutions of physical relevance must decay in time, we can see that the required separated equations with separation constant must be given by

$$S' = -\lambda^2 S, \quad \text{so } S = e^{-\lambda^2 t},$$

and

$$R'' + 2R'/r + \lambda^2 R = 0. \tag{9.8}$$

Maple gives the pair of linearly independent solutions $\sin(\lambda r)/r$ and $\cos(\lambda r)/r$ for (9.8), and the second solution is physically unacceptable because it's unbounded at the centre of the sphere.

The boundary condition at the surface, namely that $\mathcal{T} = 0$, thus requires that $\sin\lambda = 0$ and hence the eigenvalues are given by $\lambda_m = m\pi$ where m is a positive integer. In order to satisfy the initial condition (9.7) we require that

$$1 = \sum_{m=1}^{\infty} C_m \phi_m(r), \quad \text{where} \quad \phi_m = \sin(m\pi r)/r, \tag{9.9}$$

for a suitable choice of the constants C_m. In order to identify the orthogonality condition associated with the eigenfunctions ϕ_m it's necessary to rewrite the defining equation (9.8) for $R(r)$ in the standard *adjoint* Sturm-Liouville form (8.3), and thus identify the appropriate *weighting* function. In standard form the equation (9.8) for $R(r)$ is given by

$$(r^2 R'(r))' + \lambda^2 r^2 R(r) = 0,$$

as can be seen by direct comparison of the two equations, so the appropriate weighting function, see (8.5) is r^2, and thus the theory tells us that the eigenfunctions are orthogonal in the sense that

$$\int_0^1 r^2 \phi_n \phi_m dr = 0 \quad \text{for } n \neq m.$$

To obtain the coefficients C_n we thus need to multiply (9.9) by $r^2 \phi_n$ and integrate across the domain. Using the result

$$\int_0^1 \sin(m\pi r)\sin(n\pi r)\, dr = \begin{cases} 0 & \text{for } n \neq m \\ 1/2 & \text{for } n = m. \end{cases}$$

(partly implied by Sturm-Liouville theory) this gives

$$C_n = 2\int_0^1 r \sin(n\pi r)\, dr = (-1)^{n-1} 2/(n\pi).$$

Thus the solution obtained by the Fourier method is

$$T(r,t) = 1 - 2\sum_{m=1}^{\infty} (-1)^{m-1} \sin(m\pi r) e^{-m^2\pi^2 t}/(m\pi r).$$

The largest terms are

$$T(r,t) = 1 - 2\sin(\pi r)e^{-\pi^2 t}/(\pi r),$$

and since the remaining terms are negligible for moderate t this might as well be regarded as the solution. In particular the approach to uniform temperature at the centre of the sphere is given by

$$T(0,t) = 1 - 2e^{-\pi^2 t},$$

so that the unscaled time for heating the central region is $\gamma d^2/[\kappa\pi^2]$ where the multiplying number γ depends on the criterion adopted for the "close enough" approach. If "within 1%" of the external temperature is acceptable, then $\gamma = 4.61$.

A cuboid

Here we choose the length of the diagonal as our representative length scale d, see Fig. 9.3, and again adopt the time scale κ/d^2. Let the length of the sides of the cuboid be ad, bd and cd. Let xd, yd and zd be cartesian coordinates corresponding to axes aligned with the edges of the cuboid and with the origin at one corner. Again we determine the transient using the

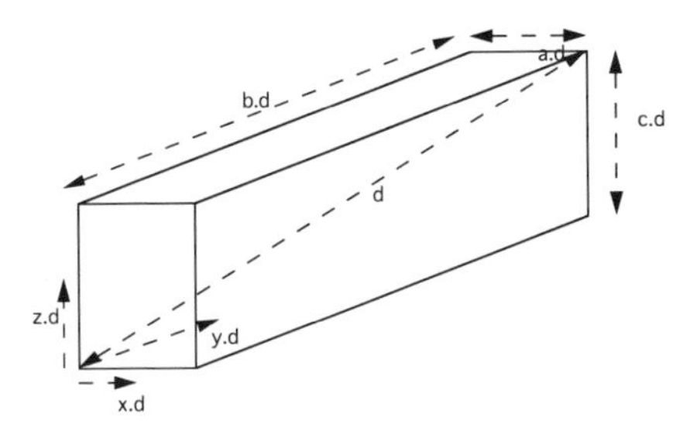

Figure 9.3: The heated cuboid.

Fourier technique. We look for solutions of the form $X(x)Y(y)Z(z)U(t)$. The scaled heat equation will be satisfied by such a combination providing

$$X''(x)/X(x) + Y''(y)/Y(y) + Z''(z)/Z(z) = U'(t)/U(t).$$

The standard argument tells us that each term in this equation must be a constant. Thus for example if x is varied and the other co-ordinates and time are held fixed then X''/X cannot vary, and so can only be a constant. We want solutions of the second order ordinary differential equation for $X(x)$ so obtained which vanish at $x = 0$ and $x = a$. The non-trivial solutions are given by $\sin(l\pi x/a)$ where l is a positive integer. Using such arguments it's easy to verify that a satisfactory solution takes the form

$$\sin(l\pi x/a)\sin(m\pi y/b)\sin(n\pi z/c)e^{-\lambda_{lmn}^2 t}$$

with

$$\lambda_{lmn}^2 = [(l\pi/a)^2 + (m\pi/b)^2 + (n\pi/c)^2], \tag{9.10}$$

where l, m and n are all positive integers. This satisfies the heat equation and all the homogeneous boundary conditions. Thus we look for a set of constants C_{lmn} such that

$$\sum_{l=1}^{\infty}\sum_{m=1}^{\infty}\sum_{n=1}^{\infty} C_{lmn}\sin(l\pi x/a)\sin(m\pi y/b)\sin(n\pi z/c)e^{-\lambda_{lmn}^2 t},$$

satisfies the initial condition $\mathcal{T}(x, y, z, 0) = 1$.

If we multiply the equation,

$$\sum_{l=1}^{\infty}\sum_{m=1}^{\infty}\sum_{n=1}^{\infty} C_{lmn}\sin(l\pi x/a)\sin(m\pi y/b)\sin(n\pi z/c) = 1,$$

by $\sin(j\pi x/a)$ and integrate with respect to x from zero to a we obtain

$$\begin{aligned}\sum_{m=1}^{\infty}\sum_{n=1}^{\infty} C_{jmn}(a/2)\sin(m\pi y/b)\sin(n\pi z/c) &= \int_0^a \sin(j\pi x/a)\,dx, \\ &= a[1-(-1)^j]/(j\pi).\end{aligned}$$

Thus only the coefficients with j odd contribute to non-zero solutions and it's sensible to write

$$l = 2l' + 1$$

and only allow l to range over values corresponding to integer l' values. The y and z variations can be handled in the same way leading to

$$C_{lmn} = 64/[\pi^3 lmn], \text{ with } m = 2m' + 1, n = 2n' + 1, \text{ for } m', n' = 1, 2, \ldots.$$

The complete solution is thus given by

$$1 - \frac{64}{\pi^3} \sum_{l'=0}^{\infty} \sum_{m'=0}^{\infty} \sum_{n'=0}^{\infty} S_{lmn}(x, y, z) e^{-\lambda_{lmn}^2 t},$$

where the shape functions S_{lmn} are given by

$$S_{lmn}(x, y, z) = \frac{\sin(l\pi x/a)\sin(m\pi y/b)\sin(n\pi z/c)}{lmn}.$$

The largest terms are

$$1 - \frac{64}{\pi^3} \sin(\pi x/a)\sin(\pi y/b)\sin(\pi z/c) e^{-\lambda_{111}^2 t},$$

and the other terms are negligible for moderate time, so an accurate estimate of the (unscaled) time required to reach steady state is

$$t_h = \gamma d^2/[\pi^2 \kappa(1/a^2 + 1/b^2 + 1/c^2)], \tag{9.11}$$

where $\gamma \approx 4.6$. Note in particular that if one of the cuboid sides is much smaller than the others (a say), then this expression is closely approximated by $\gamma(ad/\pi)^2/\kappa$. Now ad is the actual length of the small side and so the time scale is the same as if the heat transfer was taking place in just the x direction; well worth remembering for modelling purposes! For example it's clear from this that the heating time for a bowl shape (a piece of pottery for example) is the same as that for a slab of the same thickness. Further observations concerning the dependence of cooling time on shape will be made in Exercise 9.4.

9.4.2 General shapes: engine blocks

While the special cases where Fourier's ideas can be extended may yield useful solutions and provide us with a good understanding of the effects of geometry on heat transfer, they are far too restricted in application to deal with the shapes that are the subject of modelling investigations. For example, to ensure quality control and yet minimize processing time, a steel manufacturer has asked for the time required for automobile engine blocks and other shapes to heat to a prescribed temperature. The need for a reliable theoretical result arose because of the vast range of products that need to be treated. The geometry of an engine block is of course very complicated, with cylinders bored out and various other complex shapes involved. There is in fact a rule of thumb used in the metal casting industry, namely: the required time is $t_0 = \gamma d^2/\kappa$ where d is the "maximum thickness" of the body and γ is a number arrived at using site based experience. A procedure for determining d would be: for every point in the body, determine the minimum distance to points on the surface; the maximum thickness is double the largest of these values. It's clear from the cuboid work that the rule of thumb makes sense particularly for bodies that are "shell like" (so that the heat flow is basically one-dimensional), and the above work suggests how γ should be related to the initial temperature difference between the body and its environment. For bodies that are more chunky the heat flow will not be approximately 1D and the overall shape will play a role in determining t_0. Can we use any of the ideas so far considered to improve on this rule of thumb particularly for chunky objects? (For shell-like structures an iterative scheme based on local one-dimensional flow would be probably most effective.)

In general terms in the Fourier work we've extracted solutions of the scaled heat equation of the form $U(\mathbf{x})e^{-\lambda^2 t}$ satisfying the homogeneous boundary condition $U = 0$. By adding together such eigensolutions corresponding to the possible eigenvalues λ we were able to generate an exact solution to the required initial value problem. The eigenfunctions $U(\mathbf{x}, \lambda)$ satisfy (substitute the form in the heat equation)

$$\nabla^2 U = -\lambda^2 U \quad \text{in } \mathcal{D}, \text{ with} \tag{9.12}$$

$$U = 0, \text{ on } \partial\mathcal{D}, \tag{9.13}$$

where $\mathcal{D}$ is the domain defined by the body, with $\partial\mathcal{D}$ its boundary. Now for the special shapes the U's were determined exactly and in this way we were able to explicitly determine the eigenfunctions and associated eigenvalues. In general we can't do this (exact solutions are just not available), but in principle if we can determine approximations for the eigenvalues and eigensolutions associated with the above equations, then we've solved the problem. Determining these solutions would be a massive computational task; however, the special shapes work suggests that *only* the smallest eigenvalue and eigensolution are numerically significant for determining the heating time. Thus our interest turns to ways of determining estimates for the smallest eigenvalue of (9.12,9.13). For this we need a number of general variational results associated with the solutions of the partial differential equation. As indicated in the Moorings chapter, variational and Newtonian ideas represent alternative ways for describing the physical world and often it pays to go from one formulation to the other. Variational methods are particularly advantageous for computational work.

If we multiply the partial differential equation (9.12) by U and integrate over the domain we obtain

$$\begin{aligned} -\int_{\mathcal{D}} \lambda^2 U^2\, dV_{\mathbf{x}} &= \int_{\mathcal{D}} U\nabla^2 U\, dV_{\mathbf{x}} \\ &= \int_{\mathcal{D}} [\boldsymbol{\nabla}\cdot(U\boldsymbol{\nabla}U) - (\boldsymbol{\nabla}U)^2]\, dV_{\mathbf{x}} \\ &= \int_{\partial\mathcal{D}} U(\boldsymbol{\nabla}U)\cdot\mathbf{n}\, dS_{\mathbf{x}} - \int_{\mathcal{D}} (\boldsymbol{\nabla}U)^2\, dV_{\mathbf{x}} \\ &= -\int_{\mathcal{D}} (\boldsymbol{\nabla}U)^2\, dV_{\mathbf{x}} \end{aligned}$$

as the result of using Gauss's Divergence Theorem, see (6.1), and the boundary condition (9.13). Thus we can deduce that

$$\lambda^2 = \frac{\int_{\mathcal{D}} (\boldsymbol{\nabla}U)^2\, dV_{\mathbf{x}}}{\int_{\mathcal{D}} U^2\, dV_{\mathbf{x}}}, \tag{9.14}$$

an interesting but not immediately useful result for calculations (after all we don't know what U is). We will, however, be able to exploit this special relationship between the eigenfunctions and eigenvalues.

Consider the functional $\mathcal{E}(\phi)$ defined by

$$\mathcal{E}(\phi) = \int_{\mathcal{D}} [(\boldsymbol{\nabla}\phi)^2 - \lambda^2\phi^2]\, dV_{\mathbf{x}},$$

for any sufficiently smooth function $\phi(\mathbf{x})$, defined on the domain $\mathcal{D}$, satisfying the boundary condition $\phi = 0$ on $\partial\mathcal{D}$, and where λ is a parameter. We'll show that the (U, λ) combinations that satisfy our boundary value problem (9.12, 9.13) render this functional stationary—a most useful result! To do this we proceed as usual for variational problems (see Section 4.3), and examine small variations ϵv (so that $\epsilon \ll 1$, and $v = 0$ on $\partial\mathcal{D}$) about the assumed solution (U, λ). Thus

$$\begin{aligned} \delta\mathcal{E} &= \mathcal{E}(U+\epsilon v) - \mathcal{E}(U) \\ &= 2\epsilon\int_{\mathcal{D}} [\boldsymbol{\nabla}U\cdot\boldsymbol{\nabla}v - \lambda^2 U v]\, dV_{\mathbf{x}} + O(\epsilon^2) \\ &= 2\epsilon\int_{\mathcal{D}} \left[\boldsymbol{\nabla}\cdot(v\boldsymbol{\nabla}U) - v\nabla^2 U - \lambda^2 U v\right]\, dV_{\mathbf{x}} + O(\epsilon^2) \\ &= 2\epsilon\left[\int_{\partial\mathcal{D}} v[\mathbf{n}\cdot\boldsymbol{\nabla}U]\, dS_{\mathbf{x}} - \int_{\mathcal{D}} v[\nabla^2 U + \lambda^2 U]\, dV_{\mathbf{x}}\right] + O(\epsilon^2), \\ &= 2\epsilon\left[-\int_{\mathcal{D}} v[\nabla^2 U + \lambda^2 U]\, dV_{\mathbf{x}}\right] + O(\epsilon^2), \end{aligned}$$

since $v = 0$ around $\partial\mathcal{D}$ for acceptable functions. This expression will be of order ϵ^2 if and only if (U, λ) are solutions of the specified boundary value problem. Thus the problems of solving the boundary value problem and finding the function that renders the above functional stationary are equivalent. The latter is often preferred for computations.

There is an alternative way of viewing this result that's also useful. Consider the question of finding stationary values of the functional

$$\mathcal{J}(\phi) = \int_{\mathcal{D}} (\nabla\phi)^2 \, dV_{\mathbf{x}}, \tag{9.15}$$

amongst those functions ϕ which satisfy the boundary condition $\phi = 0$ and are also required to satisfy the further constraint

$$\mathcal{I} = \int_{\mathcal{D}} \phi^2 \, dV_{\mathbf{x}} = 1. \tag{9.16}$$

Then, in the standard way of dealing with such constrained stationary value problems using a Lagrange multiplier μ, one considers the variation of

$$\int_{\mathcal{D}} [(\nabla\phi)^2 - \mu\phi^2] \, dV_{\mathbf{x}}.$$

A repetition of the calculation in the previous paragraph, with a minor reinterpretation, shows that any constrained stationary value will be achieved by a function Φ satisfying the partial differential equation

$$\nabla^2\Phi = -\mu\Phi,$$

and the boundary condition $\Phi = 0$. This has solutions only if $\mu = \lambda^2$. Thus any such stationary value is associated with some eigenvalue of the original partial differential equation and its associated solution. Furthermore the least value achievable by the ratio $\mathcal{J}/\mathcal{I}$ is clearly the smallest eigenvalue, so

$$\lambda_1^2 = \min_{\phi} \frac{\int_D (\nabla\phi)^2 \, dV_{\mathbf{x}}}{\int_D \phi^2 \, dV_{\mathbf{x}}}, \tag{9.17}$$

amongst all functions satisfying the boundary condition $\phi = 0$. Thus by searching through all admissible ϕ's one can use the stationarity properties to identify the eigenvalue and eigensolution of interest, and solve the heating problem.

Aside: Higher order eigenvalues and solutions can also be extracted.

Now it would be a practical impossibility to scan through all admissible functions and in this way determine λ_1; however, one might reasonably expect to obtain a good approximation to λ_1 if one scans through a subclass of *test functions* with almost the right shape. At the same time an approximation of the eigensolution would be obtained. This technique is referred to as the **Rayleigh-Ritz method**. Of course we need some guide concerning the appropriate shape (or class) of test functions to use. Our experience with eigenfunctions suggests that the first eigenfunction has "one hump" in the solution domain, and in fact modern functional analytic techniques show that this is so. Explicitly it can be shown that the minimum value is achieved by a function which is single signed on the domain $\mathcal{D}$. It turns out that the Rayleigh-Ritz procedure is much better for determining eigenvalues than one has any right to expect, in the sense that even shapes that poorly represent the eigenfunction give excellent approximations for λ_1 and (with care) for the larger eigenvalues. To illustrate the procedure we'll examine a situation for which we know the answers.

We'll reconsider the simple cube problem (sides of length 1). We can ensure that the test functions we use satisfy the boundary conditions by using a multiplicative term $f = x(1-x)y(1-y)z(1-z)$, and then we can generate further functions satisfying the boundary condition by using

$$\phi = f[1 + a(x + y + z) + b(x^2 + y^2 + z^2)],$$

where the parameters a and b are arbitrary, and where we have taken advantage of the symmetry in x, y and z to reduce the number of parameters appearing. (No xy, xz, yz terms appear.) We thus have a two parameter family of test functions to work with. There's no special virtue in using polynomials here except that the calculations are made easy by this choice. Sine functions are another obvious choice. Clearly it's a simple matter to broaden the class of test functions by introducing higher order polynomials, and Maple is ideal for this purpose.

It is a routine matter to use Maple to evaluate the two integrals

$$A = \int_D (\nabla\phi)^2 \, dV_{\mathbf{X}} \quad \text{and}$$

$$B = \int_D \phi^2 \, dV_{\mathbf{X}}.$$

The combination A/B will be a minimum for values of a and b that satisfy the pair of equations

$$A_a B - A B_a = 0 \quad A_b B - A B_b = 0,$$

obtained by equating the derivatives with respect to a and b to zero. The solution procedure is as follows:

1. the first equation is solved for a in terms of b,
2. this value of $a(b)$ is substituted into the second equation which is then solved for b,
3. the value of a is then obtained from $a(b)$

4. both values are then used to evaluate A/B.

The set of Maple instructions for doing this is:

```
f:=x*(1-x)*y*(1-y)*z*(1-z):
g:=1+a*(x+y+z)+b*(x*x+y*y+z*z):
phi:=expand(f*g):
with(linalg,grad);
v:=grad(phi,[x,y,z]):
with(linalg,dotprod);
V:=dotprod(v,v):
int(expand(V),x=0..1):
int(",y=0..1):
A:=int(",z=0..1);
int(expand(phi*phi),x=0..1):
int(",y=0..1);
B=int(",z=0..1);
eq1:=diff(A,a)*B-A*diff(B,a):
eq2:=diff(A,b)*B-A*diff(B,b):
s:=solve(eq1,a):
s1:=s[1]:
subs(a=s1,eq2):
b1:=fsolve(",b);
a1:=subs(b=b1,s1);
evalf(subs(a=a1,b=b1,A)/subs(a=a1,b=b1,b));
```

The value of $\lambda_1^2 = 29.609$ is obtained, which may be compared with the exact value of $3\pi^2 = 29.610$—very good! This is a little misleading in the sense that for non-symmetric situations the results are not quite so good, but nevertheless the agreement is quite remarkable. We've also in the course of the above process determined (a, b), and thus an approximation for the eigensolution. It is the general experience that the Rayleigh-Ritz method gives remarkably accurate predictions for the smallest eigenvalues but nevertheless it should be stated that the test function which achieves this minimum is not necessarily a very good description of the exact solution $U(\mathbf{x})$. *To obtain a good approximation it's necessary to choose the shape functions wisely.* Experience can be gained by examining simple cases with known solutions, see Carrier and Pearson (1968), also Carrier and Pearson (1976).

Since it is the smallest eigenvalue which serves to determine the rate of approach to the final state, this method is of great value in determining the time it will take to get within some specified level of the final state of uniform temperature for awkward shapes, see Exercise 9.5. This solution process is particularly suitable when real time control considerations place severe restrictions on the available computation time.

Important Comment: Eigenvalue determinations are of special interest in wave problems and in quantum mechanics for determining the natural vibrational modes, and the Rayleigh-Ritz method is invaluable in these contexts. Variational formulations are in fact generally available for most of the equations of importance in mathematical physics (and elsewhere), so that the above procedures can be brought to bear on these problems; such procedures are particularly valuable for approximation work.

9.4.3 The finite-element method

The **finite-element method** has perhaps become the most useful (and favoured) technique for numerically determining the solution of boundary value problems. It's the flexibility and robustness of the technique that stand out. Complicated situations are handled effortlessly. We've briefly encountered the method earlier—you'll perhaps recall in Chapter 4 we indicated its use for determining the deflection of a table top under load. The technique is a standard tool in engineering and programs are available for most PC's.

The technique uses variational ideas and in fact is just a special case of the Rayleigh-Ritz method. The disadvantage of the Rayleigh-Ritz technique as presented above is that it can be *very* difficult to construct analytic expressions that are of the right form to satisfy the required boundary conditions for awkwardly shaped domains. The 2D domain shown in Fig. 9.4 is a case in point. In the context of the heating problem we need test functions that vanish on the boundary. The elementary device introduced earlier of using $f \cdot g(a, b, \cdots)$ where f vanishes on the boundary and the g is the "fitting" function might be used, but how do we get an f? For really complicated 3D shapes such as the engine block things get much worse! The finite-element method handles such situations with ease. In fact much more complicated situations can be handled—different materials and shapes all joined together etc.

The approach is simple: one simply breaks up the domain into simple subregions (elements) and uses simple approximating functions (usually polynomials) in each of the subregions. As an example consider the problem of determining the time taken for the cylindrical "block" with section shown in Fig. 9.4 to heat up. All we need is the value for λ_1, and the minimization procedure (9.17) determines this once we have useful approximating ϕ's that vanish on the boundary, which we'll call $f(x, y)$. Triangular elements seem sensible so we divide the region as shown, and number the nodes (in any order) also as shown.

Let's denote our solution approximation by $\psi(x, y)$, and let ψ_j denote the value of ψ at the jth node. For simplicity we'll use linear functions in each of the subregions, so that, if for example, $\psi = ax + by + c$ in the region bounded by nodes #1, #2 and #5, see Fig. 9.4, then we can identify

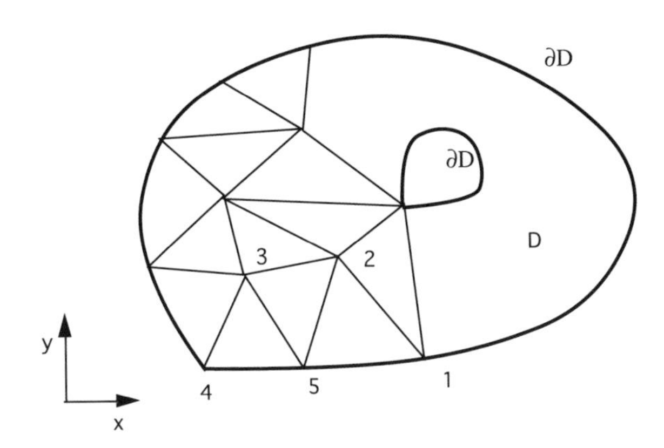

Figure 9.4: Finite elements: a 2D domain.

the coefficients a, b, c in terms of the values of ψ at the vertices. Thus the equations:

$$\begin{aligned} ax_1 + by_1 + c &= \psi_1 \\ ax_2 + by_2 + c &= \psi_2 \\ ax_5 + by_5 + c &= \psi_5 \end{aligned}$$

(where (x_j, y_j) are the cartesian coordinates of node j), uniquely determine (a, b, c) in terms of the values (ψ_1, ψ_2, ψ_5) at the nodes. This means of identifying the coefficients is particularly convenient because automatically the ψ constructed in this way will be continuous across triangle boundaries, irrespective of the actual values of the ψ_j. Now for points on the boundary f (such as #1, #5 in Fig. 9.4), the values of ψ are prescribed ($\psi = 0$ for our problem). The remaining values of ψ (eg. ψ_2) are to be determined so that the integral $\mathcal{J}$ is minimized subject to the constraint $\mathcal{I} = 1$, see (9.15, 9.16). The values of ψ_x, ψ_y are constant in each subregion and thus (summing over all triangles) the integrals are simply quadratic functions of the unknown ψ_j values. Using a Lagrange multiplier μ and setting the derivatives with respect to the unknown ψ_j's to zero we obtain enough linear equations to determine the ψ's in terms of μ, and the constraint $\mathcal{I} = 1$ completes the system. The coefficient matrix resulting from the linear equations is large but sparse and so easily solved. Simple examples are given in the exercise set. It's likely that your department has a finite-element package, so spend some time experimenting with it—the finite-element method is an essential tool for the modeller. Most packages are well set up for defining (and refining) your own elements etc. For a clear account of finite-element work see Strang (1986).

Given the ease with which such numeric packages handle complex problems, one might well ask why bother with analytic work at all? The answer is that if it's a "once off" situation then there's no point in undertaking such work! If you're a modeller you won't, however, be asked to carry out such a task—normally such standard numerical work would have been carried out before the problem reaches you. If one's interested in determining the effect of various parameters on the outcome of a process (which is normally the type of set-up presented to a modeller) then *understanding* is necessary—the real problem with straight numerical techniques is that *they just produce numbers!* Insight requires that one understand the effect of the various dimensionless parameters on the outcome, and numbers help little in this respect. As a simple example of the relative advantages of analytical as opposed to numerical work, note that in the above we obtained *very* accurate information about the heating of cuboids *of all sizes* using simple analysis. A finite-element scheme would shed little light on this general problem but would produce accurate results for special cases. On the other hand, for the engine block problem the finite-element technique is probably the only real option. In this regard the algebraic packages when used in conjunction with methods such as the Rayleigh-Ritz method represent a major advance. The real advantage of the Rayleigh-Ritz method is that it produces a *semi-analytic* result. Thus the solution is in the form of an analytic result with certain coefficients numerically determined. Using such a result one can see how the dimensionless groups affect the outcome! The inevitable algebraic complexity that arises when proceeding to higher approximations has meant in the past that the Rayleigh-Ritz method was not practical; the advent of algebraic packages has changed all this—perhaps it may be possible in the future to get the best of both worlds (analytic and numeric)!

Comment: Some experimentation with numerical evaluations may lead to suggestions about how to go about solving a problem analytically. Thus for example, after experimenting with finite-element determinations for the time to heat bowls of various shapes, one might be lead to the (correct) belief that it's the thickness of the bowl that matters, which might suggest that one should investigate the possibility of "almost" one-dimensional models. This experimental use for numerical schemes has been found to be particularly valuable for dealing with complicated fluid mechanics problems where the experiments have enabled researchers to see what aspects of such problems require careful attention.

9.5 Summary

In this chapter we examined inverse problems that arose out of a cooking context. A major difficulty associated with such **inverse** problems is that often they are not well-set, and it's not easy to tell if a particular problem is well-set. In one of the cases examined (the surface temperature problem) a unique solution existed and was constructed using knowledge about the corresponding forward problem. The other situation (the temperature distribution problem) was not well-set and an approach based on adding together a class of special solutions to generate one of the possible solutions was presented.

Fourier methods were used to obtain useful exact solutions for problems with simple geometry, and the results indicated the importance of the first eigenvalue for determining the thermal adjustment time. This observation, together with a variational formulation, led to approximation procedures that are useful for handling awkward geometries.

9.6 Exercises

Exercise 9.1 *Roasting*

One presently accepted roasting procedure is to initially raise the surface temperature of the roast (from T_0 say) to a high level T_h, and then progressively (over $0.1t_0$ say, where t_0 is the roasting time) reduce the temperature to a medium level T_l, and maintain this temperature (for $.8t_0$ say) until the final cooking stage. During the final stage of cooking the temperature is again increased to high levels T_h (say), while basting to achieve browning.

(a) Set up an appropriate mathematical description of the required temperature variation, scale appropriately, and use the result (9.2) and Maple to numerically determine the required heat input rate to achieve this outcome for a semi-infinite body. Plot your results. Check to see that your solution for q produces the required T output.

Hint: Use convenient simple functions to describe $T(t)$, for example exponentials.

(b) The above solution for $q(t)$ is acceptable only if it's robust i.e. only if the inevitable small fluctuations of heat input that arise in any practical situation don't have a major influence on the outcome; in this case the surface temperature. To investigate robustness one normally adds a fluctuation (eg. $\epsilon \sin \omega t$ where ϵ is relatively small) and observes the numerical effect on the outcome. Add in such a heat flux term with a range of ω values and see its effect on the surface temperature profile. Comment. On physical grounds why would one not expect any robustness difficulties for heat problems of this type?

Aside: Largely as a result of incorrect modelling during the early scientific investigations of cooking in the 30's, traditional roasting techniques briefly lost favour in the U. S., see Mc Gee (1984). Chefs quickly rejected the new theories.

Exercise 9.2 *Cooker efficiency*

If a forced convection oven is not very efficient, or there are too many roasts in the oven, then a constant temperature boundary condition will not be sustained at the surface of the roasts, and the cooking time will increase as a result. Of course the production gains made by "stacking" the oven may more than compensate for the increased cooking time. To understand the influence of this decreased heat transfer on cooking time we'll work with spherical stuffed chickens of radius a and assume a Newtonian boundary condition

$$-kT_r(a,t) = \beta(T(a,t) - T_0),$$

where T_0 is the oven temperature (assumed fixed), and $T(a,t)$ is the surface temperature of the chickens. The heat transfer coefficient β would be dependent on the number of roasts and the heater characteristics (air speed, heating rate etc.) in a way that would need to be largely determined experimentally.

(a) Scale the equations describing temperature changes in a typical chicken. Look for separated solutions of these equations of the form $e^{-\lambda^2 t}R(r)$.

(b) Show that the cooking time is determined by the smallest root of the equation

$$\lambda \cot \lambda = 1 - \mu$$

where $\mu = a\beta/k$, and extract relevant information (plot graphs etc.).

Exercise 9.3 *Sturm-Liouville Problems*

The Sturm-Liouville problems that arise out of the Fourier Method can be formulated in variational terms; a useful thing to do for approximation work.

(a) $\star$ Show that for functions $\phi(x)$ with $\phi(a) = \phi(b) = 0$ and with the further constraint

$$\mathcal{I} = \int_a^b r(x)\phi^2(x)dx = 1,$$

that the functional

$$\mathcal{J}(\phi(x)) = \int_a^b [p(x)(\phi'(x))^2 - q(x)\phi^2(x)]dx$$

is rendered stationary by the solutions $y(x,\lambda)$ of the Sturm-Liouville system

$$(p(x)y'(x))' + (q(x) + \lambda r(x))y(x) = 0 \quad \text{in} \quad a < x < b,$$

with $y(a) = y(b) = 0$, and that the minimum value of $\mathcal{J}$ is given by λ_1, the smallest eigenvalue of the Sturm-Liouville system.

(b) The above variational formulation of Sturm-Liouville systems is useful because it enables one to obtain accurate estimates for the eigenvalues and eigensolutions in cases when exact answers aren't available.

(i) To test the usefulness of the variational approach, determine approximations for the first eigenvalue and eigensolution in the simple oscillatory case in which $p = q = r = 1$, and $a = 0, b = 1$, using appropriate test functions. Make comparisons with the exact solutions.

Hint: Make use of the symmetry of the solution.

(ii) Determine an approximation for the first eigenvalue in the case in which $r(x) = 1 + x$ (for which no finite exact solution is available).

Exercise 9.4 *Loose Ends: Shape Effects*

Naïvely one might expect the heat transfer rate from a cooling object of a given volume to go up roughly in proportion to the exposed surface area, so that one might expect the cooling time to go down like the inverse of the surface area.

(a) Using the expression (9.11) obtained for the cooling time of a cuboid with sides (a, a, b) (i.e. with one square section) examine the dependence of cooling time on shape factor $\alpha = b/a$ for cuboids of the same volume. Using Maple plot out the cooling time vs. surface area relationship for such cuboids. Comment.

Comment: Many Nature made objects are designed to efficiently absorb a chemical species. For example the roots of plants are designed to efficiently absorb nutrients from the soil, and blood cells are designed to efficiently absorb oxygen. Thus issues of the above type are of importance. The final design adopted by Nature is almost always a compromise because other design features (for example strength) also need to be taken into account.

Exercise 9.5 *Ellipsoidal Chickens*

A spherical stuffed chicken will take longer to cook than a real stuffed chicken of the same weight because the spherical chicken will expose less surface area per unit volume. To get some idea of the effect of shape on the cooking time we'll estimate the heating time for an ellipsoidal body.

We'll use the results obtained in the text that the smallest eigenvalue λ of the system (9.12,9.13) determines the heating time, and that the Rayleigh-Ritz variational approach (9.17) provides an accurate estimate for this value. It pays to creep up on problems, so it's as well to see first if the approach produces accurate results in the spherical case.

(a) It's sensible to work with a spherical co-ordinate system and to use simple test functions. Using the class of test functions

$$f(r) = A + B - (Ar^2 + Br^4) \quad \text{for } 0 \leq r \leq 1,$$

which contains one free parameter after the integral condition on f^2 is satisfied, estimate the smallest eigenvalue for the spherical situation, and compare with the exact value. Compare the exact eigenfunction with its approximation.

Hint: The equations for (A, B) are nonlinear so be careful with branches. Make sure also that you home in on the *smallest* eigenvalue.

(b) ⋆⋆ We now look at the elliptical body case.

Formulate the problem of determining the heating time for an ellipsoidal whose surface is the surface of revolution generated by rotating the ellipse

$$(x, y) = (a \cos\theta, b \sin\theta), \quad 0 \leq \theta \leq 2\pi$$

about the x axis. Use test functions related to those used in the spherical case.

Hint: The surface in the (x, y) plane with the above boundary can be parameterized

$$(x/a, y/b) = (r\cos\theta, r\sin\theta), \quad 0 \leq r \leq 1,\ 0 \leq \theta \leq 2\pi.$$

You may need to look up the expressions for the volume and surface elements, and gradients in general curvilinear co-ordinates. Any reasonably comprehensive vector calculus book will contain this information, for example Kreyszig (1983).

(c) ⋆⋆ Using integration software available to you carry out calculations for a variety of ellipsoids with volume 1 and make comparisons.

Exercise 9.6 *Pipe insulation*

Fluids often need to be transported by pipes at high temperatures, either because the high temperatures are needed at the point of delivery, or because unacceptable chemical or physical changes occur at lower temperatures. For example, oil at high temperatures has low viscosity and as the temperature drops the viscosity increases and solidification can occur, making transport difficult. To ensure the liquid is delivered at the required temperature one can either transport the fluid at high flux levels so the transport time is reduced, or provide more pipe insulation. The aim is usually to determine the least expensive combination of these strategies.

(a) Firstly we examine the radial heat transport in a pipe with external radius r_1 surrounded by insulation material of thickness d and conductivity k. We'll assume the fluid within the pipe is well mixed and also, since pipes are normally made of a good conductor, we might as well assume the pipe's temperature is the same as that of the liquid, T_1 say. We'll assume the environment maintains the external radius of the insulator at T_2 (not so good; a Newtonian condition would be better).

Write down the equations governing the steady state radial heat transport in the insulator. Note that in cylindricals

$$\nabla^2 T(r) = T_{rr}(r) + \frac{1}{r} T_r(r).$$

Show that the steady state solution is given by

$$T = \frac{T_1 \log(\frac{r_2}{r}) + T_2 \log(\frac{r}{r_1})}{\log \frac{r_2}{r_1}},$$

where $r_2 = r_1 + d$, and show that the heat loss rate per unit pipe length is given by

$$\mathcal{L} = -2\pi r_2 k T_r(r_2) = \gamma(T_1 - T_2),$$

a Newtonian cooling relation. Identify γ and plot $\mathcal{L}(d/r_1)$. Comment.

(b) The above results will now be used to construct a "global" heat loss model. The idea is simple: at any location x along the pipe one expects the above expression to provide an accurate description of the loss if $T_1(x)$ is interpreted to be the *local* temperature of the fluid. To determine the temperature at any location x along the pipe one simply needs to account for the accumulated losses from the source to that location. Show that heat conservation requires

$$\rho_f c_f Q \frac{dT_1}{dx} = -\mathcal{L},$$

where Q is the volume flow rate in the pipe and ρ_f,c_f are the density and specific heat of the fluid. Solve the resulting equations and identify the parametric combination determining the length scale for significant temperature changes of the transported fluid. Comment on the results obtained.

Exercise 9.7 *Finite-Elements 1*

We wish to investigate how the choice of interval size affects the accuracy of the approximation obtained using a finite-element approach on a test system. The approximating functions to be used are piecewise linear (as in the main text), and so different approximations differ only in the number of domain divisions used. The test system we'll use is

$$y'' + \gamma^2 y = 1 \quad \text{in} \quad 0 < x < 1$$

$$\text{with } y(0) = y(1) = 0.$$

Firstly the $\gamma = 1$ case will be examined. Follow the following steps:

(a) Using Maple, or otherwise, solve the system exactly.

(b) Show that the functional corresponding to the above differential equation is

$$\int_0^1 [\phi'^2 - \phi^2 + 2\phi]dx$$

and explicitly specify the associated variational problem.

(c) Given the symmetry of the problem, the simplest finite-element approximation that makes sense is one based on breaking the domain into just 2 equal divisions. Either on paper or using Maple determine the resulting finite-element approximation and make solution comparisons.

(d) Double up on the number of divisions and make comparisons.

(e) $\star$ Using Maple plot out the exact solution of our problem if $\gamma = 10$. How many divisions do you think would be necessary to obtain a reasonable approximation in this case?

Hint: How many oscillations does the solution have?

(f) $\star\star$ How many divisions would be necessary in the $\gamma = \pi$ case?

Hint: Be careful!

Exercise 9.8 *Finite-elements 2*

To test the efficacy of the finite-element technique for obtaining approximate solutions we'll investigate its use for determining the steady state temperature distribution in a square plate with one of its edges maintained at temperature 1 and the others maintained at temperature 0, see Fig. 9.5. We've examined this problem in Chapter 6, see Exercise 6.1, where we used linearity and symmetry ideas to show that the temperature at the plate's centre is 1/4. Later in Chapter 8, see Exercise 8.5, we used Fourier techniques to extract exact results for the temperature distribution throughout the plate.

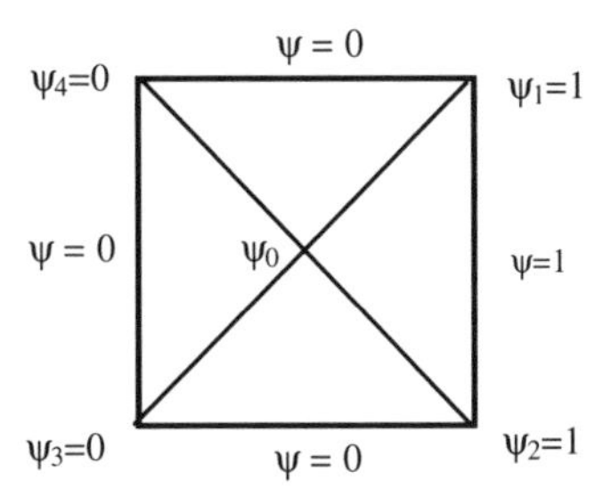

Figure 9.5: Steady temperature in a plate: a finite-element scheme.

(a) Firstly show that the appropriate functional for this problem is

$$\int_0^1 \int_0^1 [\phi_x^2 + \phi_y^2] dx dy.$$

(b) Using the finite-element set-up illustrated in Fig. 9.5 determine the finite-element approximation for $\phi(0,0)$.

Comment: The result obtained is excellent, but this is not a good test. Note that, the approximation is poor except the origin. More elements are necessary to provide a sensible test. Notice also that it's inevitable that discontinuities in the finite-element solution approximation will arise in regions adjacent to boundary value discontinuities (for example at the vertices with $\psi = 1$ in our problem).

Comment: Note that a variational approach to a problem is likely to produce a solution that's not bad on average, but is likely to be poor at particular locations in the solution domain. Thus, if one is concerned with the local behaviour near a boundary value discontinuity then a variational approach will not provide a sensible answer, whereas an approach based on a local series expansion is likely to be effective.

(c) ⋆⋆ This problem is straightforward but it's more of a project than an exercise. Using Maple determine an analytic expression for the linear approximation that realizes the (arbitrary) values ψ_1, ψ_2, ψ_3 at the nodes of an (arbitrary) triangular element. Thus produce a procedure for evaluating the above functional over this element. Having done this you're in a position to attack any 2D steady state heat problem—it's just a matter of specifying the nodes and solving the linear set of equations that result when the contributions are added and the resulting functional optimized.

(d) ⋆⋆⋆ Set up a more refined mesh for the square plate, proceed to the answer, and make comparisons. Make sure that you make use of any solution symmetries that exist. Having developed the tools you might like to determine the temperature distribution in the case in which there's a square hole in the plate with a variety of boundary conditions.

Bibliography

Abramowitz, Milton and Stegun, Irene A. (1970). *Handbook of Mathematical Functions*, Dover Publications, New York.

Carrier, George F. and Pearson, Carl E. (1968). *Ordinary Differential Equations*, p125, Blaisdell Waltham, Massachusetts.

Carrier, George F. and Pearson, Carl E.(1976). *Partial Differential Equations*, Academic Press, New York.

Jastrzebski, Z. D.(1959). *Engineering Materials*, Wiley International Edition.

Kreyszig, E.(1983). *Advanced Engineering Mathematics*, 5th ed., John Wiley.

McGee, Harold (1984). *On Food and Cooking*, Charles Scribner & Sons New York.

Morse, Philip M. and Feshbach, Herman (1953). *Methods of Theoretical Physics, Part I, p 520*, McGraw-Hill Book Company, New York.

Strang, Gilbert (1986). *Introduction to Applied Mathematics*, Wellesley-Cambridge Press, Mass.

Chapter 10

Aspects of the greenhouse problem

10.1 Introduction

It's not easy to see how to start in modelling a situation as complex and subtle as the greenhouse problem. One such way, that exposes some of the issues, is presented in this chapter. The particular questions addressed are of interest in many other applications, some of which will be examined in the main text and in the exercises.

Before one can begin to understand the changes brought about by greenhouse gases it's necessary to understand how the normal pattern of temperature variations on the earth is determined. The incident radiation from the Sun is partly reflected from clouds (30%) and partly absorbed in the atmosphere (25%). The remainder (45%) is absorbed by or reflected from the vegetation and ground or the oceans on which it falls. In each medium the heat input produces a change of temperature leading to radiative losses from that medium to the others and into space. Additionally, since the various components are affected differently by the input, there are "internal"(conductive and convective) heat exchanges between the atmosphere, the ground with vegetation, and the oceans, and also exchanges between the tropics and polar regions. All the inputs, losses and exchanges vary with location, time of the day, and time of the year. Given the complexity it's easy to lose sight of the fundamental issue: in crude terms we need to know just how much of the Earth and atmosphere is involved in the relevant heat exchanges with the Sun. If the "effective thermal mass" of the Earth and atmospheric material involved is small in an appropriate sense then relatively small changes in the heat balance (brought about for example by the addition of hydrocarbons into the atmosphere) will result in large changes in observed surface temperatures and the response time will be short. If the thermal mass is large, then the induced changes will be small in the short term. Now it's evident that much of the atmosphere, the surface layers of the oceans, and the vegetation, are involved in the daily and yearly exchanges (witness the induced motions in the oceans and the atmosphere), but just how much of the solid Earth is involved in the exchange? It's this aspect of the greenhouse that we'll mainly address in this chapter. A simple model examining the atmospheric and oceanic heat exchanges with the Sun is examined in Exercise 10.2.

Now we've found that the tims scale required for a conducting body of the size of the Earth to adjust to a changed environment is of the order of R^2/κ where R is the Earth's radius and κ its diffusivity; which gives about 94×10^6 years, if one uses the diffusivity of granite as being representitive! This calculation ignores entirely the heat exchange mechanism and so is of little direct relevance but it does clearly indicate that very little of the Earth will be influenced by, and will influence, the exchanges over the daily, yearly, and roughly 50 year (for greenhouse effects) time spans of interest. More sensibly we should turn the question around and ask how far heat input changes will penetrate the Earth over these time periods. We'd expect the heat to penetrate to a depth of order $\sqrt{\kappa t}$, and again using the diffusivity of granite this gives a depth of penetration of about 41.5 m over a 50 year period, so one would anticipate that at most a shell of this thickness is relevant for the contexts of interest. This estimate, however, ignores the fact that most of the heat input (mainly during the day) at any location on the Earth will be lost from that location at some later time (during the evening); otherwise temperature patterns would not be (almost) repeated on a daily basis. One might thus expect the depth of heat penetration (and thus the effective thermal mass) to be somewhat less than the above estimate as a result of this absorption followed by desorption process. In Section 10.2 we'll examine the effect of such temporal fluctuations of heat input and we'll then go on to examine the effect of spatial variations on penetration depths in Section 10.3.

In Section 10.4, using a crude global balance steady model, and working with an "effective" surface temperature, we'll estimate the temperature change that might be expected as a result of the increased retention rates due to greenhouse gases. The work is incomplete in that external information is necessary to determine the "effective" surface temperature

As indicated earlier, convective motions in the atmosphere and ocean are induced by differential heating, and the heat exchanges between the various components are dominated by such convective transfers. Oceanographic and meteorological issues far beyond the scope of this book need to be examined to determine these exchanges. The efficiency of these exchanges partly determines the effective radiative temperature of the Earth and in this way influences the global thermal balance.

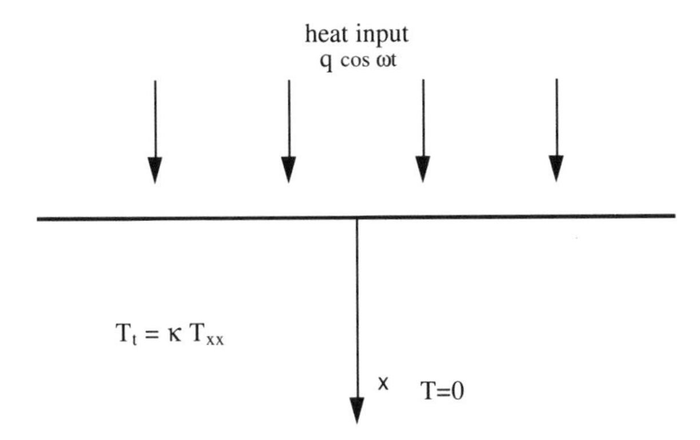

Figure 10.1: Periodic surface heating.

Greenhouse gases absorb some of the radiation emitted from the Earth and thus change the balance described above. The effect of such greenhouse gases on the overall heat balance is small compared with the effects discussed above, however, very small temperature changes are sufficient to destroy life as we know it on Earth. Thus the greenhouse problem is one that requires the examination of relatively small and subtle effects. Some of the subtleties will be briefly discussed in Section 10.4.

10.2 Temporal Fluctuations

Neglecting all consideration of the spatial variations of the heating we first examine the effect of a fluctuating heat input on the temperature distribution in the solid Earth. The fluctuating heating rate may be described by a Fourier series whose main components correspond to a daily and an annual period over most of the Earth's surface. Because the heat equation is linear these components can be handled separately. Thus we might as well simply consider the case of a purely sinusoidal supply of heat $q \cos \omega t$ to the surface of a semi-infinite body; the set-up is as shown in Fig. 10.1. Now of course the Earth has been subjected to a cyclic heat input for aeons so that any imprint of the initial thermal state of the Earth has been lost long ago. It doesn't make sense, therefore, to examine the initial value problem (as we have done for all previous heat work). One might expect that the response to a periodic input of heat would eventually also be periodic with the same period, so that we should look for a solution of heat equation with the same frequency as the heat input. In such circumstances the questions of interest are: what size fluctuations in the surface temperature are to be expected and how does the temperature vary with depth ?

Since heat takes time to penetrate to any depth there is no reason to expect that the maximum temperature occurs at the same time for all depths. Thus we will have to make provision for the possibility that there will be phase variations with depth. In dealing with purely sinusoidal fluctuations in *linear* systems in which there may be phase differences, it is always algebraically simpler to work with the complex representation of the trigonometric functions. No doubt you've used this procedure when solving ordinary differential equations. Thus we'll work with the heat input $qe^{i\omega t}$ and look for a solution of the form

$$T(x,t) = F(x)e^{i\omega t},$$

where F is expected to take complex values. To recover the real solution corresponding to $q\cos\omega t$ we can simply take the real part of the complex solution $T(x,t)$.

Substituting this guessed solution into the heat equation gives, after cancelling the common exponential factor,

$$i\omega F = \kappa F'', \tag{10.1}$$

while the surface boundary condition becomes

$$-kF'(0)e^{i\omega t} = qe^{i\omega t}. \tag{10.2}$$

Since we've focused our attention on the *forced* (particular solution) response to heating (by looking for a solution of frequency ω only) the normal initial condition is no longer relevant. Thus the only other condition to be applied is that deep within the Earth the temperature field settles down to a prescribed value which we'll use as a datum.

Aside: It's possible to determine the transient (i.e. the complementary function) using Fourier methods, but Fourier integrals rather than series are required because the domain is infinite.

Aside: Other considerations determine the temperature levels reached deep within the Earth. A balance between heating due to radioactive materials in the crust and *steady* radiative input and losses from the Earth's surface seems to be struck.

The two linearly independent solutions of the ordinary differential equation (10.1) are $e^{\pm(1+i)\sqrt{\omega/(2\kappa)}x}$, and only the selection of the negative sign leads to solutions that don't become large as $x \to \infty$. Thus the solution with the required behaviour at depth that satisfies the surface condition (10.2) is given by

$$T(x,t) = \frac{q}{(k\sqrt{\omega/\kappa})}e^{-(1+i)\sqrt{\omega/(2\kappa)}x}e^{i(\omega t-\pi/4)}, \tag{10.3}$$

where we've used $\kappa = k/\rho c$. In particular the surface temperature is given by

$$T(0,t) = q/\sqrt{\omega k\rho c}e^{i(\omega t-\pi/4)}, \tag{10.4}$$

so that the amplitude of the surface temperature fluctuations is

$$\text{amp}[T(0,t)] = \frac{q}{\sqrt{\omega k \rho c}}. \tag{10.5}$$

10.2.1 Solution features

Firstly note, see (10.3), that time averaged temperature and temperature gradient at all locations x vanishes, which implies that there is no *net* heat transfer past any location. In particular there is no net transfer into the interior of the Earth due to the fluctuating components of the heat supply from the Sun—an interesting result! In order to confirm this the transients should be examined, and of course there may be subtle effects not accounted for with our model that result in a net heat transfer.

Note, see (10.4), that the surface temperature lags behind the heating rate by $\frac{1}{8}$th of a period. Thus for example the theory suggests that the maximum temperature will be achieved three hours after the Sun is directly overhead at noon; this is in reasonable agreement in locations where temperatures are not strongly affected by such things as sea breezes or weather fronts. The delay in the yearly fluctuations predicted is about six to seven weeks, and this also seems to fit. This is comforting because it provides good evidence that the complex processes whereby heat from the Sun is absorbed and desorbed are well described by a simple heat conduction model.

The result (10.5) also tell us that the more rapid the fluctuations (i.e. the larger the ω) the smaller the surface temperature fluctuations. *Thus it's the slower variations which have most effect.* The frequency ratio for the two major Fourier components is $\omega_{day}/\omega_{year} = 365$, so the amplitude of the daily component of the heat input would have to be about $\sqrt{365} \approx 20$ *times larger* than the yearly component to have a comparable effect on surface temperature amplitudes! In high latitudes the annual variation is relatively large and it decreases gradually as the equator is approached, where there is scarcely any variation throughout the year. Thus, except for small regions close to the poles, we'd expect daily surface temperature fluctuations to be relatively small compared with yearly fluctuations. This interesting counter-intuitive result *suggests* that it *should* be possible to ignore daily fluctuations when considering questions related to global warming.

In order to examine further the variation of the temperature with depth it is convenient to rearrange the result (10.3) in the form

$$\text{amp} \cdot e^{-\sqrt{\omega/(2\kappa)}x} e^{i(\omega t - \sqrt{\omega/(2\kappa)}x - \pi/4)}.$$

The factor which involves real exponential variations in x shows that the amplitude of the fluctuations decay exponentially with depth, and that the depth of penetration is about $l_\omega = 5\sqrt{2\kappa/\omega}$. The imaginary exponential indicates that there is a wave type fluctuation with depth where the crests of the fluctuation are to be found on paths where

$$\omega t - \pi/4 - \sqrt{\omega/(2\kappa)}x = \text{constant}.$$

Thus an observer within the Earth would see a wave of heat travelling past with speed $\sqrt{2\kappa\omega}$.

The thermal diffusivity of the material at the Earth's surface varies considerably but if we again use the value for granite ($\kappa = 10^{-6}\text{m}^2/\text{s}$), then the depth of penetration of daily fluctuations obtained is 0.8 m, and for yearly fluctuations we get $20 * 0.8 = 16$ m. Thus only a surface "skin" of depth about 16 m plays a significant role in the periodic heat exchanges with the Sun. The same calculations for the oceans give a penetration distance of 1.8 m over a year, but other processes dominate diffusion in this case, see Section 10.4.1.

Not only are the surface amplitudes for higher frequencies reduced by the $1/\sqrt{\omega}$ factor, but also they're damped more rapidly with depth than lower frequencies, so that the Earth acts as a natural filter, removing higher frequency heat input components in its upper layers. Thus by placing temperature probes deep within the Earth one may be able to extract useful information about the low frequency radiation characteristics of the Sun, for example the periodic Sun spot activity which occurs on about a 11-14 year cycle.

Comment: other applications The above work is relevant in a large range of situations. One obvious question answered by the above results is how thick should one make the thermal insulation used in house construction. Normally one wishes to insulate against the daily temperature variations, so that about $\gamma\sqrt{2\kappa/\omega_{day}}$ with $\gamma \approx 4$ is the thickness required. Most insulation consists of air (a poor conductor) trapped in a fibre matrix (to prevent convection and provide thermal mass); so that the conductivity is essentially that of air, but the specific heat and density properties are determined by the properties of the fibre material and how compressed it is. Based on this model of the behaviour of such insulators and values tabled, see Table 6.1, (or on experimental values you may have access to) you might like to make some estimates for the required thickness.

10.3 Spatial Variations

As a means of estimating the effect of differential heating at different points on the surface of the Earth consider a heating rate varying as $\sin \mu y e^{\imath\omega t}$ where y is measured along the Earth's surface. We'd expect the temperature field to reflect this y variation so that

$$T(x, y, t) = T(x) \sin \mu y e^{\imath\omega t},$$

represents a sensible guess for the solution form. Substituting this expression into the heat equation leads to the ordinary differential equation

$$\mathcal{T}'' - [\mu^2 + \frac{\imath\omega}{\kappa}]\mathcal{T} = 0.$$

It's easy enough to solve this equation exactly but it is probably more instructive to examine the effects of the various terms, using results already obtained. The term ω/κ is essentially the reciprocal of the square of the depth of penetration l_ω of the fluctuating temperature, and the term μ^2 is the reciprocal of the square of the length scale l_h over which the heating rate changes. However the depth of penetration (about 16 m) is *much* smaller than the length scale over which there are significant heating variations due (directly or indirectly) to the Sun ($>$ 500 km). Thus the parameter range of interest is one in which the μ^2 term in the above ordinary differential equation is negligible in comparison with the ω/κ term. The implication of this is that the one-dimensional model of the previous section, see (10.4), will accurately describe the temperature field on the Earth's surface if the *local* value of heat input is used. The temperature variations with depth will also be described by the one-dimensional model. An iterative scheme could be used to justify this mathematically intuitive result but, in the light of experience, there's little reason to doubt it. Thus, in the absence of heat transfer by way of the atmosphere, we can see that the *local* surface temperature rise is simply related to the *local* heat input 3 hours earlier.
Comment:shape effects: Basically because $l_\omega \ll l_h$, see Fig. 10.2 in the Earth context (so that the Earth is "thermally thick"), we've seen that spatial variations of heat input could be ignored when determining local changes. The same arguments could be used for either topographic variations, variations associated with the curvature of the Earth, or conductivity variations; providing l_ω is much smaller than all the other relevant length scales the one-dimensional result will determine local temperature variations. Given the size of l_ω this is evidently the case in the Earth context. This situation occurs in many other contexts involving periodic forcing and in such cases "shape" effects can be ignored—a major simplification! (You recall the mathematical complications that arose in the cooking context when dealing with non-plane geometry.) In other contexts the forcing frequency is such that the size of the object being heated is much smaller than the diffusion length scale i.e. $l_h \ll l_\omega$. For these cases the above analysis, see (10.3), suggests that the variation in temperature through the body will be relatively small; so that one might as well think of the object as being uniformly heated by the radiation—again a major simplification! If the length scales are comparable then no simplification is possible; the complication is real in that in such circumstances geometric and diffusion effects both play a role in determining the thermal response.

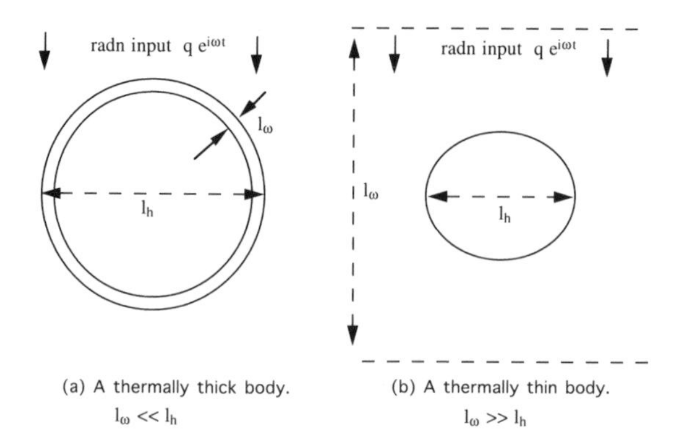

Figure 10.2: Thermally thick and thin bodies.

10.4 Adjusting to a new radiative equilibrium

Having examined the "normal" response of a solid Earth to periodic heating we now turn our attention to the question of determining how a change in average heat input due to greenhouse gases is likely to influence the surface temperatures of the Earth. The aim here is not so much to produce a complete model (a major task) but to check to see if the numbers "make sense". Such preliminary "back of an envelope" calculations are often made by modellers before attempting time consuming major calculations that may be conceptually flawed. With this limited purpose in mind we consider the effect of increasing the heat supply rate from Q to $Q + q$ on the surface temperature of a semi-infinite conducting body which is radiating according to Stefan's radiation Law. Thus (based on the earlier work) we avoid the complications associated with periodic heat input and shape, and also we assume that it makes sense to work with *a* surface temperature (an appropriate average). We assume an abrupt change in heat input occurs due to greenhouse gases. Clearly the assumptions underlying this simple picture are open to challenge, but none the less the calculations should provide crude estimates for the resulting temperature changes, and may also provide some indication of the thermal response time of the Earth to such heat input changes. The most doubtful aspect of the envisaged situation is that the Earth is treated as a conducting solid. One might well argue that it would be more sensible to model the Earth's "surface" as a well mixed fluid layer, corresponding to the atmosphere or the ocean, see Exercise 10.2.

Initially the heat supply is balanced by the radiative loss from the surface which is at temperature T_0 (where T_0 is measured in degrees Kelvin) so that

$$Q = \sigma T_0^4. \tag{10.6}$$

After the rate is increased an imbalance will occur at the surface with some of the additional heat being conducted away from the surface into the conductor. Heat conservation requires

$$-kT_x(0,t) = Q + q - \sigma[T_0 + T(0,t)]^4,$$

where $T(x,t)$ denotes the resulting increase in temperature of the body. Now in the greenhouse context the relative changes in heat supply q/Q are very small and will produce correspondingly small proportional changes T/T_0 in the surface temperature. Thus, neglecting terms of order $(T/T_0)^2$, (i.e. *linearizing*), and using (10.6), the surface condition becomes

$$-kT_x(0,t) = Q[q/Q - 4T(0,t)/T_0],$$

which you'll recognize as being of the Newtonian Law form

$$-kT_x(0,t) = \beta[T_{eql} - T(0,t)],$$

with

$$\beta = \frac{4Q}{T_0}, \quad \text{and } T_{eql} = \frac{T_0}{4}\frac{q}{Q}. \tag{10.7}$$

We've studied this Newtonian cooling problem already in Section 7.6. The heat conducted into the material causes a rise in temperature which in turn results in enhanced radiative losses. We found that eventually, after a time of order

$$t_d = \frac{\pi}{\kappa}\left(\frac{k}{\beta}\right)^2,$$

see equations (7.32,7.33,7.35) of Section 7.6, the surface temperature reaches its new equilibrium level $T_0 + T_{eql}$.

Now $T_0 \approx 290°$K so that, from (10.7) a 4% increase in heat retention rate (typical of the figures quoted) would be expected to give rise to about a 2.9°C rise in surface temperature according to the model. This figure can be checked against available data; the figures appear to be reasonable. Allowing for clouds, the rate at which heat enters the Earth's system averages at about 0.34 kw/m^2/s. Using the above expression for the time scale gives a figure of 11 days for the time for a granite surface to adjust to the changed heat input. You'll recall, however, (see Section 7.6.3 and also Fig. 7.6), that the approach to radiative equilibrium is much slower (by a factor of about 1000) than indicated by this time scale. There's considerable uncertainty associated with the above model as far as the response time is concerned so that perhaps this figure shouldn't be taken too seriously. Similar guesstimates for the atmosphere (treated as a well mixed region) give a time span of about 1 year for the exchange processes to equilibrate with the Sun. The heat transfer processes in the oceans are complex and will be discussed in the next section.

Comment: The response times calculated above suggest that the thermal adjustment time of the Earth is relatively small compared with the time span (50 years) associated with the significant accumulation of greenhouse gases. If this is the case then effectively the Earth responds immediately to the changes in greenhouse gas levels and temperatures will mirror these changes. If, however, the thermal adjustment time is relatively large (and there are good reasons for believing that this *may* be the case, see Section 10.4.1), then the effects of greenhouse gas accumulation will not be felt for some time, a potentially dangerous situation—the "environmental hangover" may not be avoidable. Often in environmental systems the response time to man's input is relatively large compared with the lifetime of a government or a generation, which is why many scientists are pessimistic about the safety of the planet.

10.4.1 Comments on the greenhouse effect

In the above work we've focused our attention on particular issues of importance in greenhouse studies. Such studies serve to provide a knowledge base so that sensible complete models can be built up. The internal exchanges (between land and ocean and atmosphere) ignored in the above work introduce complications much greater than those so far considered. Some of these will now be briefly discussed.

- One factor which presents great difficulties is the role of clouds. During the daylight hours they serve to reflect incident solar radiation; this reflected radiation doesn't enter the thermal balance of the Earth. On the other hand during the night clouds serve as a reflecting blanket for the infra-red radiation from the Earth, as is instanced by the greater tendency for frosts on clear nights. At present about 30% of the incident radiation is lost by reflection, so you can see that any change in the average cloud cover will have major implications. If warming trends (caused by added hydrocarbons) produce a greater level of cloud cover during daylight hours (because of enhanced evaporation) then the Earth may receive rather less solar radiation than expected due to this increased cloud cover; so the clouds could act as a buffer reducing the effects of any such change. Predicting cloud formation is virtually a black art.

- A second factor involving great uncertainty is the heat absorption in the oceans, including the ice caps. The oceans cover about 70% of the globe and are an enormous sink for heat, so that relatively small changes in the thermal state of the oceans can be of major importance in the greenhouse context. The heat input raises the temperature of the oceans and causes evaporation, and also melting of the ice caps. Additionally the input helps drive the oceanic and atmospheric motions, some aspects of which we will now briefly discuss. A typical temperature profile with depth is as shown in Fig. 10.3. There are several distinct layers. Pronounced seasonal variations in temperature occur in the upper 25 m, so that at the very least this top layer is involved in the thermal exchange. Estimates of the response time for thermal exchanges with just this layer give about 4 years, see Exercise 10.2. Below this seasonal ocean layer there's a thick layer (200 m to 400 m) of water at a relatively constant temperature over the globe, but of a depth that varies with location. Below this region is a region, referred to as the thermocline, in which there are rapid decreases in temperature, and rapid increases in salinity; both of which inhibit vertical water movement. Thus the deep cold lower layer below the thermocline communicates little with the upper regions and is relatively isolated from questions of interest here. Winds on the surface of the ocean drive the large oceanic gyres that play such an important role in determining our climate. Associated with these gyres are small vertical water motions (of the order or 2 m/year) that transport warm water down to the thermocline or cool water from the thermocline to the surface at locations determined by the land masses and rotational effects. The depth of the thermocline is also thereby determined. Any change in the heat input will cause immediate and obvious changes in the wind system which in turn will bring about small changes in fluxes from the thermocline. Now in the short term such changes may not be noticeable, but in the long term such effects may cause changes in the depth of the thermocline and thus may be crucial. Oceanography is a subtle subject! The disturbing feature about this phenomenon (exposed using a simple analytic model of an ocean) is that such subtleties will be inevitably missed by numerical schemes that are not explicitly designed to pick them up. The above phenomenon is reasonably well understood but one wonders what other subtleties, not yet understood, may lie hidden. Simple models enable one to see such subtleties and should thus be a necessary component of any future investigations.

Let it suffice to say that present models are restricted by available computers to very crude representations of clouds, ocean, and atmospheric effects.

- We've said virtually nothing about the greenhouse gases that are actually responsible for the present concerns. In order to predict the changes in greenhouse gas levels in the atmosphere it's necessary to understand (and model) the underlying carbon cycle. Controversial

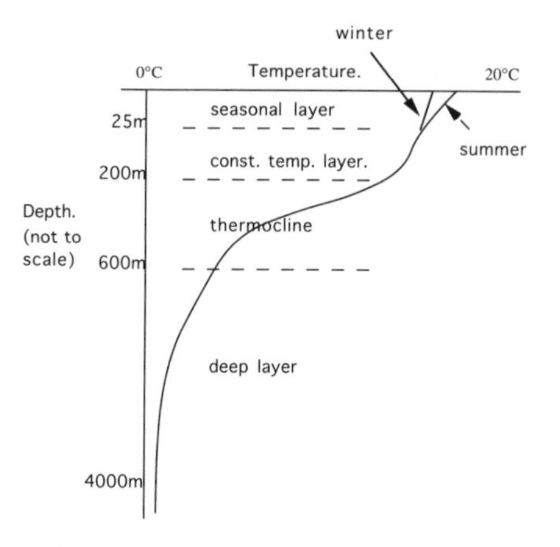

Figure 10.3: Ocean temperatures.

issues arise here also but the problems arising appear to be more straightforward. Greenhouse gas levels are expected to double in about the next 50 to 70 years. For a concise account see Schneider (1989) in the special Scientific American edition "Managing Planet Earth", which also contains other relevant articles.

The computer models currently used strain present day computer resources and are still grossly inadequate. In spite of all the difficulties and uncertainties important decisions concerning greenhouse gases have to be made based on present day calculations, however inadequate. Different models give different results and none of the investigators would claim their results are accurate. The models, however, predict temperature increases of the order of 1.5°C to 5.5°C degrees to result from the doubling of greenhouse gases expected in the next 50 to 70 years. Changes of a few degrees would have a major effect on life forms on the planet—doesn't look good! For further information on the hydrological cycle see Eagleson (1970) and Bras (1990). For more information concerning all aspects of the greenhouse problem (including models of various types) read the Scope publication on "The greenhouse effect" Bolin, Warrick, Döös, Jager (Ed.) (1986).

10.5 Summary

The above work of course just scratches the surface of the problem. However, it is hoped the work will have given the reader some appreciation of the difficulties involved in making greenhouse predictions, and also some appreciation of how to go about investigating complex modelling questions.

10.6 Exercises

Exercise 10.1 *Thermal fluctuations on the Earth's surface*

At most locations on the Earth's surface the mean surface temperature will vary over a yearly period (frequency ω_1), and there will be daily fluctuations (frequency ω_2) superimposed whose amplitude will vary over a yearly period. A model of the surface temperature of the type

$$T(0,t) = 1\cos\omega_1 t + [b + c\cos\omega_1 t]\cos\omega_2 t,$$

where the temperature scale and datum have been chosen so that the average temperature is zero and the amplitude of the seasonal variation is unity, should therefore provide useful information concerning the heat input from the Sun and the resulting heat penetration into the Earth.

(a) Choose an appropriate time scale and choose the coefficients (b,c) to reflect the seasonal and daily temperatures variations in your local area. Sketch the above profile (do not use Maple). Because of the large difference in the two relevant time scales, plotting $T(0,t)$ is a problem. One commonly used procedure to overcome this difficulty is simply to plot the upper and lower amplitude envelopes; to do this replace the high frequency term by ± 1. Plot these envelopes.

(b) Using linearity and the results (10.3,10.4) characterize the input flux (i.e. specify the amplitudes of the Fourier components) required to produce the above surface temperature variation in the case in which $c = 0$. Plot the flux envelope. Determine and plot the temperature envelope at relevant distances below the ground. Use the length scale $\sqrt{2\kappa/\omega_1}$.

(c) $\star$ In the case in which $c \neq 0$ carry out the same analysis.

Hint: One way of handling the product term in the above expression is to change it into a sum of oscillatory terms. One can then process the individual Fourier terms using complex exponentials.

Exercise 10.2 *Ocean and Atmosphere heating*

As indicated in the text, seasonal temperature changes are observed in the top 25m of the ocean. Based on the assumption that only this top layer is involved in the thermal exchanges associated with the greenhouse changes we can make crude estimates for the expected adjustment time for this layer if an additional external heat input rate q per unit area is introduced, and radiative losses are accounted for. For simplicity assume the layer is well mixed, at temperature $T(t)$.

(a) (i) Working from first principles and using the notation of the text show that heat conservation requires

$$\rho_w c_w \frac{dT(t)}{dt} = [Q + q(t)] - \sigma_w [T_0 + T(t)]^4,$$

where d is the water depth of water effected by heat input.

(ii) Approximate the radiation term and thus show that

$$\rho_w c_w \frac{dT(t)}{dt} \approx q - \frac{4Q}{T_0} T(t).$$

(iii) Obtain an estimate for the adjustment time.

(b) ⋆ Based on the same assumptions determine the time scale for the atmosphere to adjust to changed conditions. Of the three components (ocean, atmosphere and land) which has the shortest response time to external heating? What are the implications in terms of the long term exchanges between the components?

Exercise 10.3 *Linered Engine Blocks*

The following problem arose out of an MISG (Maths in Industry Study Group) meeting held at The University of South Australia in Feb 1991.

Aluminium engines are much lighter than cast iron engines, so that large cost and efficiency benefits result if aluminium engines are used. However, aluminium has inferior wear properties, so that one possible compromise is to use engine blocks with thin cast iron liners (thickness 3mm) inserted into each cylinder. Inevitably there will be a gap between the liner material and the cast iron (estimated average thickness 10μm) and thermal expansion effects will open up a larger gap under the high operating temperatures. Such gaps inhibit the flow of heat from the cylinder to the water coolant and so affect the efficiency of the engine. The temperature within the cylinder varies over a large temperature range during the combustion cycle so it's obviously important to know if significant temperature variations occur at the gap.

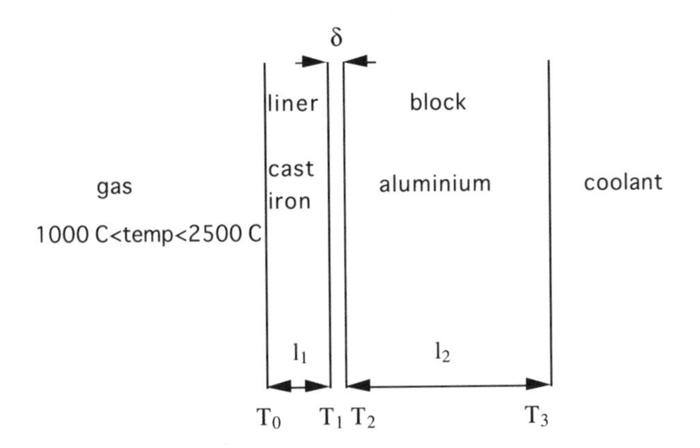

Figure 10.4: Section through a wall of a linered cylinder.

(a) Typically an engine runs at 50 cycles/s. Using the tabulated values for the physical properties (see Table 6.1), show that the penetration depth for temperature variations is of the order of 0.7mm; so that they can be ignored. The implication is that a steady state analysis using average values suffices for determining the effects of heating on the gap size.

(b) Typical dimensions are: inside radius of the liner $r = 40$mm., thickness of liner $l_1 = 3$mm, distance between cylinder wall and cooling ducts $l_2 = 7$mm, width of macroscopic gap between cylinder and liner $\delta = 0.4\mu$m to 60μm, see Fig. 10.4. By considering limiting cases assess the affect of various gap sizes on the heat flux. Assume that air fills the gap. Plot the temperature distribution. Assume the average surface temperatures $T_0 = 200°$C for the inside of the liner, and $T_3 = 80°$C for the coolant temperature.

Comment: There will not be a uniform gap between the block and the liner. The surfaces will touch at certain locations, and, given the high conductivity of metals relative to air, it's likely that the conductivity of the gap will be determined by such metal to metal contacts. Touching surfaces in fact only make contact at relatively few high points and the area of contact is affected by the normal force between the surfaces. Thus the resistance of the gap to heat flow will depend on strength of the force holding the liner in place. For further information about the interesting problems that result see the MISG report, Barton (1991).

Bibliography

Barton, N. G, (1991). *Proceedings of the Mathematics-in-Industry Study Group*, CSIRO, Australia.

Bolin, B. Warrick, R. Döös, B. Jäger. J.(Ed.) (1986).*The Greenhouse Effect, Climatic Change, and Ecosystems*, John Wiley and Sons., Chichester.

Bras, Rafael.L,(1990). *Hydrology*, Addison-Wesley Publishing Co. Reading, Mass.

Eagleson, Peter. S.(1970). *Dynamic Hydrology*, McGraw-Hill Book Company. New York.

Schneider, Stephen. H, (1989)*The Changing Climate*, Scientific American (Sept 89).

Chapter 11

Producing sheet steel?

A design feasibility study
Moving boundaries
Dealing with large parameters
Thin geometry problems

11.1 Introduction

In this chapter we present some preliminary calculations made to assess the feasibility of a possible procedure for casting steel sheets.

The standard way to produce sheet steel has been to take molten steel and pour it into large blocks (thickness 20 cm) which, after cooling, are taken to a rolling mill, **reheated** and then rolled to produce the thin sheets (thickness 1 cm or less). Obviously it would be cheaper to produce the sheet steel in one continuous operation. One way which has been considered is to pour the molten steel of near to the required thickness onto a large rotating copper drum cooled internally by circulating water, see Fig. 11.1. The problem of examining the feasibility of such a procedure was posed at a Maths-in-Industry Study Group meeting run by CSIRO Australia and held at The University of N.S.W. in 1989. A number of questions arise:

- What size drum would be necessary? Clearly the sheet can't be removed from the drum until the steel is solidified. Large drums are expensive to build and operate.
- Could the copper drum be maintained in a sufficiently cooled state? The heat taken up by the drum needs to be transferred to the water during one rotation or else there will be a build up in temperature of the drum.

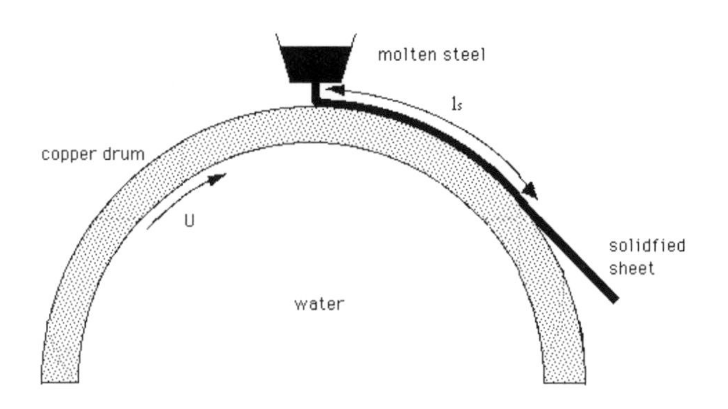

Figure 11.1: A continuous steel casting proposal.

- Are there any minor modifications of the design which would make the project more feasible or economic?

Drum sizes of the order of several metres radius, 1-10 cm thickness, operating at speeds of about 1 m/s and producing steel sheets of thickness about 1-10 mm were envisaged.

Obviously copper is under consideration for the drum because of its high conductivity; but it is expensive. Water cooling is effective only if the water is recirculated and the effectiveness improves with speed of recirculation, but the expense also goes up rapidly with the speed of recirculation. Obviously if the system is found to be feasible, optimal choices for the various components and their operation (drum radius and thickness, drum rotation speed, pump size etc.) could be selected based on a more detailed study.

Aside: alternatives? Why use a copper drum to remove the heat? For example why not simply remove the heat by spraying water onto the upper surface of the steel? Ignoring possible engineering difficulties that might arise to set up such a scheme we already have the background knowledge needed to examine this question. If a copper drum is used then we've found, see Exercise 7.5, that the surface of the steel in contact with the copper *immediately* drops to a temperature intermediate between that of the molten steel and the copper, so that the steel will immediately begin to solidify. If water is used directly then no such immediate temperature drop will occur and the cooling rate would be initially slower. After sufficient time, however, direct water cooling would realize lower surface temperatures so that a direct water removal system would be more efficient for the longer term removal of heat. Estimates, see the Appendix, suggest the copper drum system may be marginally better.

There are a couple of obvious features of the problem that lead to simplifications:

- Copper and water are much more efficient at extracting heat from a good conductor, such as the molten steel or the drum, than is air. Thus the amount of heat which would be lost from the upper surface of the molten steel by conductive/convective heat transfer would be a very small fraction of that transferred to the copper drum. The radiative losses from this steel surface will also be ignored so that a zero flux condition will be assumed on this exposed surface. Similarly the heat lost from the sides of the drum to the air will be relatively small, so that temperature variations across the steel sheet will be small and will be ignored. Of course the resulting model will lead to a (slight) overestimate for the required drum size; errors in this direction are preferable. If preliminary studies indicate that the system is feasible then iterative solution corrections could be made.

- The heat is first transferred into the copper and from there into the water. One might suspect (and crude calculations substantiate) that the amount of heat transferred to the drum *during the period of contact between the drum and the steel* will be much greater than the amount of the heat transferred to the water during this period. This being the case we can substantially simplify the initial stages of the calculations by ignoring the heat flux to the water during this stage of the process. Again this will produce a conservative estimate for drum size and corrections can be made if necessary.

The problem thus divides naturally into two independent calculations:

1. the removal of heat from the steel,

2. the removal of heat from the drum.

11.2 Solidifying The Steel

We consider first the cooling of the steel while it is in contact with the rotating copper drum. For the present we will assume that the portion of the drum coming up to the zone where the molten steel is poured onto it has been brought back to a uniform low temperature T_c. Further we assume that the steel is in contact with the drum over a sufficiently small length of its surface that the contact surface is well approximated by a plane. If this isn't the case then unacceptable bending stresses will be set up in the sheet. Let the speed of the drum surface be U and we'll take it as an initial basis for working that it's possible to pour the steel in such a way that it moves with this same speed parallel to the surface of the drum. Thus we assume there is no movement relative to the moving drum. If this isn't the case then unacceptable waves may be produced on the molten steel surface. Should the initial study suggest that the procedure is feasible as far as the freezing aspect of the problem is concerned, then flow aspects of the process would need detailed investigation.

11.2.1 Heat transfer in a shallow moving medium

In a co-ordinate system moving with the drum (x', z') (say) the usual heat equation

$$T_t = \kappa_s[T_{x'x'} + T_{z'z'}]$$

applies in the steel sheet (both liquid and solid), so that relative to the co-ordinate system (x, z) fixed in space, where

$$x = x' + Ut, \ z = z',$$

the appropriate equation for $T(x, z, t)$ is

$$T_t + UT_x = \kappa_s[T_{xx} + T_{zz}],$$

obtained by changing variables. After an initial adjustment period the temperature field is observed to be steady relative to a stationary observer. It's this steady state that we will investigate, so the T_t term will be dropped. We'll assume that the thermal diffusivities in both molten and solidified steel are the same for simplicity of presentation.

If h_s is the thickness of the steel sheet and l_s is the length of molten steel zone then one might expect $h_s/l_s \ll 1$. Now T_{xx}/T_{zz} is of order $(h_s/l_s)^2 \ll 1$, so that under such circumstances

$$UT_x = \kappa_s T_{zz} \tag{11.1}$$

should provide an accurate approximation of the heat equation. The two terms in this equation balance if the length scale in the x direction is of order Uh_s^2/κ_s, which provides us with a crude estimate for l_s, namely Uh_s^2/κ_s. The value $\kappa_s = 5 \times 10^{-6}\text{m}^2/\text{s}$ was specified by the steel maker so that with $U = 1$ m/s, and $h = 0.01$ m this gives a scale for l_s of order 20 m. With this estimate for l_s, T_{xx}/T_{zz} is of order $(h/l)^2 \approx 2.5 \times 10^{-5}$, so that the approximation to the heat equation made above looks justified—at least we have consistency.

Aside: Interpreted physically the above equation simply states that (under steady conditions) there is a balance between the net longitudinal heat transport (convected) into an element (dx, dz) fixed in space and the transverse heat lost by conduction (through its dx faces).

Note that this equation is mathematically identical with the normal heat equation, with t replaced by x/U. Thus the results from the work on one-dimensional unsteady heat transfer carry across immediately. Note that this equation will also apply in the copper drum with κ_s replaced by κ_c.

11.2.2 Solidification

The solidifying steel releases latent heat and this will have a significant effect on the process; by raising the temperature this released heat will inhibit further solidification. It will be assumed here, and this is confirmed by observation, that there is a well defined (distinct) solidification temperature T_m coinciding with the temperature of melting. Under such circumstances there will be a clearly identified surface separating the molten from the solidified steel; this is experimentally observed. This situation doesn't always occur; witness the mush (or slush) associated with melting ice or snow.

Let the position of the solid/liquid steel interface be given by $z = Z(x)$, see Fig. 11.2. To allow for solidification it's necessary to observe the changes in a steel element as it moves. Over a time interval between t and $t+\delta t$ the steel element initially at x moves off to the new location $x+U\delta t$. The change in thickness of solid steel in this steel element during this time interval will thus be $Z(x+U\delta t) - Z(x)$, so that the rate of release of latent heat by this moving element is given by

$$\rho_s LU \frac{dZ}{dx},$$

per unit sheet width and length of element, where L is the latent heat (units joules/kg), see Section 6.6. This released heat will appear at the solid/liquid steel interface (referred to hereafter as the solidification front) so that heat conservation requires that there be a flux change across this interface given by

$$k_s[T_z(x, Z(x)_-) - T_z(x, Z(x)_+)] = \rho_s LU \frac{dZ}{dx} \tag{11.2}$$

where the $\pm$ suffices indicate on which side of the interface the derivative with respect to z is to be evaluated. The other requirement at the interface is that the temperature be at the melting/solidification point temperature for steel, T_m, so that

$$T(x, Z(x)) = T_m. \tag{11.3}$$

11.2.3 The equations

We're now in a position to fully specify the problem in mathematical terms. The set-up, now to be described, is displayed in Fig. 11.2. The appropriate field equations are

$$UT_x = \kappa_s T_{zz}, \text{ for } h_s > z > 0 \tag{11.4}$$

for both the molten and solid steel regions, where h_s is the thickness of the steel zone, and

$$UT_x = \kappa_c T_{zz}, \text{ for } 0 > z > -h_c, \tag{11.5}$$

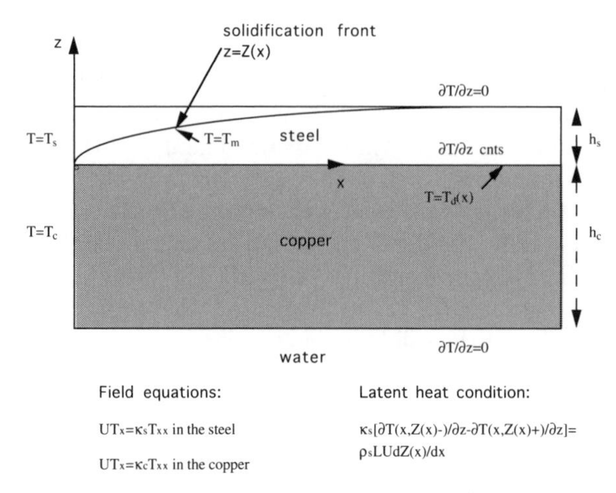

Figure 11.2: The defining equations.

where h_c is the thickness of the copper drum.

The external constraints on the drum/steel system are that there is no loss of heat from the upper surface of the steel and also from the lower surface of the copper drum, as discussed in the Introduction, so that we require

$$T_z(x, h_s) = 0 \text{ and } T_z(x, -h_c) = 0. \tag{11.6}$$

At each of the internal interfaces the temperature is continuous. We'll denote the drum surface temperature by $T_d(x)$. The temperature of the solidification front is required to be at the melting point T_m of steel, thus

$$T(x, Z(x)) = T_m. \tag{11.7}$$

The "initial conditions" of the steel and copper drum need to be specified thus

$$T(0, z) = T_s \text{ (given) for } 0 < z < h_s, \text{ and,} \tag{11.8}$$

$$T(0, z) = T_c \text{ (given) for } -h_c < z < 0. \tag{11.9}$$

Heat conservation requires that the flux of heat from the steel into the copper at the drum surface be equal to that being conducted away by the copper at this surface. This requires

$$k_s T_z(x, 0+) = k_c T_z(x, 0-), \tag{11.10}$$

Also across the solidification front we have, see (11.2)

$$k_s[T_z(x, Z(x)_-) - T_z(x, Z(x)_+)] = \rho_s LU \frac{dZ}{dx}. \tag{11.11}$$

Our knowledge of the physics of heat flow would suggest that the above specification is complete, so that a unique solution is to be expected. To press ahead with a direct onslaught on the problem at this stage would, however, be a mistake. There are a number of difficulties that need to be addressed:

- The location of the solidification front is not known a priori; in fact our primary aim is to determine this surface. The rather innocuous looking solidification conditions (11.7,11.11) introduce major technical difficulties. Basically the system is no longer linear (so that solutions can't be generated by direct addition of fundamental solutions) as a result of these surface conditions; so that standard linear techniques (Fourier series etc.) can't be used directly. Also, standard numerical techniques run into difficulties trying to keep track of the interface. Such *free or moving boundary problems* arise whenever a phase or chemical change occurs and so are of major importance. Special techniques, usually numerical, and often based on a variational formulation, have been developed to handle such problems, see Crank (1984).

- There are many physical parameters that determine the length of the molten steel zone and the relevant dimensionless combinations vary enormously in size. Let's examine a few relevant parameters. The solidification zone is long (about 20 m) and thin (0.01 m) with rapid temperature changes occurring through the sheet and *much* slower changes occurring along the sheet. However, if the longitudinal changes are not accurately accounted for then the length of the liquid zone will not be correctly determined; so the relatively small longitudinal temperature gradients are important! Additionally the thermal parameters for copper and steel are very different (eg. $k_c/k_s = 20$) so that the equations contain small and large terms that could well be important but could be ignored by any scheme that's not specially designed to recognize their importance. A direct numerical scheme would thus have to be very large (with small grid spacing and small truncation and round-off error) to avoid masking significant physical effects; even then the results would be doubtful and provide little insight.

  Comment: This difficulty is very typical of "thin geometry" problems. These problems are subtle and careful analytic processing is normally required to arrive at a useful set of defining equations. Such problems arise *very* often and are a real challenge for the modeller. For example the ocean and atmosphere are thin fluid zones on the surface of the Earth, so that oceanography and meteorology are generally concerned with motions that occur in thin zones. Virtually all such problems do not lend themselves to a direct numerical onslaught because seemingly small terms, arising because of thinness, play an important role in determining the solution.

As we've seen in earlier examples, appropriate scaling will overcome some of the difficulties, will reduce the number of number of parameters as much as possible, and could lead to useful simplifications. Of course it's important to carefully select the scales. It's evident that the temperature ranges from the low value T_c occurring initially in the copper to the high initial temperature T_s of the molten steel, so it makes sense to work with the scaled temperature, $\mathcal{T}$ given by

$$T = T_c + (T_s - T_c)\mathcal{T},$$

in terms of which the scaled value of the melting point temperature, $\mathcal{T}_m$ (clearly a major factor determining the solidification distance) is given by $\mathcal{T}_m = (T_m - T_c)/(T_s - T_c)$. The most obvious choice for the transverse length scale is the thickness of steel sheet h_s, and a longitudinal length scale Uh_s^2/κ_s based on the thermal characteristics of steel also seems sensible. This is the scale for l_s that we arrived at earlier. Note, however, that during the initial solidification stages the thickness of neither the steel sheet nor the drum is likely to strongly affect the solidification process; the thermal characteristics of copper and steel and the input temperatures are the relevant parameters. Eventually, of course, heat will penetrate the sheets and once this happens h_s and h_c will affect the process. Given that our interest lies in determining the length of the metal puddle it seems sensible to focus our attention on the early stages and choose length scales accordingly. At the very least this will provide information on how the thermal parameters (in isolation) influence solidification.

Aside: One of the interesting features of modelling is that the solution procedure is strongly dependent on the question being asked. In the above problem different choices of scales lead to different approximations, and ultimately lead to different solutions that are relevant for different circumstances. For example, if our interest were in the long term cooling of the steel sheet, h_s or h_c would be suitable transverse scales to employ depending on circumstances, and "Fourier-like" techniques would be more appropriate than the "series-like" techniques we'll find useful here.

11.2.4 The similarity solution

In line with the above observations we consider the situation in which both the steel and copper regions extend to infinity and their associated boundary conditions (at $z = h_s, -h_c$) disappear from the problem.

Given that the thicknesses of the sheet and drum don't play a role in the initial solidification process what length scale is appropriate? We're on familiar ground here. The only length scales in the problem are provided by the variables z and x and dimensionality arguments tell us that there exists a similarity solution $\mathcal{T}(\eta)$ dependent on the dimensionless combination

$$\eta = z/d_s(x) \quad \text{where} \quad d_s(x) = \sqrt{\frac{\kappa_s x}{U}}. \tag{11.12}$$

We can in fact think of $(x, d_s(x))$ as being the appropriate longitudinal and transverse length scales associated with the location x.

Aside: The recognition of the similarity character of the solution transforms the problem from an analytically intractable, numerically difficult problem to a simple problem for which exact results are available. As mentioned in earlier work, similarity techniques are very useful for handling nonlinear problems and so modellers are always very aware of potential applications.

The fact that the explicit dependence on x must be removable from the formal problem also implies that the location of the solidification front, once expressed in terms of η, can't involve the variable x, and thus must be expressible in the form $\eta_m = A$ with A constant. Thus the shape of the solidification front must be expressible in the form

$$Z(x) = Ad_s(x) = A\sqrt{\frac{\kappa_s x}{U}}. \tag{11.13}$$

Additionally the surface temperature of the drum $\mathcal{T}_d(x)$ cannot be dependent on x, because there's no other length scale available to produce a dimensionally sensible expression for $\mathcal{T}_d(x)$. The implication is the somewhat surprising result that the drum temperature must remain constant during this stage of the solidification process i.e.

$$\mathcal{T}(\eta = 0) = \mathcal{T}_d, \text{ a constant.} \tag{11.14}$$

Aside: Of course for sufficiently large x the influence of the sheet thickness will be felt and the similarity arguments will fail, so that the drum temperature will no longer be constant, see Section 11.2.5.

In terms of the similarity variable the field equations for the scaled temperature $\mathcal{T}$ defined (as before) by

$$T = T_c + [T_s - T_c]\mathcal{T}, \tag{11.15}$$

reduce to

$$\mathcal{T}'' = -\frac{1}{2}\eta\mathcal{T}' \quad \text{for} \quad \eta > 0$$

in both the solid and liquid steel zones, and

$$\mathcal{T}'' = -\frac{1}{2}\lambda\eta\mathcal{T}' \quad \text{for } \eta < 0,$$

$$\text{where } \lambda = \kappa_s/\kappa_c, \tag{11.16}$$

in the copper zone.

The (entry) boundary conditions (11.8,11.9) require

$$\mathcal{T} \to 1, \text{ as } \eta \to \infty, \text{ and}$$

$$\mathcal{T} \to 0, \text{ as } \eta \to -\infty.$$

The temperature continuity conditions and (11.7) require

$$\mathcal{T}(\eta) = \begin{cases} \mathcal{T}_d & \text{at } \eta = 0, \text{ the drum surface} \\ \mathcal{T}_m & \text{at } \eta = A, \text{ the solidification front,} \end{cases}$$

where

$$\mathcal{T}_m = \frac{T_m - T_c}{T_s - T_c}, \tag{11.17}$$

see (11.15).

In all regions the solution consists of linear combinations of an erf function solution and the constant solution.

Aside: Note that the erf and erfc functions of positive and negative arguments differ only by a multiplicative factor or a constant, see (7.1) Chapter 7, so that there's some flexibility available in representing solutions. In general terms it doesn't really matter which of the options (erf or erfc) are used, however, the clutter is greatly reduced if an appropriate choice is made. Thus the erfc function is favoured in semi-infinite regions in which the solution is required to vanish at ∞, and the erf function is often used when matching conditions at a given finite location.

The solutions are given by:

- In the molten zone $\eta > A$:

$$\mathcal{T} = 1 - (1 - \mathcal{T}_m)\text{erfc}(\eta/2)/\text{erfc}(A/2). \tag{11.18}$$

- In the solid steel zone zone $A > \eta > 0$:

$$\mathcal{T} = \mathcal{T}_d + (\mathcal{T}_m - \mathcal{T}_d)\text{erf}(\eta/2)/\text{erf}(A/2). \tag{11.19}$$

- In the copper zone $\eta < 0$

$$\mathcal{T} = \mathcal{T}_d[1 + \text{erf}(\sqrt{\lambda}\eta/2)] \equiv \mathcal{T}_d\text{erfc}(-\sqrt{\lambda}\eta/2). \tag{11.20}$$

There are thus two unknown constants $\mathcal{T}_d$ and A which need to be determined so that the remaining conditions (the flux conditions) are satisfied. In similarity variable terms, the flux condition at the drum surface (11.10, 11.11) becomes

$$\mathcal{T}_\eta(0+) = \mu\mathcal{T}_\eta(0-) \tag{11.21}$$

where

$$\mu = k_c/k_s, \tag{11.22}$$

and the flux condition at the solidification front becomes

$$A = 2\nu[\mathcal{T}_\eta(A-) - \mathcal{T}_\eta(A+)], \tag{11.23}$$

where, (using (11.13) and $\kappa_s = k_s/(\rho_s c_s)$),

$$\nu = c_s[T_s - T_c]/L. \tag{11.24}$$

After substituting in the solution forms, the first of these conditions gives (after some manipulation)

$$\mathcal{T}_d = \mathcal{T}_m/[1 + \mu\sqrt{\lambda}\,\mathrm{erf}(A/2)], \tag{11.25}$$

and the second gives

$$A = \nu\left[\frac{\mathcal{T}_m - \mathcal{T}_d}{\mathrm{erf}(A/2)} - \frac{1 - \mathcal{T}_m}{\mathrm{erfc}(A/2)}\right]\mathrm{erf}'(A/2).$$

After eliminating $\mathcal{T}_d$ in favour of A we obtain the equation

$$A = \nu\left[\frac{\gamma\mathcal{T}_m}{1 + \gamma\mathrm{erf}(A/2)} - \frac{1 - \mathcal{T}_m}{\mathrm{erfc}(A/2)}\right]\mathrm{erf}'(A/2), \tag{11.26}$$

where

$$\gamma = \mu\sqrt{\lambda}. \tag{11.27}$$

It's interesting to note that the thermal material properties only influence the location of the solidification front and the drum temperature through the combination of μ and λ given by $\gamma = \mu\sqrt{\lambda}$. This result is a consequence of mathematical structure and cannot be obtained by any simple physical argument. It's the same combination that determined the surface temperature when two bodies were brought into contact as in Exercise 7.5 Chapter 7, so (if one had been extremely observant) the result might have been anticipated.

The equation for A needs to be solved numerically for the range of values of the dimensionless groups:

- $\mathcal{T}_m$, see (11.17) which measures how "close" the input temperature is to the solidification temperature.
- γ, see (11.16,11.22,11.27), which measures the thermal effectiveness of the particular drum material for removing heat from the steel.
- ν, see (11.24) which measures the relative proportion of sensible heat to latent heat that needs to be removed from the molten steel.

Once A has been determined the location of the solidification front follows from equation (11.13). Thus according to our model the solidification front reaches the steel surface at the location given by $Z(x) = h_s$ so that the puddle length l_s is given by

$$x = l_s = \frac{U h_s^2}{A^2 \kappa_s}. \tag{11.28}$$

This represents a first estimate for the puddle length. Of course without an error check one should be cautious about the result, but nevertheless we'll proceed to examine the implications.

The results

The data provided by the steel maker were:

Input Data	
T_c=150°C	T_m= 1400°C
k_c = 400watts/m/°C	k_s = 20watts/m/°C
$\kappa_c = 10^{-4}$m²/s	$\kappa_s = 5 \times 10^{-6}$m²/s
c_c=400joules/kg/°C	c_s = 600joules/kg/°C
$\rho_c = 8.9 \times 10^3$kg/m³	$\rho_s = 7.6 \times 10^3$kg/m³
$L = 2.7 \times 10^5$joules/kg	T_s=1400°C to 1460°C
Scaled Values	
$\lambda = 0.04$	$\nu = 2.7$
$\gamma = 4$	T_m = 0.95 to 1

The following instructions to Maple define the procedure that generates the solution for A:

```
Afn:=proc(nu,gamma,Tm)                   % the right hand side of
equation (11.26)
local c;                                                   %c=A/2
   1/2*nu*(Tm*gamma/(1+gamma*erf(c))-(1-Tm)/
erfc(c))*diff(erf(c),c);
   fsolve("-c,c);
   2*";                                            % we recover A
end;
```

The solution can be obtained by entering specified values of the parameters as arguments for the **Afn** procedure. Using this we can obtain information about the effects of the various parameters on the solution. Of the parameters, γ is fixed and ν can be varied little (and with little effect), so that (for fixed thickness plate, and production rate) only the temperature of input of the molten steel can be altered to improve the operation of the envisaged arrangement. The predicted effect of an input temperature range from 60°C above the melting point (with $\mathcal{T}_m = 0.95$), to 0°C above the melting point ($\mathcal{T}_m = 1$) on the puddle length is shown in Table 11.1.

$\mathcal{T}_m$	1	0.995	0.99	0.98	0.97	0.95
A	1.61	1.59	1.57	1.55	1.52	1.47
l_s	7.74m	7.89m	8.11m	8.31m	8.6m	9.25m

Table 11.1: Solution dependence on $\mathcal{T}_m$.

It's clear from this table that the puddle length can be reduced considerably if the molten steel can be placed on the drum at a temperature close to solidification. However molten steel has to be brought to the drum and this will be difficult if the temperature is too close to solidification. Achieving a figure for $\mathcal{T}_m$ as high as 0.995 would mean pouring the molten steel at a temperature about 6°C above the solidification temperature; a potentially risky situation. If it does solidify before being poured it's costly to repair the damage. The steel maker suggested that a temperature level of about 20°C above solidification, corresponding to $\mathcal{T}_m = 0.98$ might be feasible, and later calculations will be based on this value.

A number of additional observations can be made:

- The $l_s = 7.74$ m result obtained under close to solidification entry conditions is large in engineering terms (requiring a drum size that's much too large), so it's sensible to see how this might be reduced. It's clear from (11.28) that $l_s \propto U$ so that, at the expense of production rate, a smaller apparatus could be used.

- Note that $l_s \propto h_s^2$ so that halving the sheet thickness would reduce l_s by a factor of 4—very significant! It is in fact possible to achieve an effective *halving* of the thickness by simply using two rollers as in Fig. 11.3. There are other advantages to this double drum arrangement. The molten steel surfaces would immediately freeze and this would reduce the possibility of fluid instabilities disturbing the sheet surface. Frozen-in stresses in the final product are also less likely for such an arrangement.

- Copper is expensive so what about a steel drum (assuming it won't melt)? For the $\mathcal{T}_m = 0.98$ case, if a steel drum is used ($\gamma = 1$), the result obtained is $A = 1.19, l_s = 14.1$ m—very much larger than the 8.3 metres obtained for a copper drum.

The similarity solution is in error because it ignores the fact that both the drum and the steel sheet are of finite thickness. Because the steel is introduced at temperatures close to solidification and latent heat is expelled as the front advances, one might expect the temperature variations in the molten steel zone to be relatively minute. An examination of the solution

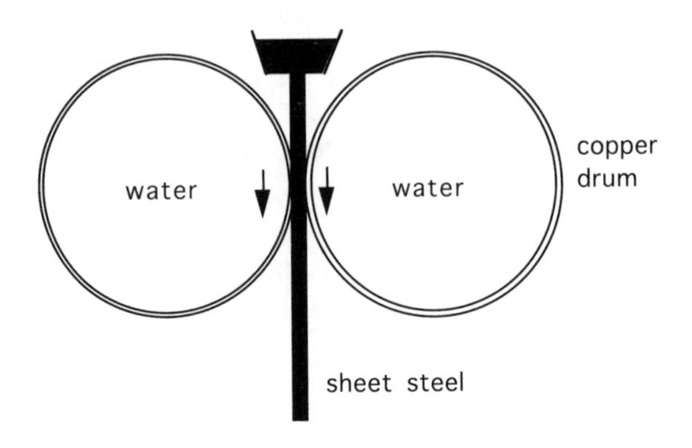

Figure 11.3: An Alternative caster.

(11.18) in this zone reveals this to be the case. Under such circumstances one might expect that errors arising as a result of treating the steel zone as infinite to be negligible. The copper drum is another matter. It's clear that economic efficiency demands that the drum be *just* thick enough to cope with the released heat, so that drum thickness effects need investigation. Ideally, from the solidification point of view, the copper drum should be sufficiently thick to soak up the heat released from the steel up to the stage at which it completely solidifies—after all water is not as efficient as copper at removing heat. A plot of the temperature profile with depth at the position corresponding to $x = l_s$ (based on the infinite model) should provide a reasonable estimate of how thick the copper should be. Such a plot is shown in Fig. 11.4 for the case in which the steel entry temperature is 20°C above freezing. The information is presented in °C and centimetres to aid visualization. The drum surface temperature, calculated using equation 11.25, is found to be $T_d = 462$°C in this case. Observe that the heat penetrates to a much greater depth in the copper than in the steel as a result of its greater conductivity. More explicitly the penetration depth will be about ($1/\sqrt{\lambda} = 5$ times the depth in steel) as can be seen from equation (11.20), which suggests that the drum has to be about 5 times the steel thickness or greater (i.e. about 5 cm) to "accept" all the heat from the steel. This is born out in Fig. 11.4. It's also evident from this figure that a 10 cm drum would be excessive and that a 2 cm drum is likely to be inadequate and thus lead to a significantly change in the puddle length. Before a rational decision can be made, however, an improved approximation is necessary. We'll turn our attention to this in the next section.

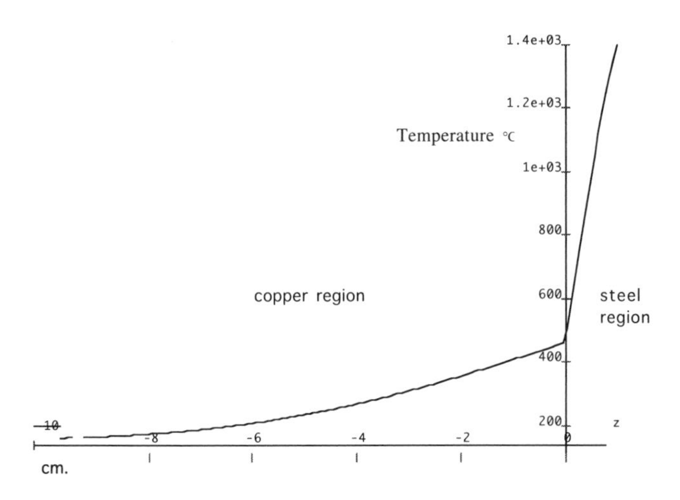

Figure 11.4: Temperature distribution through the steel and drum immediately after solidification.

11.2.5 ⋆ Drum thickness effects

The above work has given the result, see (11.12,11.19,11.20)

$$\mathcal{T}^{(0)} = \begin{cases} \mathcal{T}_d + (\mathcal{T}_m - \mathcal{T}_d)[\mathrm{erf}(\frac{z}{2d_s(x)})/\mathrm{erf}(A/2)] & \text{for} \quad z > 0 \\ \mathcal{T}_d \mathrm{erfc}(-\frac{\sqrt{\lambda}z}{2d_s(x)}) & \text{for} \quad -h_c < z < 0, \end{cases} \tag{11.29}$$

$$\text{where} \quad d_s = \sqrt{\frac{\kappa_s x}{U}},$$

in the solid steel and copper zones respectively; where we now think of $\mathcal{T}^{(0)}$ as being the primary term in an approximation scheme, and where the explicit (x, z) dependence of the solution is presented.

The solution in the copper zone is in error because the assumed zero flux condition at the internal surface of the drum $z = -h_c$ isn't satisfied. The introduction of an image solution $\mathcal{T}_d\mathrm{erfc}(\sqrt{\lambda}[z + 2h_c]/2d_s)$ symmetrically placed relative to this drum surface will cancel out the flux due to the primary solution. (Note that the defining equation (11.1) is invariant under a change in the origin for z.) This corrective term however results in an increased temperature and flux at the external drum surface $z = 0$, so that additional corrections need to be introduced. Now if the material in $z > 0$ were copper, then a second image solution in the $z > 0$ region centred on $z = +2h_c$ would compensate for the changes in both temperature and flux. This isn't the case, and in order to match both temperature and flux changes introduced by the image solution, two additional corrective terms are needed; one in each zone. Explicitly the correction (in total)

$$\mathcal{T}^{(1)} = \begin{cases} M\,\text{erfc}([z+2\sqrt{\lambda}h_c]/2d_s) & \text{for } z>0 \\ \mathcal{T}_d\,\text{erfc}(\sqrt{\lambda}[z+2h_c]/2d_s) + N\,\text{erfc}(\sqrt{\lambda}[2h_c - z]/2d_s) & \text{for } z<0, \end{cases}$$

is required, where N and M are constants to be chosen to ensure temperature and flux continuity at $z = 0$. To see the rationale behind this choice note that the corrections include the image solution and are of the right form to:

- satisfy the required field equations,
- cancel out variations of the (image solution) form $\text{erfc}(\sqrt{\lambda}h_c/d_s)$ along $z = 0$,
- ensure the boundary condition at $z = +\infty$ is unaffected.

Note that the correction involves introducing an image at $z = -2\sqrt{\lambda}h_c$ for the steel zone and images at $z = \pm 2h_c$ for the copper zone. The temperature continuity requirement across the surface $z = 0$ gives

$$M = \mathcal{T}_d + N,$$

and the flux continuity condition (11.10) gives

$$M = \gamma[\mathcal{T}_d - N],$$

see (11.22,11.27). These equations give

$$M = \mathcal{T}_d 2\gamma/(\gamma+1) \quad \text{and} \quad N = -\mathcal{T}_d(\gamma-1)/(\gamma+1).$$

With the above corrections included, the solution in the copper and solid steel regions is given by $\mathcal{T}^{(0)} + \mathcal{T}^{(1)}$, where $\mathcal{T}^{(1)}$ is given by:

$$\begin{cases} \mathcal{T}_d\{\frac{2\gamma}{1+\gamma}\text{erfc}([z+2\sqrt{\lambda}h_c]/2d_s)\} & \text{for } z>0, \\ \mathcal{T}_d\{\text{erfc}(\sqrt{\lambda}[z+2h_c]/2d_s) - \frac{\gamma-1}{\gamma+1}\,\text{erfc}(\sqrt{\lambda}[2h_c - z]/2d_s)\} & \text{for } z<0. \end{cases} \tag{11.30}$$

The procedure for satisfying continuity conditions at $z = 0$ has, however, led to the introduction of a term (the N term) with unbalanced flux at $z = -h_c$. We've seen this pattern before when dealing with image solutions; further images need to be introduced. The thing to notice is that the introduction of $\mathcal{T}^{(1)}$ has lead to an improvement; the term producing an unwanted heat flux at $-h_c$ has been reduced by the factor

$$-[\frac{\gamma-1}{1+\gamma}]\,\frac{\text{erfc}(\sqrt{\lambda}3h_c/2d_s)}{\text{erfc}(\sqrt{\lambda}h_c/2d_s)},$$

which is certainly less in modulus than one, but more importantly (because the erfc function decays rapidly as a function of its argument for moderate values of the argument) will be small for moderate values of $3\sqrt{\lambda}h_c/2d_s$. By repeating the above correction process the discrepancy at $-h_c$ can be reduced to a term of order $\text{erfc}(n\sqrt{\lambda}h_c/2d_s)/\text{erfc}(\sqrt{\lambda}d_c/2d_s)$ with $n = 5, 7\ldots$, so that *providing* d_s *is not large* at most a few corrections will be necessary to produce an extremely accurate result. Here we'll simply examine the effect of the first correction.

The Results

You'll recall that the similarity solution evaluated to a constant $\mathcal{T}_d$ along the external drum face $z = 0$. The correction term $\mathcal{T}^{(1)}(x, 0)$ is not constant along this face and the resulting temperature variation provides a good indicator for the influence of the drum thickness on the heat transfer process. Explicitly the temperature variation, see (11.29, 11.30), is given by

$$\mathcal{T}(x, 0) = \mathcal{T}_d \left[1 + \frac{2\gamma}{1+\gamma}\text{erfc}(\sqrt{\lambda}h_c/d_s)\right].$$

As x increases the thermal zone thickness $d_s = \sqrt{\kappa_s x/U}$ increases and the erfc term grows in size, and thus the influence of the drum becomes more marked. Our concern, however, is with the pre-solidification range $x < l_s$ so the image solution should be satisfactory. We're particularly interested in knowing about the influence of drum thickness on conditions at $x = l_s$. Now at $x = l_s$, $d_s = h_s/A$, see equation (11.13), so that the proportional temperature change at the external drum face produced by the correction term is given by

$$\mathcal{T}^{(1)}/\mathcal{T}_d = \frac{2\gamma}{1+\gamma}\text{erfc}(\sqrt{\lambda}Ah_c/h_s),$$

which is plotted in Fig. 11.5 as a function of copper drum thickness. Also plotted in the same figure is the proportional temperature change induced by drum thickness at the edge of the solidification zone (given by $[2\gamma/(1+\gamma)]\text{erfc}(A/2 + \sqrt{\lambda}Ah_c/h_s))$, again evaluated at l_s for a range of drum thicknesses.

You might recall that earlier, based on the results displayed in Fig.11.4, we expected that a drum thickness of about 5 cm might be necessary to soak up the required heat; however, we can see from our present figure that significant drum thickness effects on the external drum surface will occur only if $h_c < 2$ cm, and the effect at the actual solidification front will even be relatively smaller. *Thus a substantially thinner drum than expected will be effective for solidification purposes.* The rapid changes with h_c exhibited in this figure arise again because the erfc function changes rapidly with its argument for moderate arguments. The implication is:

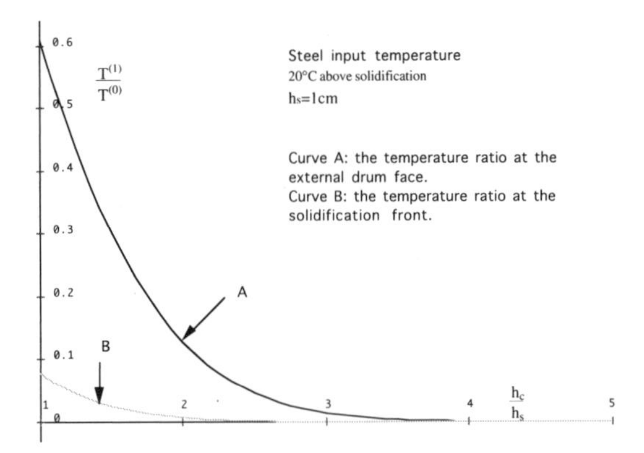

Figure 11.5: The effects of drum thickness.

- a drum thickness of between 1 and 2 centimetres is sufficient for solidification purposes.
- the results obtained earlier for the puddle length (8 m) are accurate for drums of this thickness.

Aside: As indicated the above approximation scheme, based on the primary similarity solution with introduced images, works for drums of thickness greater than about 1cm, with more image terms being necessary for thinner drums. The experience gained when dealing with the surface heating of a finite slab in Section 6.5 Chapter 7 suggests that the calculation scheme would not, however, be sensible for very thin drums. It would be better to perturb about a steady state using a Fourier-like approach for such drums.

Aside: The sources introduced to compensate for finite drum effects will cause heating throughout the steel zone and in particular at the solidification front, so that the location of the front will be changed by the presence of these terms. Images can again be introduced and a perturbation scheme can be used to correctly position the front. Given the smallness of the $\mathcal{T}^{(1)}$ at the solidification front such calculations are unwarranted, especially for a preliminary study.

11.3 The revolving drum

The design under consideration called for the interior surface of the drum to be cooled by recirculating water. We'll assume the cooling system is such that the heat transfer rate per unit area of the drum surface is given by βT where T is the temperature difference between the local interior surface temperature of the drum and the temperature of the recirculating water

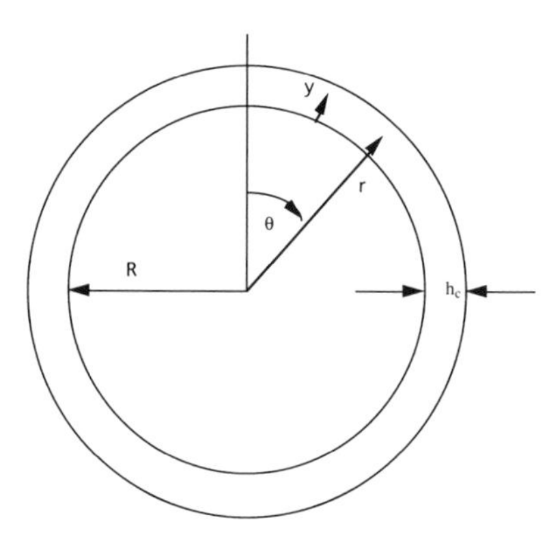

Figure 11.6: Drum geometry.

(assumed fixed), and β is a known parameter for the design configuration. A large β corresponds to an efficient but expensive cooling system. After the system has been in operation for some time one would expect a steady state situation to be set up. We'll confine our attention to this steady state situation. Relative to cylindrical polar co-ordinates, (r, θ) fixed in space and based on the centre of the drum, see Fig. 11.6, the steady state heat equation becomes

$$\Omega T_\theta = \kappa_c[T_{rr} + (1/r)T_r + (1/r^2)T_{\theta\theta}], \tag{11.31}$$

where Ω is the angular speed of rotation of the drum, and $T(r, \theta)$ refers to the drum temperature relative to that of the water. To establish this result one can proceed as in Section 11.2.1. If R is the inner radius of the drum and $R + h_c$ the outer radius, then clearly $h_c/R \ll 1$ for configurations of interest and approximations based on the smallness of this parameter are appropriate. Thus by introducing a scaled radial ordinate y based on the drum thickness, so that

$$r = R + h_c y,$$

and ignoring terms of order h_c/R and smaller, the equation reduces to the familiar one-dimensional heat equation

$$T_\theta = \mathcal{P} T_{yy} \tag{11.32}$$

with θ playing a "time-like" role, and where the dimensionless group

$$\mathcal{P} = \kappa_c/(h_c^2 \Omega) \tag{11.33}$$

provides a measure for the ratio of the time span for heat to penetrate the drum to the time for one rotation of the wheel. The sensible operating range is the one in which $\mathcal{P}$ is of unit order. (For if $\mathcal{P}$ is small the drum will heat up significantly during operation, whereas if it's large the drum size is unnecessarily large.)

As indicated in the introduction, the amount of heat lost to the water while the steel is still in contact with the wheel is relatively small. It would seem then that the appropriate "initial" condition to impose for the present problem is

$$T(y,0) = f(y), \tag{11.34}$$

where f is the temperature rise in the copper immediately after the drum loses contact with the steel, given by $\mathcal{T}^{(0)} + \mathcal{T}^{(1)}$ (see equations (11.29, 11.30)) with $x = l_s$, after variable adjustments are made. Notice that we've applied the boundary condition at $\theta = 0$ rather than at the small angle where the contact is broken.

In addition to the above we need to ensure that no heat is lost from the external surface of the drum after contact is lost, and that the heat lost to the cooling system is as prescribed earlier, so that

$$T_y(1,\theta) = 0, \quad \text{and} \quad T_y(0,\theta) = \mathcal{B}T, \tag{11.35}$$

where the parameter

$$\mathcal{B} = h_c\beta/k_c$$

is the dimensionless ratio of the thickness of the drum to the thermal thickness k_c/β associated with water cooling.

The above boundary conditions and "initial" condition (at $\theta = 0$) are typical for heat conduction problems with t replaced by the "time-like" variable θ. Experience suggests (and uniqueness theorems confirm) that with these conditions there will be a unique solution for our equation for *all* $\theta > 0$. There is, however, one unusual feature of the problem that needs further thought. In our problem the time-like variable θ is limited to the range $[0, 2\pi)$, and since there's no way that the solution as $\theta \to 2\pi$ will match onto the solution at $\theta = 0$, there is a difficulty. Clearly we need to think more carefully about the physics and mathematics of the situation.

Let's focus our attention on a particular small segment of the drum. Each time this segment makes contact with the steel an additional pulse of heat is given to its top surface, and this heat subsequently disperses into the drum. The eventual temperature reached is due to the cumulative effect of all such heat inputs and dispersals. Fortunately the system is linear so the effects are simply additive. Thus to determine the eventual temperature at any location we need first to find the temperature distribution due to single pulse at $t \equiv \theta = 0$, *treating the θ domain as being infinite* for this calculation, and then add up the effects due to an infinite number of such periodically applied pulses to determine the eventual steady state of interest. Given this new perspective it's as well to reflect back on our previous discussion. It's clear for a start that our supposition that the steady state solution should satisfy $T(0, y) = f(y)$, where $f(y)$ is the temperature distribution due to a single pass by the steel input point is incorrect; the steady temperature profile at $\theta = 0$ will be due to the cumulative input of all such passes, and this can only be determined by summing the effects of such passes. The steady solution as $\theta \to 2\pi-$ will also be determined by summation and there will be a discontinuity across $\theta = 2\pi-, 0+$, that has to be accepted as a consequence of the approximations used in our analysis. More explicitly:

- we've assumed all the heat input occurs *at* $\theta = 0$; in reality it occurs over a small θ range. This is a familiar point source type approximation so one might expect a temperature discontinuity,
- By removing $T_{\theta\theta}$ from the heat equation we've removed the possibility of diffusion in the θ direction. Diffusion in this direction would smooth out any temperature discontinuities in this direction so that the apparently small $T_{\theta\theta}$ term must be important close to $\theta = 0, 2\pi$.

Comment: Steady State Approximations: Notice that in order to resolve the above issues we needed to consider the way in which the steady state solution is set up. The steady state situation represents an idealization introduced for mathematical convenience. Often no difficulty is introduced by such a simplification but sometimes it's necessary to examine the transients to resolve contradictions that may arise.

Aside: There's a thin transition zone of thickness $\delta\theta = \kappa_c / R^2\Omega$ joining up the $\theta = 2\pi-$ to the $\theta = 0+$ regions. Such ***boundary layers*** play an important role in many areas of science. The analysis of this zone is somewhat peripheral to the questions of interest here and so will not be undertaken.

The single pass solution

Solutions satisfying the heat equation and the homogeneous boundary condition at $y = 1$ are given by

$$T(\theta, y) = E^{-\mathcal{P} p_n^2 \theta} \cos p_n(1 - y), \tag{11.36}$$

and the boundary condition at $y = 0$ will be satisfied provided

$$p_n \sin p_n = \mathcal{B} \cos p_n,$$

which generates an infinite set of eigenvalues and corresponding eigensolutions. If $f(y)$ denotes the temperature distribution resulting from one pass through the contact zone with the molten steel then the resulting temperature distribution is given by

$$T(\theta, y) = \sum_1^\infty f_n E^{-\mathcal{P} p_n^2 \theta} \cos p_n(1 - y),$$

where

$$f_n = \int_0^1 f(y) \cos p_n(1-y)\, dy \Big/ \int_0^1 \cos^2 p_n(1-y)\, dy.$$

The Steady State

The contribution from all the previous passes will be obtained by replacing θ by $\theta + 2m\pi$ and summing for m running from 1 to ∞. Because all the y variations have the same form we obtain a geometric progression for each Fourier component, so the summation can be performed exactly giving

$$T(\theta, y) = \sum_1^{\infty} f_n \frac{e^{-\mathcal{P} p_n^2 \theta}}{1 - e^{-2\pi \mathcal{P} p_n^2}} \cos p_n(1-y),$$

the required steady state solution.

Our interest is in the temperature of the surface of the copper drum as it comes into contact with the molten steel, which is obtained by substituting $(y, \theta) = (1, 2\pi)$ in the above to give

$$T(1, 2\pi) \equiv T_c - T_0 = \sum_1^{\infty} f_n \frac{e^{-2\mathcal{P} p_n^2 \pi}}{1 - e^{-2\pi \mathcal{P} p_n^2}}$$

for the temperature difference between the external drum surface and the water (at T_0). Given the separation of the eigenvalues, experience suggests (and calculations confirm) that only the first term of the series is numerically significant so that

$$T_c - T_0 \approx f_1 \frac{e^{-2\pi \mathcal{P} p_1^2}}{1 - e^{-2\pi \mathcal{P} p_1^2}} \cos p_1(1-y). \qquad (11.37)$$

Interpretation

The temperature decays exponentially in θ from an effective level given by $f_1 \cos p_1(1-y)$ just past the pouring point $\theta = 0$ to the value given by (11.37) just before the pouring point. Thus the drum will be returned to the temperature of the water after each cycle if parameters are chosen so that the argument of the exponential is greater in size than about 4, requiring

$$2\pi \kappa_c p_1^2 / (h_c^2 \Omega) \approx 4. \qquad (11.38)$$

The first thing to do is to check to see if this produces sensible numbers—it's *very* easy to make "obvious" errors when working in unfamiliar areas and end up with embarrassing large or small numbers, for example for the drum size in this case. It's comforting to note that, given the large range in size of the actual numbers to be substituted into the formula we've developed, that it's unlikely the numbers will end up the correct size *unless* the model makes *some* sense. Engineers suggest that a heat removal rate of up to $\beta = 10\text{kW/m}^2/\text{s}$ is achievable by water recirculation cooling systems. Using this as a representative value and using the parameter values tabulated we get $\mathcal{B} = 0.5$ for a drum of thickness of $h_c = 2\text{cm}$. The first eigenvalue of (11.36) is then computed to be $p_1 = 0.65$, and the required angular frequency, using (11.38), evaluates to $\Omega = 0.16$ radians per second. Thus, assuming a drum speed of 1 metres per second, the required water cooling rate will be realized for a wheel of radius $1/\Omega = 6$ metres or greater. This fits what experienced engineers expect from cooling systems, which is reassuring.

The drum size predicted is pretty large; clearly this indicates that the drum cooling aspect of the problem is not a minor consideration. The above crude calculations assume that the drum should be large enough to enable cooling of the drum to water temperature over the cycle. Obviously it's not necessary to cool to this extent providing one's willing to work with higher input temperatures T_c and accept a less efficient solidification regime. The above model of course quantifies this situation. The savings in drum size and thus in expense achieved in this way may be worthwhile.

11.4 Matching

After appropriate variable modifications the equations (11.29,11.30), and (11.37) express $T_c - T_0$ in terms of the other relevant physical parameters, and provide the connecting link between the solidification and water cooling problems. These equations need to be solved simultaneously. There are many possible arrangements that will achieve solidification and the best is the cheapest in an appropriate sense. The values of h_c, β, and drum radius R, are available for adjustment, and there may be other adjustments possible. There will be strength requirements limiting the range of h_c, and also that parameter must be large enough to ensure that the copper provides sufficient cooling, but not so much that the capital cost becomes excessive. The running costs will be reduced by having the water cooling parameter β small and so we have an optimization problem which can be tackled only when all the costings are available. It's also likely that many other practical considerations not so easily quantified will be of importance. Under such circumstances the optimization would almost certainly have to be tackled by tabulating the relevant options; a very frequent situation in practical problems.

11.5 Summary and Conclusions

The investigation has been basically an exploratory one with the basic issue being: what drum size (radius and thickness) is required to continuously cast steel sheets with the envisaged arrangement?

The length scales and parameter ranges varied over orders of magnitude, necessitating a careful analysis of the comparative importance of various terms. The circumstances were such that the solidification and drum cooling problems could be first solved independently, and then later the solutions matched. This simplified the analysis considerably. Similarity arguments were used to solve analytically the potentially difficult solidification problem.

The conclusions were:

- Based on estimates of the puddle length it would appear that for the envisaged arrangement a drum of about 8 metres radius would be required to solidify a 1 centimetre steel sheet being cast at 1 metre per second.
- The drum thickness of about 2 centimetres would be adequate to soak up the heat from the solidifying steel.
- In order to water cool the drum a radius of somewhat less than 6 metres would be required.
- The drum size required to solidify the steel varies like Uh_s^2 so that in particular the drum dimensions can be *substantially reduced* if thinner sheets are cast, and less substantial gains can be made if smaller casting speeds are used.
- A double drum arrangement would *quarter* the drum radius required to solidify a sheet compared with the arrangement suggested. Also, drum cooling could be more easily accomplished. Additionally, flow instabilities would be less of a problem for such an arrangement.
- Both the drum cooling and solidification problems lead to comparable drum dimensions which suggests that an optimal balancing of competing features could achieve substantial savings.

For further details see Barton and Gray (1985).

11.6 Appendix: an alternative cooling system

In the main text, see Section 11.1, we raised the question of removing heat from the steel by "direct" water contact with the steel (either by spraying or using a thin copper drum). Here we make crude comparative estimates.

If the copper drum system is used then the interface drops to the temperature

$$T_i = [T_s - T_c]/[1 + \gamma]$$

where T_s and T_c are the initial temperature of the molten steel and the copper respectively and γ is the dimensionless parameter dependent on the relative thermal properties of the materials, see Chapter 7 Exercise 7.5. For these materials the value of γ is close to 4 so that there is a substantial immediate reduction in the surface temperature of the steel well below the melting point, and so it begins to solidify immediately. This surface temperature drop is maintained until the finite thickness of the copper and steel layers is felt. From the solution obtained in the same exercise the temperature drop in the steel is $[T_s - T_i]\text{erfc}(x/\sqrt{4\kappa_s t})$. Thus, reducing the steel temperature at a depth h down to the melting point T_m will take a time implicitly defined by

$$\frac{T_s - T_m}{T_s - T_c} = \text{erfc}(h/[4\kappa_s t]).$$

Preliminary estimates supplied by the steel maker suggested a value of about 0.05 for the left hand side which gives the time needed to be about $0.72h^2/4\kappa_s$, i.e. 3.25 seconds.

When the cooling is achieved by the direct transfer to water then the surface temperature is as obtained in Section 7.6 using an integral equation. We can use that solution to find out how long it will take to bring the cooled surface of the steel down to the temperature achieved by the copper instantaneously. Until this temperature is reached (at the surface of the steel) the rate at which the steel is being cooled by water will be less than that being achieved by having the copper present. Since that interface temperature for copper represents 80% of the total temperature difference, we can gain a reasonable estimate of the time needed to reduce to this temperature from the asymptotic solution of the integral equation. Thus the scaled time τ is given by

$$1 - \sqrt{\tau}/\pi = 0.8,$$

which corresponds to an actual time of $0.04\pi^3 k_s \rho_s c_s/\gamma^2$, where the heat transfer parameter γ depends on the efficiency of the water cooling system. The efficiency is related to the speed at which fluid can be brought into significant thermal contact with the surface. The power required to overcome the resistance to the flow is proportional to the cube of the speed of the water and hence to the cube of the value of γ. Thus the running cost of the cooling system is proportional to γ^3; efficient cooling systems are expensive! Further there are practical limitations on the speed of water flow because if the speed is increased too much, thermal contact with the surface will be lost. At present upper limits for γ are in the range of 10 to

$20\text{Kw/m}^2/\text{s}/^\circ\text{C}$ in the very best of circumstances. For $\gamma = 10\text{Kw/m}^2/\text{s}/^\circ\text{C}$ the time required to reach the surface temperature reached by copper instantaneously is about 1 second. During this time the water would have extracted a total heat energy content of $10\text{Kw/m}^2/^\circ\text{C}$, while the copper would have extracted $20\text{Kw/m}^2/^\circ\text{C}$, and so the amount of heat penetration into the steel would be twice as great. After this period of time the water system would be more efficient. The situation looks marginal. Obviously, however, for this or other reasons the steel makers have felt that the expensive drum with cheaper water cooling system is preferable to the direct water cooling system.

Bibliography

Crank, John (1984). *Free and Moving Boundary Problems*, Oxford at the Clarendon Press.

Barton, N.G. and Gray, J.D. (1985). *Proceedings of the Mathematics-in-Industry Study Group*, CSIRO Division of Mathematics and Statistics

Part III

Vibrations and Waves

Cars, Trains, etc.

Introduction

In the Pilbarra region of Australia extremely long trains (200-300 trucks driven by up to 4 engines) are used to carry very heavy loads of iron ore over very long distances. The economics of the operation dictate that there can only be a single track, with many sidings to permit trains to pass. There are serious delays to operations if the couplings between trucks fail but it is expensive to keep replacing the couplings before failure. Despite the fact that heavier than normal couplings are used, the incipient failure rate is too high and it is desired to reduce it. This may be achieved by either changing the way the trains are driven or varying the design of the couplings. For either method to be effective it's necessary to understand the factors controlling the loadings occurring at the couplings so that strategies for reducing them may be developed. This is just one of many problems that have arisen recently concerning wave/vibration phenomena associated with trains. The problem described above is little understood and of great importance and interest; unfortunately the background required to get to grips with this challenging problem can't be attempted here, but the issues discussed in the next few chapters bear on this problem and so it serves to introduce this material.

One has only to have heard a train being shunted to be aware that when the motion of a train is varied there develops a progression of reactions at couplings which travels along the train, changing the motion of each truck. The noise associated with shunting arises from impact loadings at each coupling. The fact that the problem gets worse with increasing train length means that, despite the introduction of impact absorbing mechanisms at each coupling, there is something which causes a growth of the coupling loads as the progression of impacts proceeds. The couplings can be modelled as (highly nonlinear and damped) springs. The couplings break because an unacceptable amount of energy is transferred from the train's translational motion into "vibrations" in the couplings. The progression of impacts suggests that a wave like displacement pattern propagates along the train with increasing amplitude as it goes.

Vibrations in a system grow large either because the forcing is large or the forcing is small but *tuned* to the system. In Chapter 12 we'll develop fundamental material on the effects of forcing on a mass spring system, the archetype problem of the area. We'll then consider the application of these ideas to a vibrational problem that arises in the transmissions of cars, in Chapter 13. In Chapter 14 we'll take a very brief look at the rich area of wave phenomena. We'll examine traffic flow phenomena. This area is particularly useful because we're all experienced in its practical aspects, and also because the model equations are both simple in form and remarkably rich in a phenomenological sense. It seems likely the increasing amplitude waves observed in the train context may well be analogous to the well understood traffic jam phenomenon.

Chapter 12

Vibrations

Resonance
Damping
Nonlinear Detuning

12.1 Equilibrium and Vibrations

If a mass is carefully attached to a spring, see Fig. 12.1, the spring will stretch until the tension is sufficient to support the mass, and the resulting equilibrium will sustain until a disturbance acts on the system. The effect of an impulsive disturbance is to set the mass into oscillations about the equilibrium position at the *natural frequency* of the spring system, and the oscillations will continue until some *damping* mechanism (for example air drag or hysteresis) dissipates the energy input of the disturbance and the system settles back into its equilibrium configuration. If a sustained disturbance acts, then the system will move in a complicated way determined by the both the applied force and the natural oscillatory behaviour of the

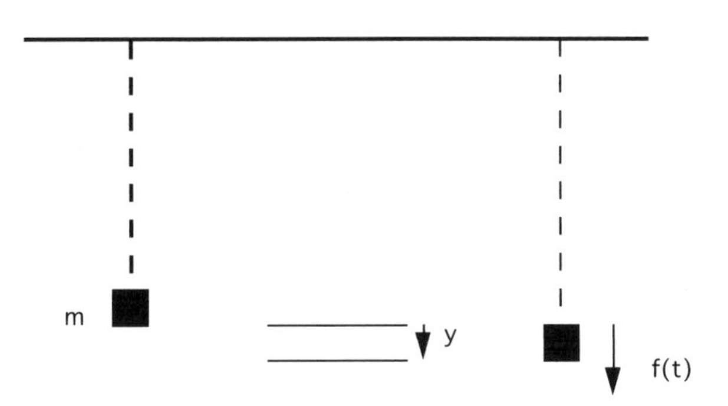

Figure 12.1: A spring mass system.

system. This simple and seemingly uninspiring spring mass system is much more important and interesting that the context suggests. (Indeed an understanding of this archetype problem is essential for the understanding of all vibration and wave phenomena.) Most stationary solid physical structures when bumped will vibrate at a frequency or frequencies determined by the *stiffness and mass* of the component parts and their couplings, and will respond to external forcing in a way that can be often characterized in terms of the simple spring mass model with an effective mass and stiffness related to the components and connections. In more complicated cases the simple spring ideas extend in a natural way. Bodies rotating about an axis also prefer to sustain this motion (as angular momentum conservation principles indicate), so that impulses will cause vibrations about this axis. The rotating shaft of an engine's transmission is such a case. The equations describing structural vibrations also model the behaviour of electrical circuits and the resulting analogy is often usefully exploited. Thus scientists often find it useful to think in terms of mechanical systems, experimentally work with electrical systems, and produce quantitative results using analytical and computational examinations of the defining equations—to restrict one's attention to the specific context out of which the problem arose is a mistake.

In the situations described above practical interest normally *does not* centre on determining a detailed description of the displacement vs. time relation for a particular mode of forcing; it is the magnitude and frequency of the response that matters. Mostly the practical aim is to ensure that large magnitude responses (and thus failures) are avoided, however, in some circumstances a very large response *is required*; for example detection devices such as a seismograph, a bar used to detect gravity waves, or a tuned radio.

In the situations described above the status of the system before the application of a disturbance was one of *stability* in the sense that forces induced within the system by the external disturbance (eg. the elastic restoring force in the spring) tend to restore the initial state. The forces in fact overcompensate, thus sending the system into oscillations. In other situations the induced forces act to carry the system away from the equilibrium situation. In such cases the equilibrium is said to be *unstable*. Thus, although it's theoretically possible to balance a ruler on one's nose, the situation is *unstable*—the floor is the preferred stable location for the ruler. This particular situation is not of great scientific interest. A more interesting situation arises when a ruler is subjected to a load applied longitudinally, see Fig. 12.2. The ruler is observed to compress until the load exceeds a particular level P_c, and then it *buckles* dramatically. The situation thus changes from one in which the straight ruler set-up is the stable configuration, to one in which the straight ruler represents an unstable configuration and the buckled ruler represents the stable state. Obviously the determi-

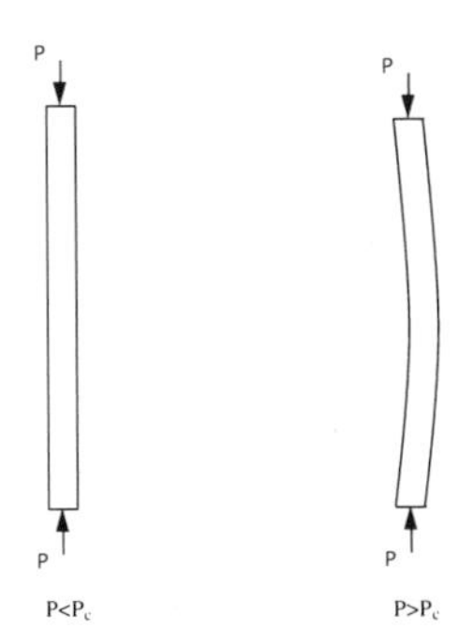

Figure 12.2: Buckling of a rod.

nation of the *critical* conditions for this changed situation is of importance. This situation in which there is a transition from one stable solution to another stable solution as the physical balance changes (as specified by an appropriate dimensionless group), occurs in many circumstances of major concern. One of the most celebrated and difficult problems of science is the problem of understanding the transitions that are observed in a turbulent moving fluid. Limited progress has been made on this problem. Here, in Section 12.3, we'll examine the transitions that are observed with nonlinear oscillators when the applied frequency passes through resonance; a similar but much simpler (archetype) problem.

12.2 The Linear Oscillator

If y is the displacement of the mass m suspended from a spring from its equilibrium position, see Fig. 12.1, and k is the spring constant (so the restoring force due to the spring is $-ky$), and the damping force is modelled by $-\iota\dot{y}(t)$ (where ι is the *damping coefficient*), and the externally applied or *driving* force given by $f(t)$ acts on the system, then

$$\ddot{y}(t) + \mu\dot{y}(t) + \omega_0^2 y(t) = f(t)/m, \tag{12.1}$$

where

$$\omega_0^2 = k/m, \quad \mu = \iota/m,$$

describes the motion. In the absence of damping the *natural frequency* of the motion is ω_0 and is thus larger for hard (i.e. large k) springs supporting small masses.

Usually to obtain solutions for linear constant coefficient differential equations one first works with an oscillatory forcing term and examines the corresponding complex oscillator

$$\mathcal{L}[x] \equiv \ddot{x} + \mu\dot{x} + \omega_0^2 x = ae^{\iota\omega t}, \tag{12.2}$$

where $\mathcal{L}$ denotes the indicated *linear* operator. If $a = \alpha e^{\imath\phi}$, then the real part of the $x(t)$ solution gives the $y(t)$ associated with the forcing term $\alpha\cos(\omega t + \phi)$, of *amplitude* α, *phase* ϕ, and *driving frequency* ω (i.e. $y(t) = \Re x(t)$). Superposition (using a Fourier representation of $f(t)$) can then be used to obtain the solution corresponding to general forcing. This procedure is useful because it significantly reduces the algebra (an advantage that is less significant if one uses computer algebra) and works because $\mathcal{R}\{\mathcal{L}[x(t)]\} = \mathcal{L}\{\mathcal{R}[x(t)]\}$; linearity is thus essential.

12.2.1 The forced undamped linear oscillator

The general solution of the above ordinary differential equation (12.2) in the undamped ($\mu = 0$) case is

$$x = Ce^{\imath\omega_0 t} + De^{-\imath\omega_0 t} + \frac{a}{\omega_0^2 - \omega^2}e^{\imath\omega t}, \tag{12.3}$$

with the corresponding solution for $y(t)$ given by

$$y(t) = A\cos(\omega_0 t + \psi_0) + \frac{\alpha}{\omega_0^2 - \omega^2}\cos(\omega t + \phi), \tag{12.4}$$

where the constants, (C, D) or (A, ψ_0) associated with the complementary function need to be determined from the initial conditions. A useful interpretation of this result is as follows:

- In the absence of an applied force (put $a = 0$), if the system is set in motion by an impulse it will vibrate about its equilibrium location $y = 0$ with its *natural frequency* ω_0, at an amplitude A determined by the energy transferred into the oscillator by the impulse.

- If forced by a disturbance at a frequency $\omega \neq \omega_0$ the response will consist of two superimposed oscillations: a *forced* response with frequency ω and amplitude determined by the amplitude and frequency of forcing (and *not* the initial conditions), and a *natural* response at the *natural frequency* with its amplitude determined by the initial conditions. The resulting characteristic *beating* pattern, see Fig. 12.3, should be familiar to readers. The fact that the solution simply consists of superimposed undamped oscillations tells us that in the long term the average energy of the system does not change. Given that the applied external force can perform work on the system, this is a non-trivial result. The work done by the external force up to time t is $\int_0^t f(t)\dot{y}(t)dt$ and can be readily calculated either from the solution or from the defining equation. Such a calculation confirms that the external force puts energy into the spring system during one stage of the solution "cycle" and extracts energy over the remaining part of the cycle, see Exercise 12.1. In the amplitude vs. frequency domain the response is much more easily characterized than in the time domain:

$$\text{amp} = \begin{cases} \alpha/(\omega_0^2 - \omega^2) & \text{at } \omega \\ A & \text{at } \omega_0 \\ 0 & \text{otherwise,} \end{cases}$$

and this amplitude vs. frequency characterization is usually the most useful way to describe vibrational or wave phenomena. Note also that the amplitude of the forced response varies with frequency of forcing and becomes very large as $\omega \to \omega_0$. It is clear from this that if a large response to external forcing is undesirable, then the structure modelled by the spring mass system should be designed to have a natural frequency significantly different from the expected frequency range of forcing. On the other hand, detection systems should be designed so that the natural frequency "matches" the frequency to be detected. The forced response due to general forcing $\sum_{i=1}^{n} a_i e^{\imath\omega_i t}$ (obtained by superposition) is given by

$$x = Ce^{\imath\omega_0 t} + De^{-\imath\omega_0 t} + \sum_{i=1}^{n} \frac{a_i}{\omega_0^2 - \omega_i^2} e^{\imath\omega_i t},$$

a complicated pattern in displacement/time space but a simple pattern in amplitude/frequency space. It is noteworthy that, because different frequencies are amplified by different amounts by the oscillator, the forced response of the oscillator will have the same shape in the time domain as the forcing function *only* in the special case of pure sinusoidal forcing. Sinusoidal forcing is thus *very special* for linear dynamic systems. This is why both theoreticians and experimentalists work mainly with oscillatory forcing. For detection systems the aim is to listen only to a particular frequency range and *filter out* other information, so such a non-uniform response is a major advantage in such cases.

- The solution given above is not well defined for the special case in which $\omega = \omega_0$ (i.e. when the frequency of forcing matches the natural frequency), and a different form of particular solution is required, namely $[-a\imath/(2\omega)]te^{\imath\omega t}$, which is called the *resonant* response of the system. Thus in this special case the forced response of the system grows indefinitely with time. The oscillator response is said to always remain *in phase* or *tuned* to the applied force in this special case so that the force continues to put energy into the system, as an examination of the integral defining the work done on the oscillator confirms. When ω is close to ω_0 the amplitude is also large but

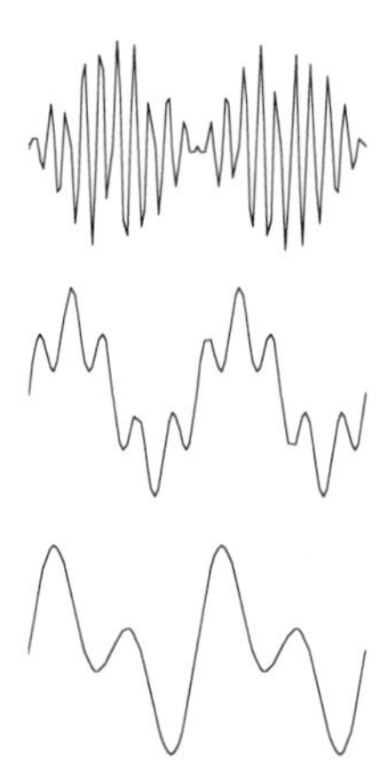

Figure 12.3: Superimposed oscillations.

does not grow indefinitely; eventually *phase coherence* is lost and the external force then starts to extract energy from the system, see Exercise 12.1. The word resonance is used to describe the phenomenon of a large response compared to the size of the forcing term when the forcing frequency is close to the natural frequency of the system. The range of frequencies in which the forced response is large is termed the *resonant bandwidth* of frequencies, and we will mainly focus our attention on this interesting frequency range.

In practical applications the amplitude will not continue to grow without limit so that the formal mathematical solution of the linear ordinary differential equation in the resonance case represents an approximation to reality that works over a limited time span; for larger times effects ignored in the simple model become important. Apart from the inevitable *damping* mechanisms (for example air resistance and hysteresis) that can remove energy from the system and thereby restrict the amplitude, the precise *tuning* required to continuously feed energy into the oscillator cannot be sustained in practice. Two detuning effects are commonly of significance in the resonant bandwidth of frequencies and thus serve to determine the amplitude of the response in this frequency range:

1. Nonlinearities that normally only slightly modify the oscillator response in frequency ranges remote from resonance cause detuning at the larger amplitude levels associated with resonance.

2. It is extremely difficult in practice to maintain a perfectly sinusoidal forcing term, so that the applied force does not remain perfectly in tune with the oscillator. Often the forcing frequency wanders randomly over a narrow frequency range. Of course the

oscillator frequency may also vary. Under these circumstances the amplitude of the forced response will depend on the statistical properties of the frequency fluctuations. Variation of parameters can be used to extract the solution corresponding to a particular $f(t)$ realization ($f(\Omega, t$ say) and statistical properties of the outcome can be obtained by appropriately averaging over the sample space Ω. The analysis can often be adapted to include the nonlinear effects that will be examined shortly . For an account of this work see Dincă and Teodosiu (1973).

12.2.2 The damped linear oscillator

All natural oscillatory systems contain energy dissipating mechanisms which lead to the decay of any natural oscillations in time, and which also effect the resonant band response. Rarely is it possible (and necessary) to accurately model the dissipative mechanism. The linear equation (12.2) represents an analytically tractable dissipative model that provides an adequate description for many such damped oscillatory systems. The relatively large amplitude resonant response cases of particular interest here will occur only if damping is relatively small.

The transient

The solutions of the homogeneous equation take the form

$$e^{-\mu t/2 \pm \imath\sqrt{(\omega_0^2 - \mu^2/4)}t}.$$

Thus for the small damping cases of interest:

- The natural frequency of the system is slightly modified

$$\omega_0 \longrightarrow \omega_0\sqrt{(1 - (\mu/2\omega_0)^2}$$

  by damping.

- The amplitude of natural unforced oscillation of the system decays to zero like $e^{-\mu t/2}$; so that after a time of order $1/\mu$ any natural oscillation of the system induced by the initiating conditions disappears. This is why the natural oscillations of damped systems are referred to as *transients* and are often ignored when considering forced phenomena. They will be ignored in the work that follows in this chapter.

Often for real systems where the dissipation mechanism cannot be identified, the damping coefficient μ is chosen so that amplitude decay predictions based on the above model match observations. Unless the damping mechanism is linked in some special way to the natural response of the oscillator one might expect this crude linear model (which removes the right amount of energy per cycle) to be adequate.

The forced solution

The particular solution describing the forced response of the oscillator (12.2) can be obtained by looking for a solution of the form $Xe^{\imath\omega t}$ and determining X so the equation is satisfied. This gives

$$x = \frac{a}{\omega_0^2 + \imath\omega\mu - \omega^2} e^{\imath\omega t}. \tag{12.5}$$

Obviously this forced solution represents a sustained response to forcing, with the amplitude of the size required to ensure that the energy input due to the applied force is just sufficient to compensate for the energy dissipated over a cycle. The coefficient of $e^{\imath\omega t}$ in this expression is complex so there will be a phase difference θ between the forcing term $\alpha e^{\imath(\omega t+\phi)}$ and the oscillator response (recall $a = \alpha e^{\imath\phi}$). Explicitly

$$x = Ae^{\imath(\omega t+\phi+\theta)}, \text{ where}$$

$$A = \frac{\alpha/\omega_0^2}{\sqrt{[1-\sigma^2]^2 + \nu^2\sigma^2}}, \quad \tan\theta = \frac{-\nu\sigma}{1-\sigma^2},$$

where

$$\sigma = \frac{\omega}{\omega_0} \text{ and } \nu = \frac{\mu}{\omega_0} \tag{12.6}$$

are the appropriate dimensionless frequency and damping parameters. Thus both the amplitude and phase of the response are strongly frequency dependent. One should note:

- The system will oscillate at the frequency of forcing with amplitudes which become large if the damping (measured by ν) is small and the forcing frequency is close to the natural frequency of the system (i.e. $\sigma \approx 1$). Under other circumstances the response will be moderate (of order α/ω^2 or α/ω_0^2). For no value of the forcing frequency will the response become unbounded. Explicitly, in unscaled terms, the amplitude of response will be of size $\alpha/(\omega_0\mu)$, if the applied frequency is in the range $\omega - \omega_0$ of order μ. Note that the smaller the damping, the larger the maximum response but the narrower the bandwidth. Some illustrative response curves are plotted in Fig. 12.4.

- For small values of σ the applied force and the oscillator response are almost in phase. The oscillator response phase always lags that of the force by an amount that increases rapidly as σ approaches and passes through the resonant frequency (1), with the phase lag being $\pi/2$ at resonance and asymptotically "reaching" π for values of $\sigma - 1$ of order ν. Thus an observer would see the apparent paradox of the applied force causing motion in the *opposite* direction to that of its application for frequencies just exceeding the resonant frequency, see Exercise 12.2 and Fig. 12.4.

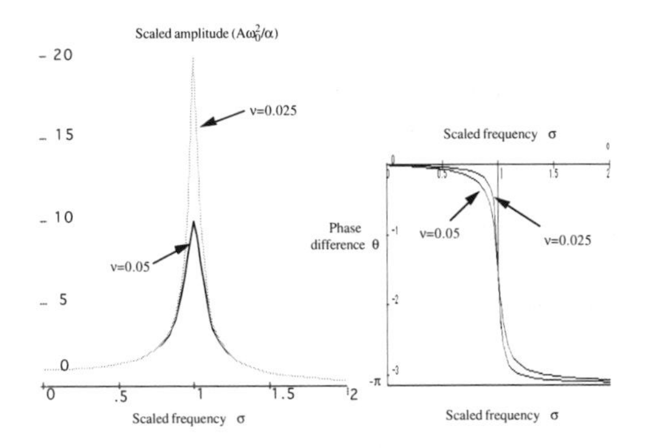

Figure 12.4: Damped linear oscillator response.

12.3 Weakly Nonlinear Resonance

In order to see the effect of nonlinearity on the near resonant response of an oscillator it is desirable to use scaled variables based on the nonresonant response of the linear system (12.1). Thus we work with the scaled variables η and τ defined by

$$y = (\alpha/\omega_0^2)\eta \quad \text{and} \quad \tau = \omega_0 t, \tag{12.7}$$

in terms of which the oscillator equation reduces to

$$\eta''(\tau) + \eta(\tau) + \nu\eta'(\tau) + \epsilon f(\eta(\tau)) = \cos(\sigma\tau + \phi), \tag{12.8}$$

where σ, ν, are the dimensionless frequency and damping parameters encountered earlier (see (12.6)), and ϵ scales the nonlinear term ϵf. The phase ϕ is included in the forcing function to enable arbitrary initial conditions to be accommodated. We'll work with the cubic nonlinearity case so

$$f = \eta^3.$$

Comment: This cubic term can be thought of as being the first nonlinear Taylor expansion term for odd (smooth) functions f of displacement y. (A quadratic term corresponds to a biased oscillator response and occurs less frequently in practice than the cubic case.) Observe that there is no loss of generality in having a coefficient of 1 for the cubic term, since any such non-unit coefficient could be absorbed into the definition of the parameter ϵ. A broad class of cases is thus covered by the above equation. For ease in presentation ϵ will be assumed to be positive.

Our concern here is with situations in which nonlinear and damping effects are relatively small under non-resonant conditions so that the dimensionless parameters ϵ and ν are small. Our primary aim in fact is to determine the amplitude of the response as a function of (ϵ, ν, σ), for frequencies σ close to the resonant frequency 1.

Under the prescribed circumstances one might expect the dominant response of the oscillator to be at frequency σ, with corrections resulting because of the nonlinearity. With this in mind it seems sensible to write the equation in the form

$$\eta''(\tau) + \eta(\tau) - \cos(\sigma\tau + \phi) = -(\nu\eta'(\tau) + \epsilon\eta^3), \tag{12.9}$$

and look for a dominant solution of the form

$$\eta = C\cos(\sigma\tau + \psi). \tag{12.10}$$

It's sensible to first tackle the undamped oscillator ($\nu = 0$) problem.

12.3.1 The undamped nonlinear oscillator

With $\nu = 0$ the oscillator equation (12.9) becomes

$$\eta''(\tau) + \eta(\tau) = \cos(\sigma\tau + \phi) - \epsilon\eta^3. \tag{12.11}$$

Unfortunately, because of the nonlinearity, it's not possible to extract the solution to this oscillator using an equivalent complex oscillator, so much of the algebraic and intuitional advantage of using complex analysis is lost. In spite of this it's still algebraically simpler to work with a complex solution representation. Thus we write the dominant solution in the form

$$\eta = Ae^{\imath\sigma\tau} + A_* e^{-\imath\sigma\tau},$$

where $A \equiv Ce^{\imath\psi}/2$ is a complex number to be determined, and the suffix $*$ denotes the conjugate complex of the variable suffixed.
Aside: Working with sines and cosines when dealing with nonlinear expressions inevitably means struggling with the multitude of cumbersome rules

$$2\cos\alpha\cos\beta = \cos(\alpha + \beta) + \cdots \text{ etc.}$$

Such complications are *just algebraic,* and so one might think they are largely minor, especially if one uses a computer algebra package. However, even for a computer, the size of the computation can become unmanageable with problems of the present type, so that it *usually* pays to work with the simpler exponential functions.

A substitution of the expression for η into the correction term gives

$$\epsilon f = \epsilon[A^3 e^{3\imath\sigma\tau} + 3A^2 A_* e^{\imath\sigma\tau} + 3A_*^2 A e^{-\imath\sigma\tau} + A_*^3 e^{-3\imath\sigma\tau}].$$

It may appear that nothing has been gained by using a complex representation, but it should be observed that it is not really necessary to write out all the terms involving the negative exponentials because these can be inferred from the positive exponentials by taking the conjugate complex of those terms—after all the complete solution must be real. Thus we can write the correction term in the form

$$\epsilon f = \epsilon[A^3 e^{3\imath\sigma\tau} + 3A^2 A_* e^{\imath\sigma\tau} + *],$$

where the $*$ denotes taking the complex conjugate of all that has been written explicitly on the same side of the equation. The correction term thus introduces the third harmonic $e^{3\imath\sigma\tau}$ into the system of size ϵ which will be reflected in the corrected solution. Thus we anticipate a solution of the form

$$\eta = Ae^{\imath\sigma\tau} + \epsilon Be^{3\imath\sigma\tau} + *,$$

and substitute this into the defining equation (12.11) to give

$$\{A[1-\sigma^2]+3\epsilon A^2 A_* - e^{\imath\phi}/2\}e^{\imath\sigma\tau} + \{B[1-(3\sigma)^2]+\epsilon A^3\}e^{3\imath\sigma\tau} + O(\epsilon^2) + * = 0.$$

Thus, equating the coefficients of the two oscillations to zero we see that the choice of coefficients A and B given by

$$A[1-\sigma^2] + 3\epsilon A^2 A_* = e^{\imath\phi}/2 \tag{12.12}$$

and

$$B[1-(3\sigma)^2] + \epsilon A^3 = 0, \tag{12.13}$$

hopefully ensures the equation is satisfied to $O(\epsilon^2)$. The first of these equations determines the amplitude of the dominant response and so is of primary importance. Although they have not been written explicitly the other equations which serve to determine A_* and B_* are obtained by writing out the complex conjugate forms of the above two equations.

Aside: The approach employed here is a slight variant of the perturbation and iterative techniques used in earlier chapters. The present approach focuses attention on the modes of oscillation, which our earlier experience with the linear oscillator suggests are of primary physical (and therefore mathematical) significance. Had we proceeded using the usual approach we would have reached the same equations via a more circuitous route. Often slight variations of approach can lead to useful clarifications, so you should familiarize yourself with the alternatives and understand what it is that makes them effective in particular circumstances.

Comment: A physical explanation Although the above mathematical scheme will be seen to provide useful answers it's not particularly illuminating in the sense that it provides little understanding of either the physics of the situation or why the process works. As a modeller such understanding is the essence—any ritual can be looked up in an appropriate handbook. A way that you *may* find useful to interpret the above results is as follows. The nonlinear term can be thought of as an additional forcing term acting on the linear oscillator whose size is approximately given by $\epsilon\eta_0^3$ where η_0

is a first estimate for η. Thus (working with the real solution (12.10)), $-\epsilon C^3 \cos^3(\sigma\tau + \psi)$ should provide a first estimate for the effective forcing induced by the nonlinearity, see (12.11). Now you'll recall from elementary trigonometry that

$$\cos^3(\sigma\tau + \psi) = \frac{1}{4}[\cos 3(\sigma\tau + \psi) + 3\cos(\sigma\tau + \psi)],$$

and in this form one can more easily see the effect of the nonlinearity. Of the two oscillatory components of η_0^3 it's only the component with frequency σ (close to the resonant frequency) that can strongly affect the linear oscillator. If the size of this component (determined by C, or equivalently A) is sufficient to cancel that of real external force $\cos(\sigma\tau + \phi)$, then the net energy transfer into the linear oscillator due to these two "forces" (one real and the other "induced") will vanish, and equilibrium will be realized. The equation (12.12) for A expresses this physical requirement, and thus determines A. Thus, concentrating on the "resonant" components and choosing A so that these components balance makes physical sense.

The solution

Firstly notice that the equations for A and B are weakly coupled; meaning that both equations don't have to be simultaneously solved. Thus we first solve for A using (12.12), and B then follows from (12.13). Secondly note that unless $1 - \sigma^2 \approx 0$, the nonlinear term in the equation for A appears to be relatively small (of relative order ϵ) so that

$$A \approx \frac{1}{2(1-\sigma^2)} e^{\imath\phi} \text{ for } \sigma \not\approx 1$$

should provide a good solution estimate. An iterative procedure could be used to check this out, but the experience we've already gained in other contexts enables us to confidently make this claim. Thus, unless the situation is almost resonant, the nonlinearity results in only minor modifications in the amplitude of the primary mode.

Aside: Under "non-resonant" circumstances the main effect of the nonlinearity is to introduce higher order harmonics, which changes the look of the response (a superficial effect) but more importantly spreads the energy into other modes. Given that such modes may be resonant modes for systems coupled into the oscillator, this is a much more important effect. This spread of energy into harmonics and sometimes *subharmonics* is characteristic of vibration and wave propagation systems and many engineering disasters have resulted because the slow feed of energy from the fundamental mode to the other modes has resulted in unanticipated resonances. The historic collapse of the Washington State Tacoma bridge in the 50's is a case in point. Exercise 12.3 examines these issues.

For σ close to 1 the linear term in equation (12.12) is almost zero, so the nonlinear term can no longer be ignored by comparison. If $A = \mathcal{A}e^{\imath\alpha}$ then equation (12.12) becomes

$$\{\mathcal{A}[1-\sigma^2] + 3\epsilon\mathcal{A}^3\}e^{\imath\alpha} = e^{\imath\phi}/2,$$

so the magnitude and phase of both sides match if

$$\alpha = \phi \quad \text{and} \quad \mathcal{A}[1-\sigma^2] + 3\epsilon\mathcal{A}^3 = 1/2. \tag{12.14}$$

Thus the oscillator response is in phase with the forcing (to $O(\epsilon^2)$) and the amplitude of the motion is governed by a cubic. Notice that at the resonant frequency $\sigma = 1$ the solution is given by $(1/6)^{1/3}\epsilon^{-1/3}$. Thus we have a large but *finite response* at the resonant frequency; it looks as if we're getting somewhere! To explore further we need to examine the cubic in detail. By squaring the amplitude equation (12.14) we obtain the following equation for the scaled energy $\mathcal{E} = \mathcal{A}^2$ in the dominant mode:

$$\mathcal{E}[1-\sigma^2 + 3\epsilon\mathcal{E}]^2 = 1/4, \tag{12.15}$$

which is a little more convenient to work with. This equation can be derived more directly by multiplying equation (12.12) by it's complex conjugate. The presence of the small parameter in the above equation indicates the need for further processing. Thus any normal numeric scheme is likely to incorrectly assess the apparently small term; in fact we know that the essence of the mathematics lies in the correct assessment of this (apparently small) nonlinear term.

Now for $\sigma < 1$ the left hand side of the $\mathcal{E}$ equation is monotonic in $\mathcal{E}$ so there will be just one solution, and it will be close to the linear oscillator solution. The obvious question is: how small does $(\sigma^2 - 1)$ have to become before the nonlinear term (i.e. the ϵ term) in this equation becomes significant? There is a remarkably simple way to determine the extent of this *critical* frequency zone which we'll now describe. The linear approximation neglects a term of size $\epsilon\mathcal{E}$ in the $[\cdot]$ term of equation (12.15). Now just outside this critical zone $\mathcal{E}$ is well estimated by the linear solution $1/(4[1-\sigma^2]^2)$; so, based on this estimate for $\mathcal{E}$, the neglected term in $[\cdot]$ is of size $\epsilon/[1-\sigma^2]^2$, which is fairly neglected in comparison with the other term $[1-\sigma^2]$ only if $[1-\sigma^2]^3 < \epsilon$ i.e. only if $(1-\sigma)(1+\sigma) < \epsilon^{1/3}$. Given that $\sigma \approx 1$ we can thus see that the approximation breaks down when $(1-\sigma)$ is of order $\epsilon^{1/3}$. This suggests that one should look for nonlinear resonance effects in the frequency range of order $\epsilon^{1/3}$ around $\sigma = 1$, and that $\mathcal{E}$ is of order $\epsilon^{-\frac{2}{3}}$ in this region. This argument may seem ad hoc and crude, but it works.

Aside: This procedure is generally not presented in papers and books, perhaps because such intuitive approaches seem to be an embarrassment

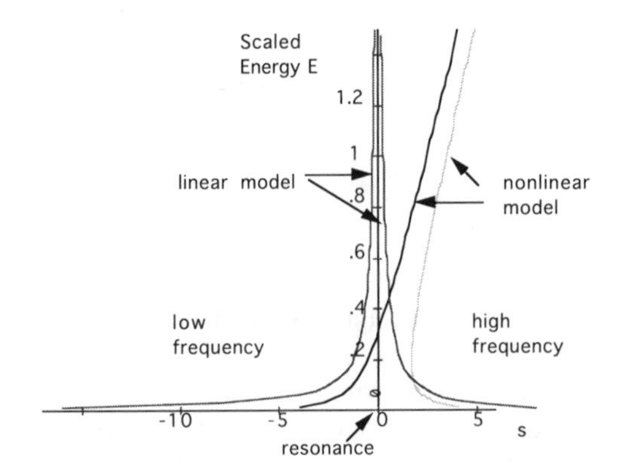

Figure 12.5: The forced nonlinear oscillator.

to mathematicians. How do you justify such a procedure except by saying that it works! This procedure is particularly useful for delineating the correct parameter range in which to seek interesting effects when making approximations involving more than one small parameter, as we'll see later.

In view of the above estimates it's appropriate to scales the variables thus:

$$\sigma^2 - 1 = s\epsilon^{1/3} \quad \text{and} \quad \mathcal{E} = E\epsilon^{-2/3},$$

in terms of which equation (12.15) becomes

$$E(-s + 3E)^2 = 1/4, \tag{12.16}$$

which is now in a suitable form for solving numerically (all the coefficients are of unit order). Maple has difficulty determining $E(s)$ but can handle $s(E)$. The energy vs. frequency results are shown in Fig. 12.5, plotted using Maple.

Response Size As the applied frequency increases from relatively small values $\sigma = 0+$ (corresponding to $s = -1+$) through resonance $\sigma = 1$ (corresponding to $s = 0$), to relatively large values $\sigma \gg 1$ (corresponding to $s \gg 0$) the above theory predicts:

- For frequencies below resonance there's just one solution with growing amplitude that departs significantly from the linear solution within a frequency band of width $\epsilon^{1/3}$ around the resonant frequency. The theory predicts a finite (large) amplitude response at the resonant frequency of the linear oscillator of magnitude $\epsilon^{-1/3}$. Thus the nonlinearity *detunes* the oscillator, thereby limiting the amplitude of the response to a size $\epsilon^{-1/3}$ determined by the strength of the nonlinearity.

- For frequencies just above resonance *three* solution branches exist, see Fig. 12.5. That is, there are three possible values of E corresponding to a specified σ in this range. Of the three branches, two have unbounded energy levels as $\sigma \to \infty$. One might expect that the two unbounded solutions are spurious, however, note that one of the infinite branches joins smoothly onto the "acceptable" branch for small σ values; life is much more interesting! Obviously the nonlinear model raises as many questions as it answers. Damping effects are likely to prohibit infinite responses, so we need to examine the damped nonlinear oscillator equation (12.9).

12.3.2 The damped nonlinear oscillator

The scaled oscillator equation of interest, repeated here for convenience, is

$$\eta''(\tau) + \eta(\tau) - \cos(\sigma\tau + \phi) = -(\nu\eta'(\tau) + \epsilon\eta^3).$$

You'll recall that the time and displacement scales used here were chosen to reflect the linear oscillator response under non-resonant conditions. Now if the damping is relatively heavy, the amplitude of the oscillator response will never be sufficient for nonlinear effects to play a significant role, and the damped linear oscillator solution is an appropriate model. The interesting parameter range is therefore the one in which *both* nonlinear and damping effects limit the amplitude of the response. From the previous results on the undamped case we expect the scale of η in the resonant band to be of order $\epsilon^{-1/3}$, so that we might expect the damping term $\nu\eta'$ to be of order $\nu\epsilon^{-1/3}$ in this band. The nonlinear term is of order $\epsilon\eta^3$ i.e. of unit order. We expect the nonlinear and damping effects to be comparable if their corresponding terms in the above equation are of the same size, and this will be the case for values of ν of order $\epsilon^{1/3}$. Thus we write

$$\nu = \upsilon\epsilon^{1/3},$$

and also it's sensible to anticipate the solution scale in the resonant band by writing

$$\eta = \epsilon^{-1/3}\bar{\eta}, \tag{12.17}$$

in terms of which the equation becomes

$$\bar{\eta}'' + \bar{\eta} + \epsilon^{1/3}(\upsilon\bar{\eta}' + \bar{\eta}^3) = \epsilon^{1/3}\cos(\sigma\tau + \phi). \tag{12.18}$$

Following the same pattern as in the undamped case we look for a solution of the form

$$\bar{\eta} = \bar{A}e^{\imath\sigma\tau} + \epsilon\bar{B}e^{3\imath\sigma\tau} + *,$$

and thus arrive at the corresponding equation, see equation (12.12)

$$\bar{A}[1 - \sigma^2] + \epsilon^{1/3}(\upsilon\imath\sigma\bar{A} + 3\bar{A}^2\bar{A}_*) = \epsilon^{1/3}e^{\imath\phi}/2,$$

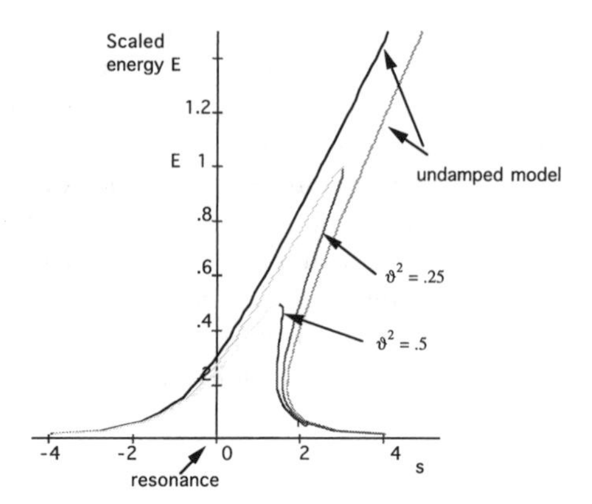

Figure 12.6: The forced nonlinear damped oscillator.

for $\bar{A} \equiv \epsilon^{-1/3}A$.

Comment: The calculations here are getting to be somewhat demanding and certainly tedious. Computer algebra is *especially suitable* for problems of the present type. One advantage of the approach is that, once one has set up a sensible scheme, it's a simple matter to add in additional terms, or to extend the calculations to higher orders. In this way one can easily check the validity of first estimates, improve the approximation, or more importantly (in more complex cases) search for effects that may be not be obvious. In Exercise 12.4 such a scheme will be examined.

Again we focus our attention on the resonant band by writing

$$\sigma^2 - 1 = \epsilon^{1/3}s,$$

and, by writing $\bar{A} = \bar{\mathcal{A}}e^{\imath\alpha}$, and equating the amplitudes of the two sides of the $\bar{A}$ equation, we arrive at the following equation for the scaled energy $E = \bar{\mathcal{A}}^2$ in the dominant mode:

$$E[(-s+3E)^2 + \upsilon^2(1-\epsilon^{1/3}s)] \approx E[(-s+3E)^2 + \upsilon^2] = 1/4, \qquad (12.19)$$

which should be compared with the corresponding undamped result (12.16). Solutions for $E(s)$ corresponding to a range of values of υ are plotted (using Maple) in Fig. 12.6.

Solution behaviour

It can be seen from the figure that the main effect of the additional damping term in the energy equation is to close off the solution branches that go off to infinity. The implication is of course that, as anticipated, damping and nonlinearity effects together limit the amplitude of response of the oscillator. Note, however, that there are still three possible solutions in a frequency range just above resonance! Which if any of these is chosen in practice? This is a *stability* question of the type briefly discussed in Section 12.1 (remember the ruler balancing act). Although all three solutions are legitimate equilibrium solutions, such solutions will only be observed in practice if they are *stable* to the inevitable buffeting that occurs in the real world. Mathematically one can examine such stability questions by adding in a perturbation and checking to see if the effect of the perturbation dies out in time (the stable case) or if it grows so that the solution drifts from equilibrium (the unstable case); thus effectively one simulates the buffeting. *Comment:* In special cases (eg. simple unforced systems) elementary mathematical techniques (eg. phase plane or graphical techniques), or simple physical arguments (eg. energy arguments) can be employed to establish the nature of the equilibrium but in general these questions, which are of great general interest, are difficult to answer.

In the present case a stability analysis reveals that the upper and lower solution branches are stable, and that the intermediate branch is unstable, and so is not observed. This still leaves us with two possible stable solutions over portions of the frequency range. Interestingly enough, both solutions are viable; the actual solution realized in a particular circumstance depends on past history. For example, see Fig. 12.7, if the applied frequency is gradually increased through resonance the response follows the obvious branch ABC until the point C of vertical tangency is reached, then the response jumps to the second stable branch DE. If the applied frequency is then reduced gradually through resonance the branch EDF is followed again until vertical tangency is reached at F, and then the response jumps to the other branch GBA. Usually to determine the state realized under prescribed circumstances it's necessary to follow the temporal development of the system. Techniques for doing this will be developed in the next chapter. Generally speaking, however, the unstable branch separates out regions of attraction for transients. Thus if the initial oscillator state lies in the region BGCF, the oscillator is likely to gravitate towards the upper equilibrium branch BGC etc.

The above situation with a non-unique (history dependent) steady state solution corresponding to a given parameter combination is an example of a phenomenon known of as *hysteresis*.

Response Shape

So far our interest has centred on the amplitude and phase of the response of the dominant mode and we've ignored the accompanying change in the total $\eta(t)$ shape. Information concerning this aspect of the problem is contained in the higher order harmonics. In particular equation (12.13) explicitly determines the third harmonic in the undamped case. The shape is displayed in Fig. 12.8 for a range of ϵ values at the resonant frequency. An

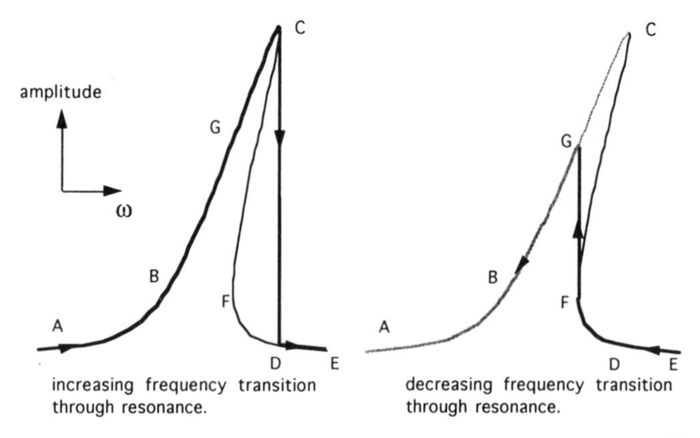

Figure 12.7: Hysteresis behaviour of a forced nonlinear oscillator.

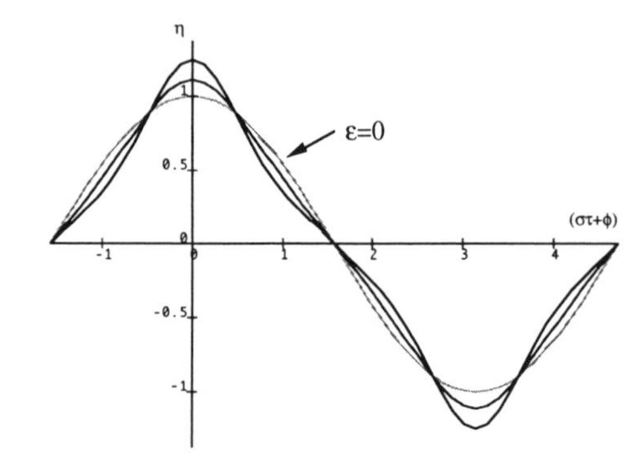

Figure 12.8: Shape changes due to nonlinearity.

iterative scheme can be used to determine the higher order harmonics (the 5th etc.) whose amplitude will grow in magnitude with increased ϵ or A. Thus the energy input from the applied force into the oscillator leaks into more and more modes, the response shape becomes more complex, and the iterative computations more and more cumbersome and less useful; a different solution scheme is required. No one knows how to deal effectively with situations such as this. Often in such cases the motion ceases to be steady with random or almost random (*chaotic*) superimposed time dependent fluctuations occurring. A numerical output in such cases is interesting but not particularly informative. What's required is a characterization of the response of a different type eg. an energy vs. frequency characterization would be useful for many practical purposes.

12.3.3 ⋆ Dissipation modelling

The above work gives a good idea of the major issues involved in analysing resonance situations but is misleading in that real world situations are usually much more difficult to analyse. In particular the above handling of dissipation is normally inadequate for ***resonant*** situations. We'll briefly indicate how to handle such complications here.

Whereas under non-resonant circumstances a linear damping model should suffice to describe solution behaviour, under almost resonance conditions the modelling of dissipation is likely to be a much more important issue; after all it's the energy input/dissipative loss balance over a cycle that determines the equilibrium amplitude. The most commonly occurring dissipative mechanisms in the real world (air resistance and solid to solid frictional resistance) are nonlinear and of an awkward form that makes for quite difficult analysis. Physically speaking such mechanisms *both* remove mechanical energy and detune the system.

We'll consider the case of a body suspended by a linear spring mass and subject to purely sinusoidal (almost resonant) forcing in the case in which the major resistance to motion is the effect of air resistance, typically proportional to the square of the speed and acting in the opposite direction to the instantaneous motion. The relevant equation for the motion about the equilibrium location is

$$m\ddot{x} = -kx - \alpha\dot{x}^2\text{sign}(\dot{x}) + f\cos\omega t.$$

After introducing (the usual) scaled variables defined by

$$x = \frac{f}{m\omega^2}X, \quad \tau = \omega t,$$

the defining equation becomes

$$X'' + (1+\delta)X = \cos\tau - \iota X'^2\,\text{sign}(X')\ , \tag{12.20}$$

where

$$\iota = \frac{\alpha f}{m\omega^2}, \text{ and } 1+\delta = (\frac{\omega_0}{\omega})^2,$$

where $\omega_0 = k/m$ is the natural frequency of the oscillator. The prime denotes differentiation with respect to τ. Clearly the dimensionless parameter δ defines how close the scaled frequency of forcing is to the scaled resonant frequency of the oscillator ($\sqrt{1+\delta}$), and ι provides a measure of the relative size of the air resistance force to the restoring force of the oscillator. For small values of δ and ι the response is expected to be large and this is the range of interest here. Given the experience of earlier work we know what to look for: we expect the solution to be of the form $A\cos(\tau+\phi)$ with A and ϕ determined so that the damping and forcing terms *at the forcing frequency* balance, and with the interesting δ range being such that the detuning term also is of the same size. With the solution as prescribed the resistance term is given by $\pm\iota A^2\sin^2(\tau+\phi)$ or $\pm(1/2)\iota A^2(1-\cos 2(\tau+\phi))$, where the positive sign applies when $\sin(\tau+\phi)$ is positive and otherwise the negative sign applies. The resistance is therefore a periodic function with period 2π. Furthermore it's an odd function of τ and so admits of the Fourier representation of the form

$$(1/2)\iota A^2 \sum_{m=1}^{\infty} R_m \sin m\tau,$$

where the coefficients are given by

$$R_m = \frac{2}{\pi}\int_0^{\pi} (1 - \cos 2\tau')\sin m\tau'\, d\tau'.$$

The even coefficients vanish and the first of the odd coefficients is given by $R_1 = 16/3\pi$. The contribution of the resistance term at the forcing frequency is $(R_1/2)[\iota A^2 \sin(\tau + \phi)]$. Thus the dominant contributions in equation (12.20) are given by

$$A\delta \cos(\tau + \phi) = \cos\tau - \frac{R_1}{2}\iota A^2 \sin(\tau + \phi),$$

and the anticipated balance of terms will be realized for amplitudes A of order $1/\sqrt{\iota}$ (which provides us with an estimate for the amplitude of the response at the resonant frequency) and for values of δ of order $\sqrt{\iota}$ (which gives us the resonance frequency band width). Writing

$$A = \bar{A}/\sqrt{\iota}, \text{and } \delta = d\sqrt{\iota}$$

gives

$$\bar{A}d\cos(\tau + \phi) = \cos\tau - \frac{R_1}{2}\bar{A}^2 \sin(\tau + \phi).$$

To solve for $(\bar{A}, \phi)$ we can expand $\sin(\tau + \phi)$ in terms of $\sin\tau$ etc., and if we compare the coefficients of $\sin\tau$ and $\cos\tau$ on both sides of the equation we obtain the pair of equations

$$\bar{A}d\cos\phi = 1 - \frac{R_1}{2}\bar{A}^2 \sin\phi,$$

$$-\bar{A}d\sin\phi = -\frac{R_1}{2}\bar{A}^2 \cos\phi.$$

These equations can be solved by first treating them as a pair of linear equations for $\cos\phi$ and $\sin\phi$ and then using the fact that the sum of their squares is unity. In this way the response amplitude can be determined; a straightforward but tedious process even with Maple. Note that at $d = 0$ the solution is given by $\phi = \pi/2$, with $\bar{A} = \sqrt{2/R_1}$.

Sliding Friction The solid friction situation is even more awkward to handle. Imagine an oscillator with the mass in contact with a wall on which it can slide. During that phase of the forcing cycle in which the net force (external plus elastic) is insufficient to overcome the frictional force, the oscillator mass simply sticks to the surface of contact waiting for the applied force to increase to the required level to cause motion. Over the rest of the cycle the frictional force is of the form $-F_0\text{sign}(\dot{x})$. Although it's easy to numerically integrate the defining equations over a few cycles, computing the resonant amplitude (i.e. the amplitude after *many* oscillations) is a major task. For more information on these problems and vibrations generally see Dincăand Teodosiu (1973).

12.4 Summary

A relatively large response results if an oscillatory system is forced at or near one of its natural frequencies. Under such circumstances damping and nonlinear effects often limit the response and thus determine the magnitude of the response and the bandwidth. Linear damping effects simply remove mechanical energy from the system so that the equilibrium amplitude is the amplitude at which the dissipation per cycle balances the work done by the external force per cycle. No significant change in the shape of the response results. Nonlinearities detune the system and thus reduce the response and also produce a shape change in the response.

12.5 Exercises

Exercise 12.1 *Oscillator Beats*

The aim of this exercise is to build up an understanding of the type of displacement vs. time patterns that result when an external force acts on a linear oscillator.

(a) Using Maple determine the solution of the oscillator

$$\ddot{x} + x = \sin \varpi t, \text{ with } x(0) = \dot{x}(0) = 0,$$

and plot displacements over an appropriate time interval for $\varpi = 0.2, 1, 1.1, 2, 8$. Comment especially on the near resonance and resonant behaviour.

(b) The work done by the external force on the oscillator up to time t is

$$\mathcal{W}(t) = \int_0^t \sin \varpi t' \dot{x}(t') dt'.$$

Plot $\mathcal{W}(t)$ for $\varpi = 2, 1.1, 1$ and make relevant observations. Especially comment on the near resonance and resonant situations.

(c) Introduce a small amount of damping into the system and plot out the displacement and the work done on the oscillator by the applied force for one value of ϖ.

Exercise 12.2 *Resonance*

Using Maple plot out the amplitude and phase of the forced response of a linear damped oscillator as a function of frequency for a variety of levels of damping. Use appropriate scales. Especially focus your attention on near resonant behaviour. Comment.
Hint: Be careful when plotting the phase function.

Exercise 12.3 *Rattling Cars*

⋆ An oscillatory force (or voltage) applied to one part of a mechanical (or electrical) system will result in vibrations (or voltage oscillations) of other components coupled to it. Often the coupling is weak (and unexpected) and it's of obvious importance to know how rapidly energy can be transferred by the coupling, and how large the induced vibration (or oscillation) is likely to be under such circumstances. Large vibrations may be expected when the frequency of the input matches the natural frequency of any one of the components, so this is the situation that is of most interest. Automobile and aircraft manufacturers are particularly interested in such vibrations. The following model examines some of the relevant issues.

An oscillator with natural frequency ω is weakly attached to an oscillator with natural frequency $\omega_0 \neq \omega$ which has an external force acting on it of frequency ω (i.e. the natural frequency of the first oscillator). We'll model the situation by the equations

$$\ddot{X} + \omega_0^2 X = \epsilon(X - x) + e^{\imath\omega t},$$

$$\ddot{x} + \omega^2 x + \nu\dot{x} = -\epsilon(X - x),$$

with $\nu, \epsilon \ll 1$, and assume zero initial displacements. Maple can be used to extract exact results to the above equations but the results thus obtained are cumbersome, so hand calculations based on the smallness of the coupling parameter are more revealing. Use whichever suites you.

(a) What's the time scale for energy feed into the oscillator with displacement specified by x? Comment on the dependence of the result on the strength of coupling, as specified by ϵ.

(b) Determine the eventual amplitude of the response of the oscillator associated with x. Identify the parametric combination that primarily determines the response size and make appropriate observations.

Hint: Again a perturbation procedure is likely to lead to more revealing results.

(c) ⋆⋆ Given that the response can be large even though the coupling is weak, it's sensible to design systems so that the natural frequencies are remote from any of the possible driving frequencies and the major harmonics that can be generated by nonlinearities. It is thus of major importance to determine the natural frequencies of such systems. In the car case the major 'external' forces are associated with the motor, the suspension, and the movement along the road surface. Such external forces produce vibrations in the skeletal structure with attached 'plates'.

It's the natural vibrational modes of the system *as a whole* that are relevant, rather than the isolated components. Assuming the damping is relatively small, determine the characteristic equation associated with the above system and thus determine the natural frequency of the system mode that's close to that of the component associated with x, and comment on the effect of the coupling. Again avoid algebraic complications if possible.

Exercise 12.4 *Computer Algebra*

As seen in the main text, the algebra that arises when considering nonlinear effects in a vibration context can become daunting, and computer algebra becomes an essential tool when dealing with many real world problems of this type. Apart from the initial observations to determine the solution form, the procedures required to determine the mode amplitudes and phases for such problems are quite mechanical, and so are ideally suited for algebraic computation. Present day packages are however still crude, and it turns out that in particular Maple (at the time of writing) is not yet well set up to deal with problems of this type. Because of these limitations we'll work with sines and cosines (rather than exponentials) for the calculations that follow. Explicitly we'll use Maple to re-determine (see Section 3.1) the equations governing the amplitudes and phases of the largest vibration modes in the cubic oscillator case. To do this we look for a solution representation of the form

$$A \cos \sigma t + B \sin \sigma t + C \cos 3\sigma t + D \sin 3\sigma t,$$

and determine A, B, C, D so that the resonant $\sin \sigma t$ and $\cos \sigma t$ terms in the defining equation vanish etc.

(a) Use the following Maple instructions to set up the relevant equation and solution form, see (12.11) (the $\phi = 0$ case is considered):

```
eqnexp:=diff(eta(t),t,t)+eta(t)+ep*(eta(t))∧3-cos(sigma*t);
soln:=A*cos(sigma*t)+B*sin(sigma*t)+
C*ep*cos(3*sigma*t)+D*ep*sin(3*sigma*t);
```

Now `subs` the solution form into the equation expression.

(b) The coefficients A, B, C, D need to be chosen so that the $\sin\sigma t$ terms in this expression vanish. Also the $\cos\sigma t$ must vanish. To extract the coefficients of $\sin\sigma t$ and $\cos\sigma t$ from the equation make use of the orthogonality of the Fourier terms over the $(0, 2\pi/\sigma)$ domain.

(c) The resulting equations are unwieldy! To make them manageable use `simplify`, and examine the largest terms (of order ϵ) using `collect`. Check to see if the results you obtain match those obtained in the main text, see (12.12,12.13).

(d) $\star$ Extend the analysis to deal with the damped oscillator case.

Exercise 12.5 *Rod Buckling*

The sudden buckling of a ruler that results when the applied longitudinal load exceeds a critical level is an interesting example of the type of "solution switching" situation that arose in the nonlinear oscillator problem; in fact it's often regarded as the archetype problem of this type. We'll briefly examine the ruler situation here. For a more complete discussion of this problem see Dym (1974).

The equations governing the *small* lateral displacement of a rod of length L subjected to a longitudinal force of P per unit sectional area (see Fig. 12.2) are

$$y'''' + \frac{P}{EI}y'' = 0, \quad \text{with} \tag{12.21}$$

$$y(0) = y''(0) = y(L) = y''(L) = 0,$$

if the rod is pinned at both ends. E is the Young's modulus and I is the second moment of area about the axis of buckling.

(a) Show that the above (eigenvalue) problem has only the trivial solution $y = 0$ unless $P = P_n = n^2 P_c$, for $n = 1, 2\cdots$, where $P_c = \pi^2 EI/L^2$. If $P = P_n$ show that the solution is given by

$$A_n \sin\frac{n\pi x}{L},$$

with A_n arbitrary.

Given the above results and what one knows about the behaviour of compressed rulers one might anticipate that P_c is the buckling load, and that the zero displacement (straight ruler solution) is the only solution for $P < P_c$. This is the case. When $P = P_c$ the above solution prescribes the shape, $\sin\pi x/L$ (the fundamental mode shape we normally observe in a buckled ruler) but fails to determine the size A_1 of the displacement. Nonlinear terms ignored in the approximating equation (12.21) serve to determine the amplitude of the buckled

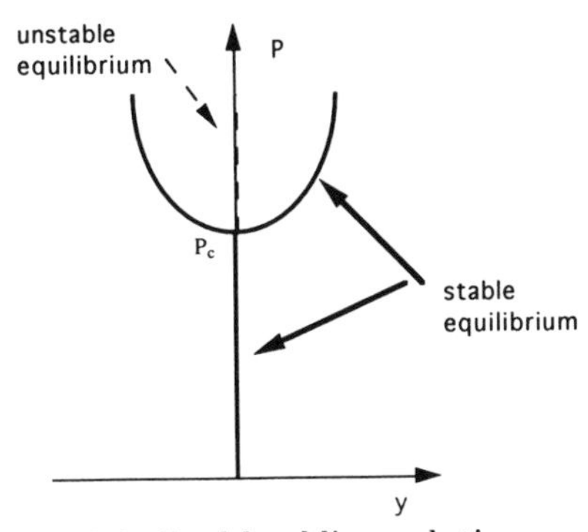

Figure 12.9: Rod buckling solutions.

solution. The procedure for determining A_1 in $P_1 < P < P_2$ is similar to that used for the oscillator. The equilibrium diagram, see Fig. 12.9 is similar in general form to that of the oscillator, with two stable solutions (the symmetric buckled solutions) and one unstable solution (the unbuckled solution) for values of $P_1 < P < P_2$. Note especially that the stable branches come in at right angles to the P axis, which indicates an abrupt change in displacement as P passes through P_c.

Exercise 12.6 *Pendulum Oscillations*

The forced oscillations of a simple pendulum are described by the differential equation

$$\theta'' + (l/g)\sin\theta = a\cos\omega t,$$

so that relatively large oscillations are expected if $\omega = \sqrt{l/g}$.

(a) Assuming that a is sufficiently small so that the angle through which the pendulum is swinging remains small even when the forcing frequency is close to the natural frequency of the pendulum, determine the scale of the angular deflection θ which occurs when the forcing is at or near resonance, and identify the associated frequency bandwidth.
Hint: Neglect the terms higher than third order in the Taylor series expansion of $\sin\theta$ in the first instance and check later that this involves a small effect.

Exercise 12.7 *Quadratic Nonlinearity*

⋆⋆ Consider the system described by the weakly nonlinear ordinary differential equation

$$x'' + (1+\delta)x + \epsilon x^2 = \cos t.$$

The two parameters ϵ and δ are both small, so a relatively large response is expected. We wish to determine response size and resonant bandwidth. Proceed as in the cubic case in Section 3.1. Thus:

(a) Show that if a dominant solution $Ae^{\imath t} + *$ is assumed then the appropriate solution form to handle the quadratic correction term is given by

$$x = Ae^{\imath t} + \epsilon[Be^{2\imath t} + C] + *.$$

On physical grounds is the displacement bias represented by the non-zero average constant term to be expected? Explain.

(b) ⋆⋆ By substituting this form into the equation and determining the coefficients of $(e^{\imath t}, e^{2\imath t}, 1)$, obtain three equations for (A, B, C) and thus show that the resonance band is of "width" $\epsilon^{2/3}$, that the response near resonance is size $\epsilon^{-2/3}$, and that the amplitude of the response is governed by a cubic. You may find it useful to use Maple to deal with the algebra.

Hint: You'll need to approximate the equations for A,B,C. A priori we don't know the appropriate scale to use for δ, so we don't know for example if $\delta \ll \epsilon^2$ etc. We do know, however, that $\delta\epsilon \ll \delta$ etc., and by making use of such evident size relationships one can simplify the equations greatly.

(c) ⋆⋆ Investigate the cubic, plot the amplitude vs. frequency diagram and speculate on the stability of the various branches.

Bibliography

Dym, C. L. (1974). *Stability Theory and its Applications to Structural Mechanics*, Leyden: Noordhoff International Publishing.

Dincă, Florea and Teodosiu, Christian (1973). *Nonlinear and Random Vibrations*, Academic Press, Inc. London.

Chapter 13

Bumpy ride?

with
David Gates
CSIRO Division of Maths and Stats

Averaging
Multiscaling

13.1 Introduction

A car is a compromise of engineering design. For example a light car with strong suspension will accurately respond to the road surface, whereas a heavier car with soft suspension will ride over bumps, producing a more comfortable but perhaps less safe ride. All this we might expect from our knowledge of spring mass systems; for vibration purposes we can think of a car as being just a complicated spring mass system. Simple models can be used to confirm our suspicions, see Exercise 13.1. We also know from our earlier work that, immediately we design a system to be responsive (in the sense of producing a significant modification of an exernal input), we open up the real possibility of resonant interactions—such is life! An interesting case in point occurs on outback roads in Australia where (in response to suspension systems or cars) the gravel roads develop corrugations of the right wavelength for the normal speed of travel to "shake cars to bits", see Exercise 13.1. Damping can of course wipe out transients and generally reduce the amplitude of the responses, but if the forcing is sustained, then it's normally best to design the car so that the natural vibrational frequencies are (as much as is possible) remote from the expected forcing frequencies. The most important of the externally applied forcing frequencies as far as the car body is concerned are those that result from movement along the road

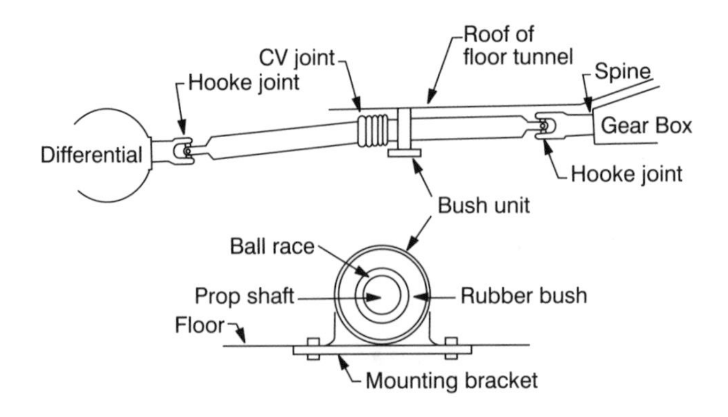

Figure 13.1: The propshaft.

and from engine operation. Of course harmonics are generated by nonlinearities, and couplings between components produce (often unanticipated) responses that may also be damaging to the machinery or uncomfortably close to human seasickness response "frequency"(a major concern for manufacturers). All told it's a complicated system that's designed partially using science and partially using knowledge gained through experience. Often a modeller is brought in to explain an unexpected response. The present problem is a case in point. It was presented to a Mathematics-in-Industry Study Group meeting held at Monash University in 1989 and this particular problem was moderated by David Gates from CSIRO Australia. The present account follows closely the work of this meeting, and provides an interesting insight into how a working group attacks such a problem. For a more complete description the reader is referred to the proceedings, see Barton and Ha (1989).

13.2 The Situation

Cars with rear-wheel drives have a propshaft (or tail-shaft) transmitting rotation from the engine to the differential and from there through to the rear wheels. Many modern cars have a two-piece rather than the traditional one-piece propshaft. The set up is as shown in Fig. 13.1. The cars include Holden Commodores, Nissans and many European makes. The two-piece shaft requires less floor clearance and hence a smaller floor tunnel, and also does not suffer from the high speed twisting resonances to which the one-piece shaft is susceptible.

The two-piece shaft is, however, susceptible to a lower speed, high load (high torque) phenomenon in which the propshaft whirls with increasing amplitude until its restraining rubber bush is fully compressed in its mounting bracket, see Fig. 13.1, causing a **thump** sound—very annoying! The whirling amplitude builds up periodically, so that a regular series of **thumps** is heard at a frequency of about 5 per second.

The effect is very sensitive to the properties of the rubber bush and to its temperature. Different designs of bush are used by different makers, each with the purpose of minimizing both the whirling problem and the transmission of other vibrations to the car body. GMH has a particular design which it would prefer not to vary because of retooling costs. It can, however, vary the properties of the rubber material used in the bush. By trial and error it has achieved rubber properties which control the whirling problem to an acceptable level under normal conditions. GMH posed the question to the group of *whether one can give theoretical guidelines for optimal rubber properties.*

GMH were also anxious to gain a better understanding of the whirling phenomenon, and felt that a mathematical description of the phenomenon was a prerequisite for the determination of the appropriate rubber properties. As well as possibly leading to an improved control of the whirling problem the resulting understanding might also be useful for the design of new or modified propshafts. Possibly the tedious and expensive trial and error process presently used could be avoided if an accurate model could be developed.

The thumping phenomenon is observed at both 40 and 80 kilometres per hour approximately, although the phenomenon has a different quality at these two speeds. The 2:1 ratio between these speeds suggests that a fundamental resonance and a harmonic resonance are responsible. Such multiple resonances are typical of nonlinear oscillating systems as we've seen, so the group looked for a nonlinear driving mechanism.

13.2.1 Possible mechanisms

We need to know more about the propshaft arrangement. Since the drive shaft of the engine and the differential are not aligned and the propshaft is broken, three joints are needed to connect up the arrangement as shown in the figure. For the arrangement of particular interest to GMH two Hooke joints are used to connect the propshaft to the differential and to the engine, while a constant velocity (CV) joint joins the two components of the propshaft. The Hooke joints are the normal joints used to connect shafts, so if you've seen a joint connecting two shafts it's most likely been a Hooke joint; the joint looks like two coupled claws. Of course the aim of all joints is to transmit torque and rotation through an angle. The joints are necessarily such that a full rotation of the driveshaft will result in a full rotation of the propshaft, however the angular displacements (and velocities) of the connecting shafts are not necessarily the same at all stages of the rotation cycle—the average just has to be the same. For a Hooke joint the relation-

ship between these angular displacements depends on the angle between the two connecting shafts, with the amplitude of the angular differences increasing as this angle increases. In Fig. 13.2 the difference in angular displacement of two connecting shafts $(\beta - \phi) - \psi$ is plotted as a function of the angular displacement of the drive shaft, for various angles between the shafts as specified by the angle θ, see Fig. 13.3. The CV (meaning constant angular velocity) joint is a remarkable piece of engineering that manages to transmit angular displacement (and thus angular velocity) without change across the joint, irrespective of angle between the shafts, providing the angle between the shafts is not too great. Such properties enable this joint to smoothly transmit torque even under circumstances when the orientation of the shafts varies. Clearly it's a useful joint to use under steering circumstances! The joint is not used more often because it's expensive and is also subject to heavy wearing. No attempt will be made here to describe the arrangement of ball bearings and machined surfaces that achieves this seemingly impossible task. For a simple explanation of the theory and its practical implementation see McGraw Hill (1992), which also has a useful list of references on automobile dynamics. Now, because the Hooke joints rotate with an angular velocity that varies, oscillatory torques will act on the shafts; so these joints are a prime candidate for exciting oscillations in the system. In fact Fig. 13.2 clearly shows that the Hooke joint introduces oscillations into the system with double the driving frequency.
Aside: The torques introduced by Hooke joints were also known to excite torsional elastic oscillations in one-piece tail-shafts at higher speeds and frequencies, see Zahradka (1973). For the shorter broken tailshafts of interest here such elastic oscillations are not a problem.

In our present set-up the propshaft can move within its restraining rubber bush (of course if it was rigidly constrained, breakages would result), so the natural frequency associated with this connection is extremely relevant. The tail-shaft has a natural frequency of oscillation in its rubber bush of about 40 cps, observed by simply displacing the propshaft in its bush and watching it vibrate when released. Since the rotational speed of the engine varies from 0 to about 120 cps, it would appear that a possible mechanism for generating an instability has been identified.

The participants at the study group set up a miniature model using Lego Technic to see if vibrations could be induced by the Hooke joint, and to try to rule out other possible mechanisms. The configuration employed resembled that shown in Fig. 13.1; however, in the model, Hooke universal joints were used for *all* joints, and rubber bands were used to approximate the rubber bush. The shaft was driven by an electric motor. The shaft-cum-rubber-band system had a natural vibration frequency of about 20 cps. It was found possible to excite resonance at this frequency by tuning the speed of the motor precisely. When the shafts were all in a straight line, no such resonance was found. This rather primitive experiment showed

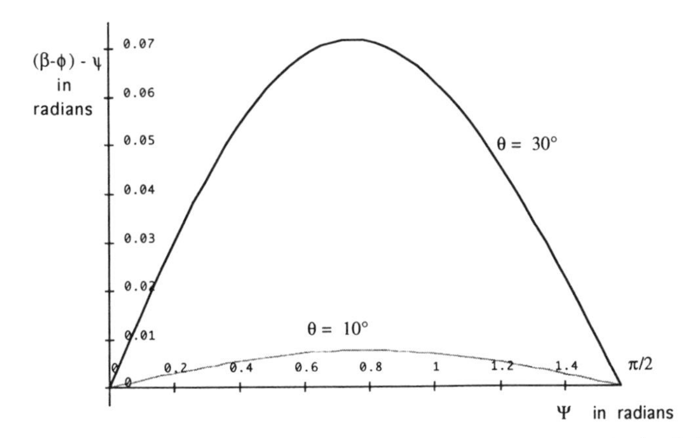

Figure 13.2: Relative angular displacement across a Hooke joint.

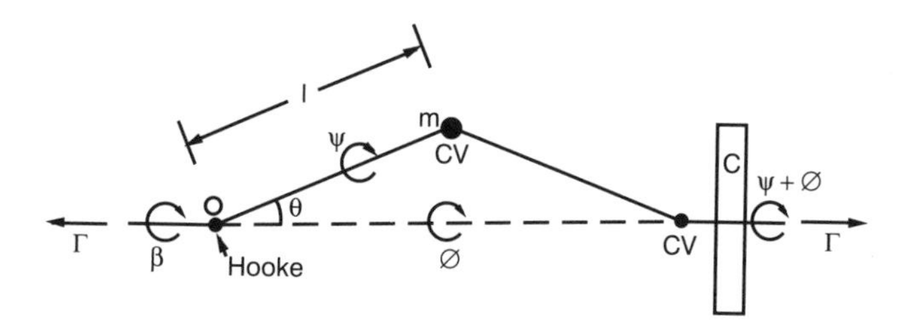

Figure 13.3: The model.

that Hooke joints between angled shafts could be the driving mechanism for instabilities in the real system, and so the dynamics of this effect was studied. Other possible driving forces were discussed and eliminated to the group's satisfaction.

13.3 A Simplified Mathematical Model

In the GMH arrangement the front half of the shaft is nearly parallel to the gearbox shaft so that the front Hooke joint behaves like a CV joint and for simplicity was treated as such. (The real angle at this joint was given as no more than 1°, while the other joints were at varying angles of around 4°, so this approximation seems satisfactory). Thus the situation considered is shown in Fig. 13.3. A single Hooke joint transmits torque through an angle θ. The other joints are assumed to be CV joints; thus they faithfully transmit angular velocity. Also a single point mass m coincident with the CV joint was introduced to model the tailshaft. Very crude! Real tail-shafts have moments of inertia and exhibit gyroscopic effects. There is no great difficulty in including such effects, but the primary aim was to focus on the crux of the resonance phenomena, and the inclusion of all sorts of additional but inessential effects might cloud the issue. The rubber bush was taken as coincident with the mass m.

The angles θ, ϕ, ψ in Fig. 13.3 are the Euler angles of rod Om (specified by O and the location of m) relative to a fixed frame through O. Thus ψ measures the angle of rotation of the shaft Om about its axis (so $\dot{\psi}$ is the angular velocity of the shaft), and ϕ and θ specify the orientation of the axis of this shaft. The angle θ measures the angular displacement of the shaft relative to the drive shaft as shown in Fig. 13.3; and ϕ measures the angular displacement of the plane of these shafts with the vertical i.e. into the page on the figure. Thus $\phi = 0$ under static conditions and changes only when the CV joint moves in the bush; it's this (very small) movement (mms) that is of interest. The angle θ was taken as fixed because this was known to be the case under operating circumstances. A flywheel of moment of inertia C was located as shown in the model. This was intended to represent the inertia of all the rotating car components connected to the tail-shaft. Thus C would include the rotating parts of the engine, gear box, differential, axles and the wheels etc. It's not easy to attach an actual value to C because it's not clear how the various components contribute to the inertia of the system; hopefully this is not an important matter.

Now the torque applied by the motor causes the car to move along the road, so the work done by the torque enables the car to increase its kinetic energy, go uphill, or simply compensate for frictional losses. Of course it makes sense to avoid dealing with such "global" issues if this is at all possible, even though one recognizes that the initiation of the instability or even it's evolution will be influenced by such dynamics. It's often not easy to see the best way to decouple one particular aspect of a problem from the problem as a whole, and still retain the essential physics. It was known that the propshaft instability is generated sporadically under conditions in which the propshaft is "loaded", so with this in mind the group decided to investigate the situation in which (known) equal and opposite torques Γ are applied at the two ends of the system envisaged in Fig. 13.3, with the shafts rotating. In this way global dynamic considerations of the above type were avoided. Also damping forces were not included—such forces are unlikely to affect the onset of the instability. If necessary in later models one might hope to study the influence of the road, the motor, and the driver's response on the propshaft instability by inputting various $\Gamma(t)$'s.

Under the circumstances considered the system (with Γ's included) is a self-contained system, and energy can be exchanged between the mass m, the flywheel, the torques, and the rubber bush. The whirling phenomenon would correspond to small amplitude oscillations in ϕ, that would be induced by the couplings in the system. The question of interest is: with this particular set-up can a significant amount of energy be channelled into this whirling motion? If so, then it's likely that under realistic operating circumstances such an energy transfer will also occur.

We've learned earlier in the Moorings Chapter that a Lagrangian formulation is advantageous when dealing with complex situations, especially when the forces involved are not of direct interest, and when generalized co-ordinates are being used. The present situation is a case in point. Specifying the forces and torques acting at the joints that cause the vibration would be something of a nightmare in the present case; the Lagrangian formulation avoids this difficulty entirely.

Only the forces and torques that do work on the system as a whole (the Γ's and the forces on the shafts due to the rubber bush) need to be considered. The system has only 2 degrees of freedom, described by the 2 generalized coordinates ϕ and ψ. The Lagrange equations for the propshaft including the bush take the form (dots denote time derivatives).

$$\begin{aligned} \frac{d}{dt}\left(\frac{\partial T}{\partial \dot{\phi}}\right) - \frac{\partial T}{\partial \phi} &= -\frac{\partial V}{\partial \phi} \\ \frac{d}{dt}\left(\frac{\partial T}{\partial \dot{\psi}}\right) - \frac{\partial T}{\partial \psi} &= -\frac{\partial V}{\partial \psi} \end{aligned} \tag{13.1}$$

where T is the kinetic energy and V is the potential energy of the system, see equation (4.21) in Chapter 4. The flywheel has angular velocity $\dot{\psi} + \dot{\phi}$ as can be seen from Fig. 13.3. The mass m has linear velocity $\dot{\phi}\ell \sin\theta$ with ℓ being the distance from O to m as shown in Fig. 13.3. Thus

$$T = \frac{1}{2}m(\dot{\phi}\ell\sin\theta)^2 + \frac{1}{2}C(\dot{\psi} + \dot{\phi})^2. \tag{13.2}$$

We suppose for simplicity that the rubber bush is ideally elastic and exerts a force ks where k is a stiffness constant and $s = \phi\ell\sin\theta$ is the displacement of m along its circular path. Thus the potential energy stored in the rubber bush is $\frac{1}{2}ks^2$, see (4.17). The potential energy due to the torques Γ is given by the negative of the work done by the torques in reaching the prescribed arrangement. Since the net angular displacement across the propshaft is $\psi+\phi-\beta$, where β is the angular displacement of the shaft on the differential side of the Hooke joint, see Fig. 13.3, we have

$$V = \frac{1}{2}k(\phi\ell\sin\theta)^2 + \Gamma(\psi + \phi - \beta), \tag{13.3}$$

where the small change of gravitational potential energy due to the movement of m is ignored.

A standard relation connecting the angular displacements $\beta - \phi$ and ψ on the two sides of a Hooke joint is, see Wagner and Cooney (1979),

$$\beta = \phi + \arcsin\left[\frac{\sin\psi}{(1 - \sin^2\theta\cos^2\psi)^{\frac{1}{2}}}\right]. \tag{13.4}$$

Aside: Modellers are not expected to have such technical information at their fingertips. Modelling is almost always a joint effort with the modeller relying very heavily on the experience and knowledge base of the customer.

The above equations completely specify the model and we have the right number of equations for the number of unknowns, however, the equations are not in a useful form and will need further processing before an attempt is made to solve them. Any attempt to process the equations numerically at this stage would be pointless. The chances of homing in on the instability would be almost zero, and if one "lucked in" what use would the results be? Maple could be used to assist in the simplification process, but sometimes it's best to just work by hand; algebraic packages are accurate but clumsy. Although it's nice to see a complex situation clarified, it doesn't make for easy reading and so this work is relegated to the Appendix. Apart from minor rearrangements, the simplifications make use of the fact that θ is small, and the moment of inertia of the flywheel is much greater than that of the propshaft about the driveshaft axis (i.e. $C \gg m\ell^2\sin^2\theta$). This processing eventually leads, see (13.38), to the single equation

$$y'' + y = \alpha\cos(y - \eta t), \tag{13.5}$$

for the whirling motion. Here $y = 2\phi$, and a time scale based on the natural frequency ω of the propshaft/bush connection has been adopted. Explicitly

$$y = 2\phi, \quad t' = \omega t, \text{ where } \omega = \sqrt{k/m}, \text{ and } ' \text{ denotes } \frac{d}{dt'},$$

$$\text{then } t' \Longrightarrow t,$$

purely for notational convenience. The dimensionless groups that arise are

$$\eta = 2N, \text{ where } N = \dot{\psi}(0)/\omega, \text{ and } \alpha = -\Gamma/k\ell^2. \tag{13.6}$$

The "tuning parameter" $N \equiv \eta/2$ is the ratio of the initial frequency of rotation of the drive shaft to the natural freqeuncy of the bush/propshaft connection, and we expect it to be the critical parameter of the problem. The "coupling parameter" α is related to the torque Γ and the propshaft bush coupling. Based on data given for VN Commodores $\alpha \approx .7 \times 10^{-4}$— very small!. The physical picture that fits the above equation is as follows: In the absence of an applied torque Γ, if the propshaft is displaced in its bush and then released, it will simply vibrate about its equilibrium

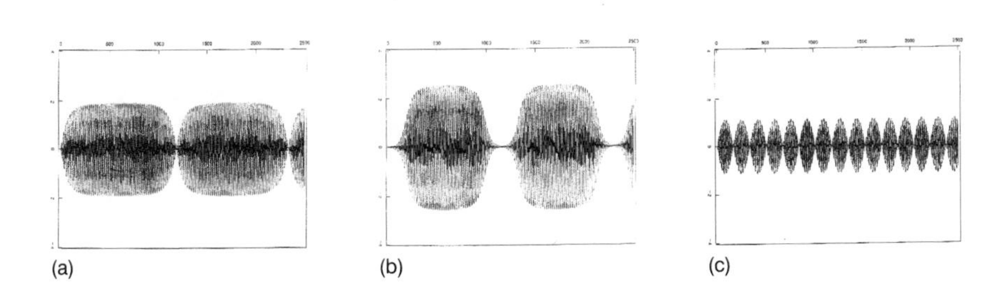

Figure 13.4: Resonant and near resonant oscillations. Case (a) is the fundamental resonance with $\eta = 1$. Case (b) is the first overtone with $\eta = 2$ Case (c) has $\eta = 0.98$ i.e. just below resonant speed.

location with the natural frequency of propshaft/bush system (1 with our scaling). Under operating conditions with Γ non-zero the applied torque will influence the propshaft/bush vibration in a complicated way that's specified by the coupling term. Our objective is to determine the conditions under which a small initial propshaft displacement

$$y'(0) = 0, \; y(0) = Y_0(\text{say})$$

will be magnified by the coupling. This is a *stability* question.

The mathematical nature of the problem is also reasonably clear from the defining equation (13.5). Since α is small, for small values of Y_0 the right hand side of this equation approximates to $\alpha \cos \eta t$, so that the forcing is small and is likely to produce a small response (of expected size α) *unless* $\eta \approx 1$; in which case the forcing is resonant. Under such resonant circumstances one might expect the amplitude to increase until the awkward cosine nonlinearity detunes the system, thus limiting its size. All this we might anticipate from our earlier experience with resonant systems. Of course we need to justify and quantify these intuitive ideas and in this regard a little preliminary numerical experimentation is in order.

13.4 The Resonant Behaviour

Some indication of the behaviour of the solutions can be obtained using numerical integration routines. Fig. 13.4 gives a suite of solutions of (13.5) for a several values of η and with $y(0) = 0$; obtained by numerical computation using IMSL routine IVPRK. Notice that the solutions obtained consist of amplitude modulated vibrations with the modulation occurring

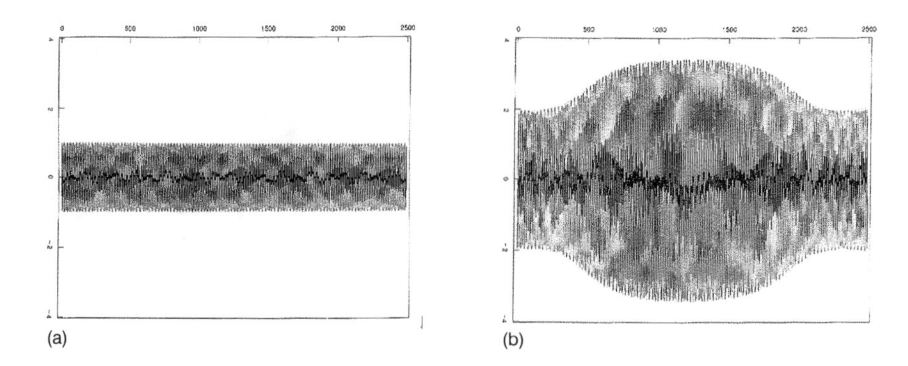

Figure 13.5: Fundamental resonant oscillations ($\eta = 1$), obtained by using different values of $y(0)$.

over a slow time scale. A value $\alpha = 0.025$ (somewhat larger than its true value) was used for these calculations so that the individual oscillations could be seen. Note the expected resonance at $\eta = 1$ (Case (a)). Note also that there are large responses at $\eta = 1/2$ and $\eta = 2$ (see Case (b)). In all cases the large amplitudes do not persist, but rather they grow and die away in a periodic fashion. As η moves away from a resonant value, the period of this **envelope** decreases quite rapidly as does the amplitude (see Case (c)). Changing the initial condition markedly affects the shape of the envelope, see Fig. 13.5. The periodic surging in the resonant behaviour described above seems quite close to GMH's description of the thumping phenomenon. The peak amplitudes are potential thumps, since the peaks occur at a frequency much less than η.
Important Comment: Although the results of the numerical calculations are encouraging one should place no great reliance on them. In order to pick up the amplitude modulations of interest it's necessary to numerically integrate the differential equation over a very large number of solution cycles. Even with small truncation error routines, and small time steps, the error will build up and could swamp the subtle effects that are of concern. *It is in the very nature of resonance phenomena that very small phase or frequency shifts can change the response drastically!* It's thus necessary to process the equations further.

13.4.1 A preliminary analysis

Encouraged by the above numerical experiments we now proceed with an analytic investigation. The present problem differs from the problems examined in the last chapter in that a fixed amplitude oscillation doesn't persist. Here the interest lies in describing the relatively slow modulations of amplitude that occur, and determining how they are influenced by the parameters of the problem. The approach to the problem is, however, basically the same: the amplitude and phase need to be selected so that the resonant terms in the defining equation balance. There are two techniques based on this underlying theme that have been used to solve such problems; **averaging** techniques that are closely related to those used in the last chapter, and **multiscaling** (sometimes called **two-timing**) techniques. In cases in which they're both applicable the two techniques seem to always produce equivalent results. It's useful, however, to first proceed with a normal perturbation analysis. Such an analysis both serves to identify the instability bands and to determine the appropriate solution form to adopt to capture the slow variations. For $\alpha << 1$ and Y_0 small we can obtain a solution $y(t)$ of equation (13.5) in powers of α via the recursion

$$\begin{aligned} y_0'' + y_0 &= 0 \\ y_{j+1}'' + y_{j+1} &= \alpha \cos(y_j - \eta t) \end{aligned} \tag{13.7}$$

for $j = 0, 1, 2, \ldots,$ with

$$y_j(0) = Y_0, \text{ and } y_j'(0) = 0.$$

Thus

$$y_0 = Y_0 \cos t.$$

Neglecting terms of order (αY_0) by comparison with terms of order α the equation for y_1 becomes

$$y_1'' + y_1 = \alpha \cos(\eta t), \tag{13.8}$$

with the solution

$$y_1 = \frac{\alpha}{1 - \eta^2}[\cos(\eta t) - \cos t] + Y_0 \cos t. \tag{13.9}$$

Aside: If the term of order αY_0 is retained the analysis is more complicated but the implications are unaltered.

The result intuitively anticipated earlier is now expressed mathematically. Thus for η not close to 1, the amplitude of the whirling is expected to remain of size Y_0, and the effect of the nonlinearity is simply to introduce a slight modification (of size α). If η is close to 1, however, then the anticipated *fundamental* resonance is evident, and it's clear that y_0 no longer represents a good first estimate for the solution behaviour.

So what happens for $\eta \approx 1$? The fact that the normal recursive scheme fails indicates that the first approximation y_0 is (at the very least) not accurate for $\eta \approx 1$, but more importantly that the response size is likely to be larger than that suggested by this first term. So how to proceed?

The simple ad hoc procedure used to determine the instability bandwidth for nonlinear oscillators in the last chapter extends. One simply identifies the conditions under which the normal perturbation fails. The solution given above doesn't apply when $\eta = 1$ so we'll focus our attention on the response at this frequency. If $\eta = 1$ the solution of the approximate equation (13.8) for y_1 is given by

$$y_1 = Y_0 \cos t + \alpha \frac{1}{2} t \sin t,$$

and the thing to notice is that the zeroth order term $Y_0 \cos t$ fails to adequately represent the solution for values of t of order $1/\alpha$. In spite of the fact that the expansion fails if t is of this order the result *suggests* that there will be a growth in amplitude of the resonant mode over a time scale of order $1/\alpha$; a simple observation, but we'll see that this leads to the appropriate solution form.

⋆ Overtone and undertone resonances

The above analysis exposes difficulties associated with describing the fundamental resonant mode and suggests an expected time span for amplitude growth in this case. The analysis to date does not, however, indicate a large response in other frequency bands. You'll recall that the numerical experiments displayed large responses at values of $\eta = 1/2, 2$. To expose these difficulties the recursion (13.7) needs to be carried through to the $j = 2$ stage. The calculations are tedious by hand and the resulting expressions long. It's appropriate to use Maple which produces an equation for y_2 of the form

$$y_2'' + y_2 = \{\alpha^2 \sin(2\eta t), \alpha^2 \sin(\eta - 1)t, \alpha \cos(\eta t), \cdot\},$$

where the functional dependence on t and the size (in terms of α) of relevant terms that arise are indicated, and where the dot indicates terms that do not result in "resonant" behaviour. The indicated terms are of frequency 1 (and thus produce unbounded solutions for y_2) if

$$\eta = 1, \ 2\eta = 1 \text{ or } \eta - 1 = 1,$$

so that (apart from the $\eta = 1$ situation already identified) a large amplitude resonant behaviour may be anticipated in the solution for y_2 when $\eta = 2, 1/2$, which fits in with the numerical work. The values $\eta = 1/2$ and 2 are the first undertone and the first overtone of the fundamental resonance.

For both cases the right hand side of the equation for y_2 will contain a term of the form $\alpha^2 \sin t$, and the associated particular solution corresponding to this forcing term is $-\alpha^2(1/2)t \cos t$. Thus it can be seen that the corresponding solution for y_2 will contain terms like $\alpha^2 t \cos t$. This suggests that the amplitude of the response at the undertone and overtone frequencies will develop over a time span of order $1/\alpha^2$; a result that's consistent with the numerical experiments shown in Fig.13.4(b). Again all this may seem to be unacceptably speculative—but the procedure has been found to be remarkably successful!

Summary The above preliminary analysis suggests that large responses may occur when $\eta = 1/2, 1, 2$. It seems likely that the fundamental ($\eta = 1$) and first overtone ($\eta = 2$) resonances correspond to the observed thumping phenomena at 40 kilometres per hour and 80 kilometres per hour. To justify the above mathematical speculations we need to obtain valid solution representations for frequencies in the resonant bands, which we now proceed to do.

13.4.2 Averaging

We'll restrict our attention to the fundamental $\eta = 1$ instability band. The other bands may be examined in the same way. Our expectation from the above work is that the dominant mode is of the form

$$y_0 = A(t)\cos[\eta t + \Theta(t)] \tag{13.10}$$

where the amplitude A and phase Θ vary significantly over a time scale of order $1/\alpha$. Under such circumstances $\frac{dA}{dt}, \frac{d\Theta}{dt}$ are of order α; $\frac{d^2A}{dt^2}, \frac{d^2A}{dt^2}$ of order α^2, so $\frac{d^2A}{dt^2} \ll \frac{dA}{dt} \ll 1$ etc.; so that, substituting the dominant solution form into the defining equation (13.5), and neglecting terms of order α^2 we obtain

$$\{A(1-\eta^2) - 2A\eta\Theta'\}\cos[\cdot] - \{2\eta A'\}\sin[\cdot] = \alpha\cos(A\cos[\cdot] - \eta t), \tag{13.11}$$

where for clarity of presentation the notation

$$[\cdot] \equiv [\eta t + \Theta] \tag{13.12}$$

is used to indicate the argument of the dominant mode. To ensure accuracy it's useful to use Maple here. As in the nonlinear oscillator problems of Chapter 12 the amplitude A and phase Θ should be chosen so that the dominant mode terms of this equation balance. The term on the right hand side of the equation is a little more awkward than we've come across before. In the examples of the last chapter you recall that we extracted the Fourier components that were "tuned" to the natural frequency. In the present case the nonlinear term that arises is not a periodic function (of period $2\pi/\eta$) because A and Θ are not constants; so that strictly speaking this term can't be represented by a Fourier series expansion. However, for small values of α, we expect that A and Θ will vary by so little over the period $2\pi/\eta$ that if we ignore this change (and thus treat the A and Θ

and their derivatives as constants) the Fourier representation thus obtained should provide an accurate estimate for the dominant forcing components associated with the nonlinearity. Hopefully any introduced error can be accounted for in later expansion terms. This is the idea underlying the method.

The upshot of the above rationalization is that to isolate out the sin[·] mode from the above equation (13.11) we simply multiply the equation through by sin[·] and integrate over a period of the variable [·], treating A, A' and Θ, Θ' as being constants for the calculation. Effectively we filter out the other Fourier components by this procedure. In the same way the cos[·] component can be extracted. This approach to our problem represents a neat trick based on sound physical insight that has been successful for handling subtle *singular perturbation* problems arising in many contexts. For obvious reasons the procedure is referred to as *averaging*. We obtain in this way the following equations

$$-2\eta A' = \alpha \mathcal{I}_s \quad \text{the } \sin[\cdot] \text{ terms} \tag{13.13}$$

$$A(1-\eta^2) - 2A\eta\Theta' = \alpha \mathcal{I}_c \quad \text{the } \cos[\cdot] \text{ terms,} \tag{13.14}$$

where

$$\mathcal{I}_s(A,\Theta) = \int_0^{2\pi} \{\cos(A\cos[\cdot] - \eta t)\} \cdot \sin[\cdot] d[\cdot], \tag{13.15}$$

and

$$\mathcal{I}_c(A,\Theta) = \int_0^{2\pi} \{\cos(A\cos[\cdot] - \eta t)\} \cdot \cos[\cdot] d[\cdot], \tag{13.16}$$

are the required Fourier components associated with the nonlinear term. Ignoring for the moment potential difficulties one might have evaluating these integrals, we can see that the above equations are simply two ordinary differential equations for the phase and amplitude of the dominant mode. A balance will be realized between the terms of these equations over a time scale of order $1/\alpha$, a result anticipated earlier, and over a bandwidth given by $(1-\eta)$ of order α. Thus with

$$\tau = \alpha t, \quad (1-\eta) = \alpha\sigma \tag{13.17}$$

the equations become (to first order in α)

$$-2A_\tau = \mathcal{I}_s, \tag{13.18}$$

$$2A\sigma - 2A\Theta_\tau = \mathcal{I}_c. \tag{13.19}$$

These two equations determine the amplitude and phase modulations of interest. Before proceeding to analyse them we'll examine an alternative approach to such problems, so that comparisons can be made. *Both* approaches are indispensable to the modeller. For an extensive account of averaging techniques see Bogolyubov and Mitropolsky (1955) (the originators of the technique) or Nayfeh (1973).

⋆ Multiscaling

Although the averaging procedure makes physical sense and has been strikingly successful, the procedure is somewhat unsatisfactory from the mathematical point of view—it seems somewhat ad hoc to treat $A(t)$ as being a constant for one stage of a calculation (when integrating over a period) and then take into account the variation with time at another stage (when solving the amplitude and phase equations). There's no doubting the correctness of the results obtained in most cases using this approach; the procedure produces verifiable results. This has prompted many theoretical attempts to validate the procedure, and also has led to the introduction of the variants that seem more mathematically acceptable, or more appropriate for handling particular problems. The multiscaling technique now to be described is more mathematically satisfactory. For a more complete account of multiscaling see Cole (1968). For the purposes of clarity of exposition we'll restrict our attention to the determination of the response *at* the fundamental resonance frequency $\eta = 1$. Minor modifications are required to determine the response for frequencies in the resonant band surrounding $\eta = 1$.

The multiscaling idea is simple and very pretty. We've seen that there are two *almost* separate processes going on: a basic oscillation (over a time scale of unit order), and a process that determines the amplitude variations (with a "long" time scale $1/\alpha$). Under the circumstances it seems sensible to introduce two time scales and correspondingly two time variables: the oscillation time t, and a "slow" time variable $\tau = \alpha t$. Mathematically it's legitimate to introduce as many time-like variables as we wish, although it seemingly adds unnecessary complications. Why convert an ordinary differential equation in t to a partial differential equation in the two variables t and τ?—partial differential equations are normally *much* harder to solve! The point is that this is a convenient vehicle for isolating out the two processes. We thus introduce the two time variables t and τ and for the moment forget that they are related in an obvious way. Later, of course, we'll remember the connection. Thus the solution $y(t)$ is no longer treated as a function of the single variable t, but as a function of the two variables (t, τ). Thus we write $y(t, \alpha) \equiv Y(t, \tau, \alpha)$, in terms of which we have

$$\frac{dy(t)}{dt} = Y_t(t, \tau, \alpha) + \alpha Y_\tau(t, \tau, \alpha),$$

and a second differentiation gives

$$\frac{d^2y(t)}{dt^2} = Y_{tt}(t, \tau, \alpha) + 2\alpha Y_{t\tau}(t, \tau, \alpha) + \alpha^2 Y_{\tau\tau}(t, \tau, \alpha).$$

Thus in terms of the two time variables the equation for $y(t, \alpha)$ becomes

$$Y_{tt} + Y = \alpha\{\cos[Y - t] - 2Y_{t\tau}\} - \alpha^2 Y_{tt}. \tag{13.20}$$

The perturbation expansion

$$Y(t,\tau,\alpha) = Y^0(t,\tau) + \alpha Y^1(t,\tau) + \cdots \tag{13.21}$$

suggests itself, and the equations

$$\begin{aligned} Y^0_{tt} + Y^0 &= 0, \\ Y^1_{tt} + Y^1 &= -2Y^0_{t\tau} + \cos(Y^0 - t), \\ \cdots &= \cdots, \end{aligned}$$

result. The first of these equations is effectively an ordinary differential equation in t with τ occurring as a parameter so it integrates to give

$$Y^0 = A(\tau)\cos[t + \Theta(\tau)], \tag{13.22}$$

where it should be noted that *arbitrary* amplitude $A(\tau)$ and phase functions $\Theta(\tau)$ of the *slow* time scale τ have arisen naturally. The relationship with the earlier work is evident. We'll again adopt the notation $[\cdot]$ for the argument $[t + \Theta]$ of the dominant mode, see (13.12) (in the present work we have $\eta = 1$). To next order we get

$$Y^1_{tt} + Y^1 = -2\{-A_\tau \sin[\cdot] - A\Theta_\tau \cos[\cdot]\} + \cos(A(\tau)\cos[\cdot] - t). \tag{13.23}$$

Again the similarity to the earlier work should be noted. This is of course effectively a non-homogeneous ordinary differential equation in the t variable, so it's a simple matter to write down an explicit particular solution. The particular solutions corresponding to the resonant $\cos[\cdot]$ and $\sin[\cdot]$ terms will be of the form $t\sin[\cdot]$ and $t\cos[\cdot]$ and such terms render the expansion $Y^0 + Y^1$ useless for αt of unit order—because the perturbation term αY^1 ceases to be small compared to the first term Y^0 for such t values. If, however, the (as yet undetermined) functions $A(\tau)$ and $\Theta(\tau)$ are chosen in such a way that the resonant terms are eliminated from the right hand side, then αY^1 will be in fact of order α, and thus uniformly small compared with Y^0. Using, as before, the orthogonality result for the Fourier terms we see that the removal of the resonant $\cos[\cdot]$ and $\sin[\cdot]$ terms requires

$$\begin{aligned} \int_0^{2\pi} \mathcal{RHS} \cdot \cos[\cdot] d[\cdot] &= 0, \\ \int_0^{2\pi} \mathcal{RHS} \cdot \sin[\cdot] d[\cdot] &= 0, \end{aligned}$$

where $\mathcal{RHS}$ denotes the right hand side of equation (13.23). The resulting ordinary differential equations for $A(\tau)$ and $\Theta(\tau)$ are the same as obtained earlier using averaging.

As you can see the approach is rather more satisfactory than the averaging technique but sometimes the averaging technique is easier to apply. Higher order expansion terms can be extracted using either method but the algebra becomes tedious unless an algebraic package is used.

13.4.3 The amplitude and phase functions

It would be difficult to proceed without detailed analytic information about the integrals $\mathcal{I}_c, \mathcal{I}_s$. Unfortunately Maple fails to recognize the integrals so it's necessary to search through one of the standard references on integrals to see if evaluations in terms of standard functions are available. One finds, see Prudnikov, Brychkov and Mariechev (1986) or Gradshteyn and Ryzhik (1965), that the integrals are closely related with standard integrals that arise in association with the Bessel functions.
Aside: The Bessel functions arise when one uses separation of variable procedures to determine solutions of the heat and wave equations in cylindrical geometry, and so are well documented.
Explicitly the tables give:

$$\int_0^{2\pi} dx \left.\begin{array}{c}\sin(z\cos x)\\ \cos(z\cos x)\end{array}\right\} \sin(nx) = 0,$$

and

$$\frac{1}{2\pi}\int_0^{2\pi} dx \left.\begin{array}{c}\sin(z\cos x)\\ \cos(z\cos x)\end{array}\right\} \cos(nx) = \left.\begin{array}{c}\sin(n\pi/2)\\ \cos(n\pi/2)\end{array}\right\} J_n(z),$$

for all integers n, where the J_n are Bessel functions. Our knowledge of the expansion formulae for the oscillatory functions suggests that it should be possible to manipulate the integrals $\mathcal{I}_c$ and $\mathcal{I}_s$ of interest, see (13.15,13.16), into combinations of the above form, and so identify them. Of course Maple can assist. The following instructions lead to the required forms:

```
% In the following Icexp refers to the integrand of I_s,
Icexp:=(cos(y0-etat))*sin(brak); % y0 ≡ y_0, etat ≡ ηt, brak ≡ [·]
expand('');
subs(etat=brak-theta,''); % theta ≡ Θ
expand('');
subs((cos(brak))∧2=(1+cos(2*brak))/2,'');    % change to
multiple angles
subs(sin(brak)=sin(2*brak))/(2 cos(brak),''); % change to
multiple angles
```

The integrand is then in the required form to enable identification. In the same way $\mathcal{I}_c$ can be identified. The results obtained are given by

$$\mathcal{I}_s = \pi[J_0(A) + J_2(A)]\sin\Theta \equiv 2\pi[J_1(A)/A]\sin\Theta \qquad (13.24)$$

$$\mathcal{I}_c = \pi[J_0(A) - J_2(A)]\cos\Theta \equiv 2\pi[J_1'(A)]\cos\Theta, \qquad (13.25)$$

where the standard results

$$J_0(A) - J_2(A) = 2J_1'(A) \text{ and}$$

$$J_0(A) + J_2(A) = 2J_1(A)/A$$

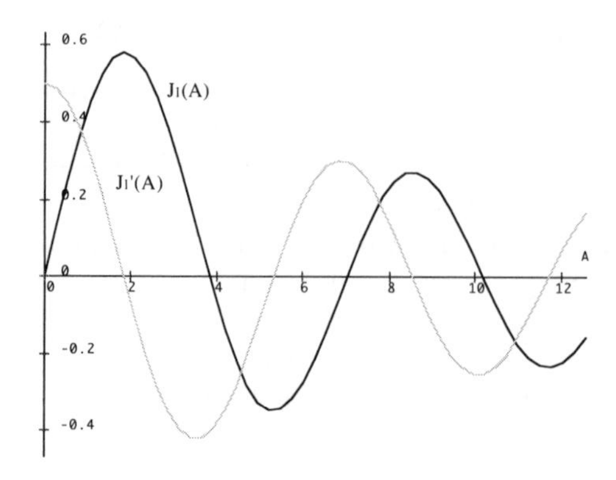

Figure 13.6: The Bessel functions $J_1(A), J_1'(A)$.

connecting the Bessel functions have been employed, see Abramowitz and Stegun (1965).

So, after that irritating diversion (mathematics isn't always fun), using (13.15,13.16) we obtain the following equations for the amplitude and phase of the vibration:

$$\begin{aligned} \frac{dA}{d\tau} &= -\pi[J_1(A)/A]\sin\Theta \\ \frac{d\Theta}{d\tau} &= \sigma - \pi[J_1'(A)/A]\cos\Theta, \end{aligned} \tag{13.26}$$

where $\tau = \alpha t$, the long time variable.

In a real sense the problem is essentially solved at this stage. The point to notice is that the small but important parameter α that made integration of equation (13.5) for $y(t)$ difficult is no longer present in this equation set—so that any reasonable numerical scheme would produce accurate results. Essentially, by averaging, the underlying oscillation has been removed, exposing the subtle processes that determine changes in amplitude of this oscillation. Of course, if possible, one should check to see if the higher order terms of the solution scheme are in fact small so that the improvement isn't just illusory. Often it's difficult to do this, and the results are judged by their effectiveness in describing the phenomenon of interest.

We'll first examine the $\sigma = 0$ case. The describing equations (13.26) for (A, Θ) are autonomous, so a phase plane procedure is appropriate for determining the behaviour of the solution. Thus, dividing the above equations we get

$$\frac{d\theta}{dA} = \frac{J_1'(A)\cos\Theta}{J_1(A)\sin\Theta}. \tag{13.27}$$

Maple recognizes the Bessel functions which are displayed in Fig.13.6, using

```
plot({BesselJ(1,A),(BesselJ(0,A)-BesselJ(2,A))/2},A=0..4*Pi);
```

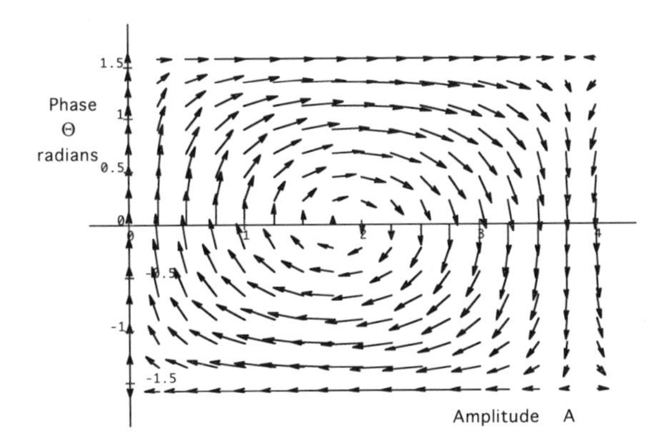

Figure 13.7: A phase plane plot of $\Theta(A)$ ($\sigma = 0$ case).

The solution trajectories in the (Θ, A) phase plane, are then obtained using:

```
with(ODE);
exp2:=(t,A,theta)→ ((BesselJ(0,A)-BesselJ(2,A))/2)*cos(theta);
% the numerator
exp1:=(t,A,theta)→ BesselJ(1,A)*sin(theta);% the denominator
phaseplot([exp1,exp2],0..4,-Pi/2..Pi/2);
```

and are shown in Fig.13.7. It's not clear from this plot whether the trajectories are closed (in which case the amplitude will display a purely cyclic behaviour), or if the trajectories spiral (in or out). Clearly this is an important practical matter. To settle the issue one needs to examine the nature of the singular points located at the points where $J_1'(A) = 0$ and $\sin\Theta = 0$. This analysis shows that the trajectories are closed, so the solutions for (Θ, A) exhibit periodic behaviour.

All the above could have been bypassed by noting that the equations (13.26) imply

$$[J_1(A)\cos\Theta - \frac{1}{2\pi}\sigma A^2]' = 0, \quad \text{so that,}$$

$$J_1(A)\cos\Theta - \frac{1}{2\pi}\sigma A^2 = K, \tag{13.28}$$

for constant K. The resulting family of curves in the (A, Θ) plane is shown in Fig.13.8 for various values of K in the $\sigma = 0$ case. The particular cyclic path followed by the system depends on the choice of $y(0)$, and there are various modes possible as displayed in the figure. However, the trajectories of interest in this problem start from small values of A corresponding to small values of the initial bush displacement $y(0)$; so mode 1 trajectories commencing close to $A = 0$ are relevant. Note that the amplitude $A \equiv 2\phi$ increases from small values to a value close to the root A_0 of the equation $J_1(A_0) = 0$ ($A_0 \approx 3.8$), as can be seen from equation (13.26), and then

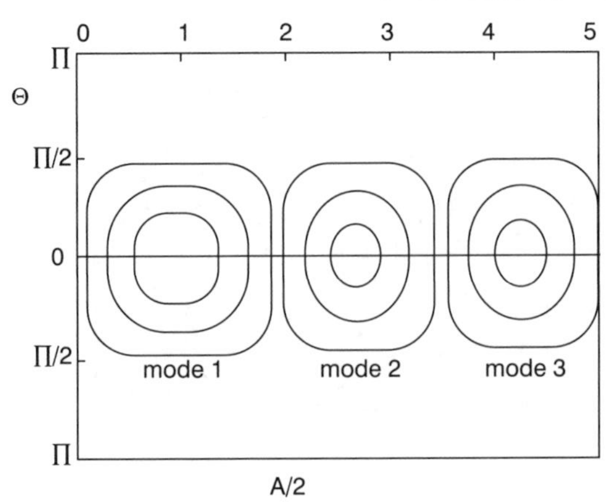

Figure 13.8: An accurate phase plane plot.

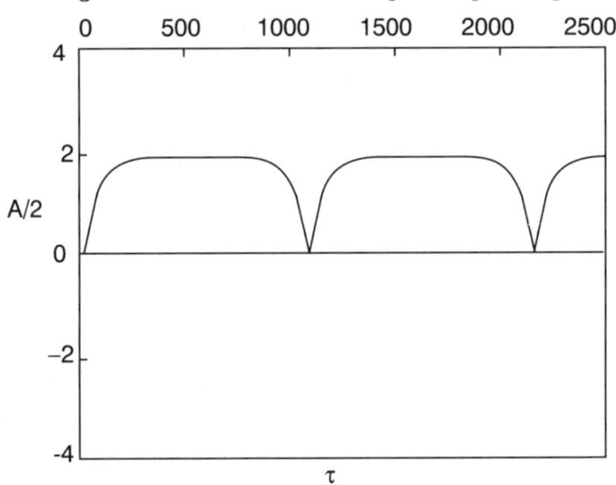

Figure 13.9: Amplitude variations.

decays back to zero. The sitution then repeats. This gives a maximum amplitude of vibration $\phi = A/2$ of about 1.9 radians. Fig. 13.9 shows a numerical solution of the envelope equations (13.26) with $A(0) = 0.01$, $\Theta(0) = 0$ in the $\sigma = 0$ case. The resulting envelope is evidently very close to the simulation shown in Fig. 13.4(a), which increases one's confidence in the numerical simulations. Thus the theory predicts that the amplitude of vibration will build up from any initial $\phi_0 > 0$ value, *however small.* Note also that a change in the initial state of the oscillator will affect the trajectory in the phase plane chosen; the simulations shown in Fig. 13.5 make sense.

Interpretation

The model shows that Hooke's universal joints can cause unit order whirling (ϕ oscillations) in the propshaft at the natural frequency of the bush propshaft system, and that the amplitude can build up from very small initial values. Similar results can be shown for the overtones and undertones. The model predictions can't be believed once the amplitude of whirling gets to be moderate, because the assumed linear elastic response of the bush rubber can't be believed and the bush housing will restrict the amplitude of movement to small values. In practice one would expect the whirling amplitude to grow over the time scale $1/\alpha$ predicted by the model until the CV joint thumps the housing; the resulting energy losses will cause a reduction of amplitude, and then again the above mechanisms will cause a build up of the whirling motion etc. There seems to be little point in trying to understand the thumping dynamics, since the aim is to avoid the onset of such whirling motion. The important feature is that the mechanisms examined in the model calculations will cause a build up in the amplitude of whirling from initially small values, so that the periodic thumping observed is consistent with the model results. Now there are a number of objections to the present model that need further investigation, for example:

- Rotating bodies (because of their angular momentum) resist changes in the direction of the axis of rotation so it's likely the rotation of the shafts could act to stabilize the vibration (a better expression for kinetic energy T, see (13.2),is required).

- The geometric effects associated with our propshaft are relatively small $(\frac{\partial\beta}{\partial\psi})\sin^2\theta$ with θ small, see equation (13.30), so that unbalanced torques associated with global car motion (that have been ignored in the model) could well influence the behaviour of the instability.

In spite of the above limitations (and others) the model seems to fit observations quite well, with the observed thumping at 40 kph and 80 kph corresponding to the two largest resonances; the fundamental (natural) frequency and its first overtone. A few implications of the model results will be examined.

The rubber property in our model is the linear elastic stiffness constant k of the bush. There is the obvious dependence of the resonant engine speeds $\frac{1}{4}\omega$, $\frac{1}{2}\omega$ and ω (radians per second) on k through $\omega = (k/m)^{\frac{1}{2}}$. Increasing stiffness pushes up these speeds. Further the thumping frequency (i.e. the envelope frequency) is proportional to $1/\alpha$, and hence to Γ/k. Thus the model suggests that increasing stiffness causes the frequency of thumping to decrease. Note that the model predicts a frequency band-width of order α so the phenomenon is very sensitive to departures from resonant frequencies. The numerical simulations, see Fig. 13.4 Case (c), display this sensitivity. These results can be confirmed using equation (13.28).

13.5 Summary

The problem examined in this chapter is very typical of problems generally arising out of practical vibration contexts. Almost inevitably a large vibration indicates the presence of a natural vibrational mode in tune with the forcing, and it's a matter of searching for this mode. As in the situation examined here, it's also often difficult to model the forcing mechanism, and the detuning mechanism that eventually limits the motion. This is, however, not normally a critical issue. In the present model, having developed reasonable equations describing the behaviour, we examined the small amplitude motion to determine circumstances under which large amplitude responses might result. The resonant bands were thus identified. Averaging and multiscaling techniques were then used to determine the amplitude variations that arise as a result of nonlinear detuning effects.

13.6 Exercises

Exercise 13.1 *Car Suspensions*

Fig. 13.10 represents a simple model of a car and its suspension that can be used to investigate the effect of different suspensions on the motion of the car. We'll assume the suspension can be modelled by a spring with elastic constant k, and the damping system can be modelled by a dash pot with damping coefficient ν, so that

$$F_s = -k\delta l, \quad \text{and} \quad F_d = -\nu\dot{\delta l},$$

are the forces exerted on the car body, where δl is the increase in length of the spring. The wheels will be assumed to be rigid.

(a) Write down the equation of motion for the car (of mass M) and, by specifying the spring and damping forces in terms of the car displacement $x(t)$ above a datum, determine the equation describing the vertical motion of the car.

(b) For present purposes assume the road is flat and horizontal. Determine the displacement x_0 of the car when stationary, and thus determine the equation governing the displacement $x(t) - x_0$ from this equilibrium location.

The effect of a sharp bump on the road surface may be modelled by assuming that it causes an instantaneous change (h say) in x. Determine the subsequent motion of the car. Comment on the effect of M, k, ν on the car's motion. How might one design the suspension to achieve a smoother ride?

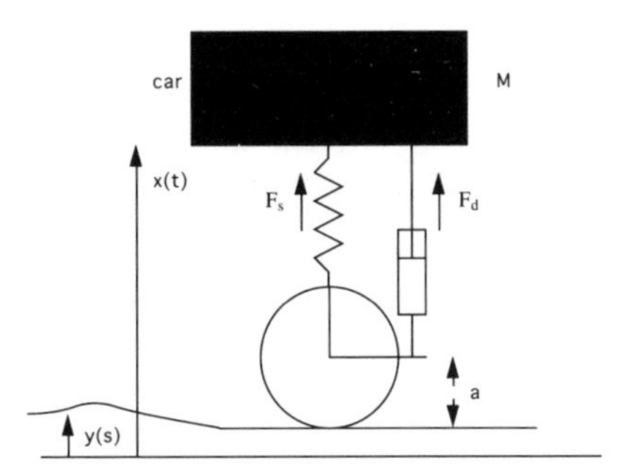

Figure 13.10: Car suspensions.

A ride is likely to be uncomfortable if the car encounters a sequence of bumps. This will be especially true if the average bump spacing l_0 "matches" the motion of the car along the road. If V is the car speed, roughly speaking at what l_0 would you expect an especially rough ride? Why might one expect a gravel (dirt) road to develop corrugations of precisely this awkward spacing after a period of useage?

(c) $\star$ A more complete description of the car's motion can be obtained if one specifies the height of the surface $y(s)$ above the datum as a function of distance s along the road. Equivalently one can specify the vertical displacement of the road/tyre contact surface as a function of time. For the model described above one can obtain a good understanding of the car's response to different road conditions by simply determining the displacemet $x(t)$ due to $y(t) = \sin \omega t$, for general ω. Why? Scale the equations appropriately and using Maple determine the required solution and plot out curves that you feel are informative in context.

(d) $\star\star$ How might one model the effect of the tyres on the situation?

Exercise 13.2 *Loose Ends*

(a) Using Maple or otherwise reproduce the amplitude vs. phase plots for the amplitude of vibration of the tailshaft in the $\sigma = 0$ case, see equation (13.27).

(b) By examining the trajectories close to the relevant singular points establish that the trajectories are in fact closed as shown in Fig. 13.8.

(c) $\star$ Examine the phase plane trajectories in the non-zero σ case, and thus determine the instability bandwidth.

(d) ⋆⋆ In order to simplify the defining set of equations as much as possible the approximation $A/B = 1$ was made, see equation (13.33) (*not* the same A as above). Carry out the analysis in the $A \neq B$ case and produce results. Interesting things happen!

Exercise 13.3 *Two-timing*

⋆⋆ Consider the unforced and damped oscillator described by

$$v''(t,\epsilon) + \epsilon(v'(t,\epsilon))^3 + v(t,\epsilon) = 0, \text{ in } 0 < t < \infty,$$

with $\epsilon \ll 1$ and initial conditions given by

$$v(0,\epsilon) = 1, \quad v'(0,\epsilon) = 0.$$

The long term decay of the oscillator is required.

(a) Determine the first two terms of a normal perturbation expansion

$$v = v_0(t) + \epsilon v_1(t) + \cdots$$

for $v(t,\epsilon)$ and thus determine a useful form for describing long term variations.

(b) Using the multiscaling technique show that a useful first estimate for the long term behaviour of the system is given by

$$v(t,\epsilon) = A_0(\tau)\cos t + B_0(\tau)\sin t \text{ where } \tau = \epsilon t,$$

and where A_0 and B_0 satisfy the equations:

$$B_0' + \frac{3}{8}B_0(B_0^2 + A_0^2) = 0, \text{ and}$$

$$A_0' + \frac{3}{8}A_0(B_o^2 + A_0^2) = 0.$$

(c) Thus show that

$$A_0 = 0, \quad B_0 = (1 + 3/4\tau)^{-1/2}.$$

Exercise 13.4 *The Cubic Oscillator*

⋆⋆⋆ Using either multiscaling or averaging techniques obtain the appropriate equations for determining the long term amplitude and phase changes of the undamped weakly cubic oscillator considered in Section 3.1 Chapter 12, and discuss the solutions which start from zero initial conditions.

Hint: Examine the equations you obtain in the phase plane. Keep in mind the equation (12.12) determining the steady amplitude of the system.

13.7 Appendix: Propshaft equations

Here we simplify the equations (13.1,13.4, 13.2, 13.3) describing the propshaft dynamics.

First we need to gain a useful understanding of the geometric tail-shaft relationship (13.4) (which can be derived using elementary vector methods). Note that if $\phi = 0$, then the angular displacements on the two sides of the Hooke joint, i.e. β and ψ, differ by an amount that's dependent on the tail-shaft geometry as specified by θ *and the specific angular displacement* ψ. If $\phi \neq 0$, so that m has moved in its bush, then the effective origin for β changes to $\beta - \phi$. The rotational angular velocities of the shafts $\dot{\psi}, \dot{\beta}$, and $\dot{\psi} + \dot{\beta}$ are large and *almost* equal, but the small differences are of course important because the whirling motion of interest (specified by $\dot{\phi}$) is associated with such differences.

Examining the geometric relationship in more detail we can see that $(\beta - \phi) = \psi$ when $\psi = \pi/2, 3\pi/2, \cdots$, and in fact $(\beta - \phi) - \psi$ is a periodic function of ψ with half ψ's period. This relationship is plotted (using `plot`) in Fig. 13.2 for a range of θ values. Now the θ values relevant to the propshaft (about 4°) are small, so $\sin\theta$ is small, and it makes sense to simplify the $\beta(\psi, \theta)$ relationship based on the smallness of $\sin\theta$. The Maple commands

```
simplify(taylor(beta,st=0,3);        %st ≡ sin t,
```

yield the result

$$\beta - \phi \approx \psi + \frac{\sin^2\theta}{4}\sin 2\psi, \tag{13.29}$$

so that

$$\frac{\partial\beta}{\partial\psi} \approx 1 + \frac{\sin^2\theta}{2}\cos 2\psi. \tag{13.30}$$

In this form we can now see more clearly the distortion in transmission of rotation introduced by the Hooke joint.

Substituting (13.2), (13.3) and (13.30) in (13.1) gives

$$\begin{aligned} (m\ell^2\sin^2\theta)\ddot{\phi} + C(\ddot{\psi} + \ddot{\phi}) &= -(k\ell^2\sin^2\theta)\phi, \\ C(\ddot{\psi} + \ddot{\phi}) &= [\frac{\sin^2(\theta)}{2}\cos 2\psi]\Gamma. \end{aligned} \tag{13.31}$$

These equations can then be combined to give

$$\begin{aligned} \omega^{-2}\ddot{\phi} &= -\phi - A\cos(2\psi), \\ \omega^{-2}\ddot{\psi} &= \phi + B\cos(2\psi), \end{aligned} \tag{13.32}$$

where $\omega = (k/m)^{\frac{1}{2}}$ is the *natural frequency* associated with the rubber bush/propshaft connection, and

$$\begin{aligned} A &= \Gamma/(2k\ell^2), \\ B &= A(1 + m\ell^2\sin^2\theta/C). \end{aligned} \tag{13.33}$$

The dimensionless constants A and B summarize the physical regime in which we are interested. For VN Commodores

$$\begin{aligned} k &\approx 10^6 \mathrm{N/m} \\ \Gamma &\approx 290 \mathrm{Nm} \\ \ell &\approx 1 \mathrm{m} \end{aligned}$$

whence

$$A \approx 1.5 \times 10^{-4},$$

very small ! Note that B/A measures the ratio of the moment of inertia of flywheel plus m to the flywheel alone. Since C represents many inertial parts, we expect $C >> m\ell^2\theta^2$, so that

$$A \approx B.$$

Using this approximation and introducing a time scale based on the natural frequency of the bush/propshaft so that

$$T = \omega t$$

the equations (13.32) reduce to

$$\begin{aligned} \phi'' &= -\phi - A\cos(2\psi), \\ \psi'' &= \phi + A\cos(2\psi), \end{aligned} \tag{13.34}$$

where $(') = (d/dT)$. The above equations encapsulate the dynamics of the propshaft driven by the applied torque Γ. We need to append appropriate initial conditions to determine the solutions.

It's evident from the physics and the equations that a specification of $\phi(0), \phi'(0), \psi(0), \psi'(0)$ will suffice to determine the solution. One expects the scaled initial angular velocity

$$N = \psi'(0) \equiv \dot{\psi}/\omega \tag{13.35}$$

i.e. the ratio of the initial angular frequency of the drive shaft to the natural frequency of the bush/propshaft connection, to be the critical parameter of the problem. The value will of course depend on the speed of car motion so a range of values should be considered. The additional initial conditions

$$\psi(0) = 0, \phi'(0) = 0, \phi(0) = \phi_0 \text{ (prescribed)}, \tag{13.36}$$

should provide sufficient flexibility to enable one to examine the effect of different disturbances on the system. Remember that our interest is in determining if the dynamics of the propshaft is such that a small initial disturbance (i.e. a small ϕ_0) will grow in time. In other words we'd like to identify circumstances (more explicitly parameter ranges) under which the system is ***unstable***. First we process the equations further.

Adding equations (13.34) gives

$$\phi'' + \psi'' = 0$$

which integrates to give

$$\phi + \psi = Nt',$$

where

$$t' = T + \phi_0/N,$$

after applying the prescribed initial conditions (13.35,13.36). It's clear from this that a change in the initial condition on ϕ can be accomplished by simply redefining the time origin (to t'). Eliminating ψ in favour of ϕ in (13.34) and changing to the time variable to t' we get

$$\frac{d^2\phi}{dt'^2} = -\phi - A\cos(2Nt' - 2\phi). \qquad (13.37)$$

It's convenient to strip the equation of the factors of 2 and to change to the (scaled and origin shifted) time variable t'. Also we'll drop the primes, so with

$$t' \to t, \; y = 2\phi, \; \eta = 2N, \text{ and } \alpha \; = -2A,$$

we get

$$y'' + y = \alpha\cos(y - \eta t), \qquad (13.38)$$

which is in a useful form for subsequent work.

Bibliography

Abramowitz, M. and Stegun, I.A. (1965). *Handbook of Mathematical Functions*, 5*th* edition. New York: Dover.

Barton, N.G. and Ha, J. (1989). *Proceedings of the 1989 Mathematics-in-Industry Study Group*, CSIRO Australia.

Bogolyubov, N. N. and Mitropolsky, Iu. A. (1955). *Asymptotic Methods in the Theory of Non-linear Oscillations,* Gordon and Breach, New York.

Cole. J.D.(1968). *Perturbation Methods in Applied mathematics*, Ginn and Company, Boston.

Gradshteyn, I.S. and Ryzhik, I.M. (1965). *Table of Integrals, Series, and Products*, p 400, Academic Press New York.

McGraw Hill Encyclopedia of Science and Technology, (1992). Vol 19 p 63, New York.

Nayfeh, A. H. (1973). *Perturbations*, John Wiley and Sons, Inc, New York

Prudnikov, A.P. Brychkov, Yu. A. and Marichev, O.I. (1986). *Integrals and Series Vol I,* Gordon and Breach, New York.

Wagner, E.R. and Cooney, C.E. (1979). *Cardan and Hooke universal joint,* Section 3.1.1, p 39 in Universal Joint and Driveshaft Design Manual, Advances in Engineering, Series 7, Society of Automotive Engineers, New York.

Chapter 14

Traffic flow

Signal Speed
Shock waves

14.1 Introduction

Heavy traffic is one of the more common and less pleasant aspects of modern urban living. Every driver is faced with balancing the conflicting objectives of reducing travel time and avoiding accidents. It's not clear what policies drivers actually follow to achieve this balance, but they clearly acquire habits which keep the traffic moving but make accidents quite rare. Individuals however are not able intuitively to drive so as to realize smooth and efficient running conditions for all drivers on the road network, or even for themselves. Engineers are given the task of designing, controlling and managing the road system to achieve such ends. Mathematical models can provide the understanding necessary for their purposes.

14.1.1 Situations of particular interest

As the afternoon peak hour approaches on a main road out of a city centre the speed of travel reduces with the increase in the number of cars using the road. This suggests that there is some relationship between the number of cars on the road and their speed of travel. What can be learned about the nature of this relationship? Also under such heavy traffic conditions relatively minor disruptions can result in major traffic hold ups (traffic jams) that can last for a considerable period after the disruption has disappeared, and affect the flow well away from the location of the disruption. Why should this be so and can the situation be improved? In this chapter a simple model, due to Lighthill and Whitham, is examined which explains these phenomena and most of the phenomena commonly observed in heavy traffic. The model also provides a useful component for use in traffic control contexts and accident models. For a more detailed account consult Lighthill and Whitham (1955).

The ideas that arise in the traffic context are of broad applicability. Thus, for example, the traffic equation also models the movement of a pollutant in a flow field. Also the propagating traffic jam is analogous to the propagating shock waves that cause noise pollution near airports, the travelling detonation and combustion fronts associated with explosives and fires, and the wetting water fronts observed in soils after rainfall. The clanking observed in shunting trains may also be such a phenomenon. Some of these situations will be examined in the exercises.

In Section 14.2 we'll examine steady flow situations, and then we'll go on to examine unsteady behaviour in Section 14.3.

14.2 Steady Traffic Flow

Since human behaviour is variable it's clear that there is no single exact relation describing the driving behaviour of all individuals. It's also clear that the behaviour of a single driver can have a major influence on a traffic stream. Thus, for example, on a single carriageway where passing is impossible, the slowest driver determines the maximum average speed of travel of the cars behind that driver. At best, therefore, a traffic model can either describe what will happen in some sort of average sense, or else attach probabilities to various events happening. When one is dealing with heavy traffic conditions the amount of variability is much reduced so a model that only considers averages makes sense; we'll confine our attention to such a model. Such heavy traffic conditions are of most interest to the engineer. Note that such a model is likely to be more useful on multilane highways because under such circumstances the occasional slow car can be passed, and so has limited influence on traffic flow.

14.2.1 Basic variables

Primarily our concern here is with phenomena that are observed in streams of traffic, rather than for example phenomena associated with the motion of individual cars. For such circumstances it's usually convenient to describe the traffic situation as seen by an observer on the side of the road. Such a description is referred to as an *Eulerian* description, as opposed to a *Lagrangian* description, which describes the motion (position, velocity etc.) of individual cars. A Lagrangian (or particle following) scheme, is normally used in mechanics, whereas the Eulerian scheme is usually more appropriate for describing fluid motion.

The variables will be defined initially for uniform, steady, single lane traffic flow circumstances. All cars are assumed to be the same size and travelling at the same speed. Such circumstances are of course highly unrealistic but in an appropriate average sense the definitions extend. The variables we'll work with are:

- The Traffic Speed, V: units km/hr,

  which simply corresponds to the driver's speed.

- The Traffic Density, N: units cars/km,

  is defined to be the number of cars per unit road length. Of course the spacing between cars is of more concern to the individual drivers. If l is the car length, then the bumper to bumper spacing is $(1 - Nl)/(N - 1)$.

- The Traffic Flux, or flow rate F: units cars/h,

  is the number of cars passing a given point on the road in one hour. The performance of the road system is often gauged in terms of the flux "through" the system.

14.2.2 The flux, speed, density relation

The above variables are not independent of each other but satisfy the relationship

$$F = NV. \tag{14.1}$$

This result is evident if one observes the following: If cars are traveling at a speed V, then after a time T the marked car initially at A, see Fig. 14.1, will cover a distance $AB = VT$. Of course during the same time interval the $N \cdot VT$ cars initially occupying AB will have all moved on, passing B in the process. Thus by definition the flux past B is $NVT/T = NV$, as indicated above.

Of course to restrict one's attention to identical cars travelling with the same speed and sharing a single lane is not necessary; the variables simply need to be defined in an appropriate average sense. In forming this average it is necessary to choose suitably the sample for which the average is defined, and the appropriate sample is the one that preserves the above flux/speed/density relation. A little thought about the basis of the above result should convince the reader that the speed and density should be defined as average values over a given stretch of road; while, for the flux, the average should be taken over the sample of cars passing a given point in a selected interval of time. If these choices are made the flux/speed/density relation carries through i.e. $F_{av} = N_{av}V_{av}$, see Exercise 14.1.

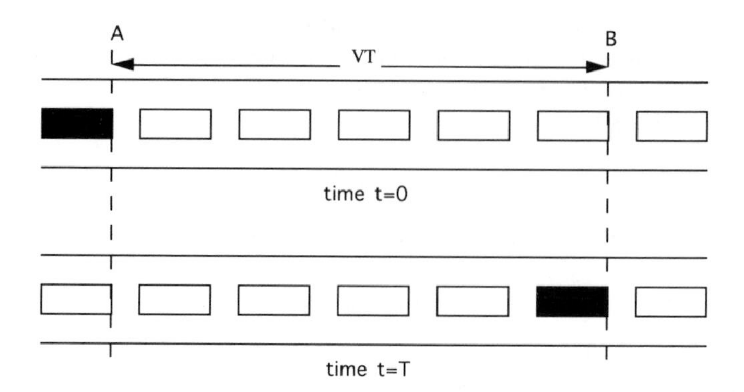

Figure 14.1: The flux, density, speed relation.

14.2.3 The flux vs. density relation

There are a few obvious statements that one can make about the driving behaviour of individuals which have implications as far as the assumed relationship $V = V(N)$ connecting the number of cars on the road and the speed at which they travel is concerned:

1. Prudence demands, and observations confirm, that drivers leave a larger bumper to bumper spacing between cars as the speed of travel increases. Thus V must be a strictly monotonically decreasing function of N. The implication is that there will be fewer cars on a given length of road if the cars are travelling faster. Observations of traffic streams confirm this.

2. On a clear highway drivers will on average travel at the maximum allowable speed V_{max}. Thus

$$V \rightarrow V_{max} \text{ as } N \rightarrow 0. \tag{14.2}$$

3. When cars are bumper to bumper drivers are reluctant to move at all. Thus

$$V \rightarrow 0 \text{ as } N \rightarrow N_{max}, \tag{14.3}$$

where N_{max} is the maximum traffic density.

The $V(N)$ relationship is thus of the form displayed in Fig. 14.2 and the linear relationship

$$V = V_{max}[1 - \frac{N}{N_{max}}] \tag{14.4}$$

provides at the very least a useful idealization to experiment with.

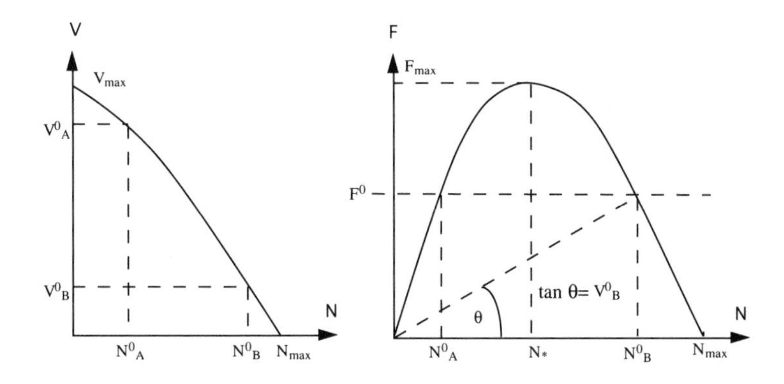

Figure 14.2: The velocity vs. density and flux vs. density relations.

Aside: Some traffic authorities suggest that drivers should allow a minimum safety distance to the car in front made up of the sum of two terms. One term is proportional to the speed, and this term allows for the distance the car will travel in the reaction time interval of the driver (typically 0.3 seconds.). The second term (based on a constant deceleration assumption) is proportional to the square of the speed, and allows for the distance the car will travel after the brakes are applied. Different numbers are usually given for different road conditions, for example wet or dry road conditions. Although the tables based on such minimum stopping distance models serve to emphasize the increased hazard at high speeds it's unlikely that they provide useful working models for drivers.

The flux/density relationship

$$F = F(N) = V(N)N, \tag{14.5}$$

can be inferred from this speed/density relationship and equation (14.1), and the relationship will have the form displayed in Fig. 14.2. In the linear $V(N)$ case this gives

$$F(N) = V_{max}N[1 - \frac{N}{N_{max}}], \tag{14.6}$$

which we'll refer to in future as *the parabolic model.* It is in principle possible to determine experimentally the flux/density relationship using aerial photography to determine the traffic density and pressure or magnetic sensors to record flux levels. There are, however, evident features of this relationship that follow directly from the $V(N)$ relationship:

- $F \to 0$ as $N \to N_{max}$ *and* as $N \to 0$. The physical explanation for this is clear: Under light density conditions, although the cars on the road travel at high speed, there are very few of them. Under heavy traffic conditions, although there are many cars on the road, those cars are travelling slowly, so again flux levels will be small.

- There will be a maximum flux level F_{max} at $N = N_\star$ (say). Flux levels above F_{max} are not achievable. Clearly F_{max} will depend on the width of the particular road, the lighting, the weather conditions etc.

- There will be *two* possible traffic densities with corresponding speeds compatible with any flux level $F^0 < F_{max}$, see Fig. 14.2; a high speed/low density solution (V_A^0, N_A^0), and a low speed/high density solution (V_B^0, N_B^0).

Of the two possible solutions corresponding to any flux level $F^0 < F_{max}$, clearly both engineers and individual drivers would prefer the high speed solution A. The motorist is happier with shorter travel times, and the engineer is happy if *all* drivers are happy. But which of the two solutions is chosen in practice? General experience provides a fair guide to the answer. If previously the traffic had been light, then drivers arriving at the road would find the traffic flowing well and would travel at the higher speed. On the other hand, if there had been heavy conditions previously, then (unless something unusual were to happen) drivers would be forced by prevailing conditions to accept the less preferred solution. *Here is clearly a situation in which the intervention of the engineer (using perhaps traffic lights) could well affect the solution adopted and so improve the situation for all.* The implication of all this is that in order to determine the steady flow in a network, some understanding of transient traffic behaviour is necessary.

Aside: You'll recall that in order to determine the amplitude of motion of a forced nonlinear oscillator it was also necessary to discuss the manner in which the oscillation was set up. It's a characteristic of nonlinear problems that several solutions satisfying the conservation principles of physics can exist, and external considerations need to be invoked to determine the unique solution observed. Thus for example a flame front can often travel through an explosive material at a low speed (a conflagration) or at a high speed (a detonation). Obviously knowing which solution is adopted under prescribed conditions is likely to be important. In the flame front case an abrupt pressure change in the material usually triggers the detonation (explosive) solution while gradual heating usually results in a conflagration.

It's convenient to note for future purposes that the speed of travel corresponding to a given (F, N) state is given by the gradient of the chord joining the state point to the origin on the (F, N) diagram, see Fig. 14.2.

14.2.4 A working flux vs. density relation

For present and future purposes we need an explicit flux/density relation to work with. The simplest parameters to observe are the V_{max}, N_{max} values for the road. Given that our aim at this stage is to understand traffic behaviour rather than make accurate predictions, it makes sense to use the simplest possible relationship meeting the evident requirements. Later of course the model can be refined. The parabolic model

$$F = V_{max}[N(1 - N/N_{max})]$$

introduced earlier has the right behaviour in that

1. $F \to 0$ as $N \to 0$,
2. $F'(N) \to V_{max}$ as $N \to 0$,
3. $F \to 0$ as $N \to N_{max}$,

   and also has the right general (convex) shape. It is unsatisfactory, however, in that it suggests that maximum flux levels will be reached at $N = N_{max}/2$ with speeds of $V_{max}/2$. Although this is a possibility it does seem highly unlikely that the conditions associated with open road travel (or imposed speed limits) would have much to do with the maximum flux conditions for the road. Since there's no immediately obvious way to predict theoretically the flow behaviour under maximum flux conditions, a reasonable model should fit observed conditions. The simplest observation to make (see later) is the speed at which maximum flux is achieved. If maximum flux levels occur when $V = V_\star$, then the additional requirement that

4. $F'(N_\star) = 0$ when $F_{max}/N_\star = V_\star$, will ensure that maximum flux levels are reached at $V_\star$. Here $F_{max} \equiv F(N_*)$ and $N_\star$ are the unknown conditions associated with the observed $V_\star$. We'll see later that $V_\star$ is an especially important flow characteristic as far as the dynamics is concerned; so it's important to get it right. A still better model would also match the observed F_{max} level—we will not attempt this here.

To handle the additional condition a cubic $F(N)$ can be used. Explicitly the cubic

$$F(N) = V_{max}N(1 - AN - BN^2),$$

satisfies conditions 1. and 2. and the remaining conditions are satisfied if A, B and $N_\star$ are chosen so that

$$\begin{aligned} 0 &= V_{max}N_{max}(1 - AN_{\max} - BN_{max}^2), \\ 0 &= V_{max}(1 - 2AN_\star - 3BN_\star^2), \\ V_\star &= V_{max}(1 - AN_\star - BN_\star^2). \end{aligned}$$

The organization of the solution of these equations for A, B and $N_\star$ calls for some care. The difficulty arises because the last two equations are quadratic in $N_\star$ with coefficients involving the unknowns. Thus utilization of either of these equations to solve for $N_\star$ will lead to high degree expressions which Maple cannot handle. (Often it's necessary to provide assistance to algebraic packages.) The apparent degree of the system can be reduced by eliminating $BN_\star^2$ between the last two equations. The resulting linear equation can be solved for $N_\star$ in terms of A. The first equation can also be solved for B in terms of A, and these results can then be substituted into either of the quadratic equations in $N_\star$, leading to a quadratic equation in A; the positive root is selected. Before all this is undertaken it's useful to simplify the calculations by introducing scaled variables. The obvious scales to use are N_{max} for N, V_{max} for V, $N_{max}V_{max}$ for F. Thus with

$$n = N/N_{max}, \quad n_\star = N_\star/N_{max}, \quad a = N_{max}A, \quad b = N_{max}^2 B \text{ and}$$

$$f = F/(N_{max}V_{max}), \quad v_\star = V_\star/V_{max},$$

we have

$$f = n(1 - an - bn^2),$$

and the unknowns $a, b, n_\star$ are to be chosen so that

$$0 = 1 - a - b, \tag{14.7}$$

$$0 = 1 - 2an_\star - 3bn_\star^2, \tag{14.8}$$

$$v_\star = 1 - an_\star - bn_\star^2. \tag{14.9}$$

The Maple instructions for solving this set are as follows:

```
flux:=n*(1-a*n-b*n*n);
eq1:=1-a-b=0;
eq2:=1-2*a*ns-3*b*ns*ns=0;    % ns ≡ n⋆
eq3:=1-a*ns-b*ns*ns=vs;    % vs ≡ v⋆
bv:=solve(eq1, b);    % an intermediate b evaluation
nsv:=solve(eq2-3*eq3,ns);    % an intermediate ns evaluation
eq4:=subs(b=bv,ns=nsv,eq2);
av:=solve(eq4,a);    % av=av[2] was the positive root
f:=subs(b=bv,a=av[1], flux);
```

Appropriate values for the traffic parameters $N_{max}, V_{max}, V_\star$ can now be entered and f determined. Observations by the authors on two-lane Perth highways gave $V_\star = 65$km/h, $V_\star = 35$km/h, $N_{max} = (1/2.5)$ cars/m (corresponding to an average front bumper to front bumper spacing of 5 m on a 2 lane highway). The resulting expression for f obtained is given by

$$f = n\left(1 - 0.726\, n - 0.274\, n^2\right). \tag{14.10}$$

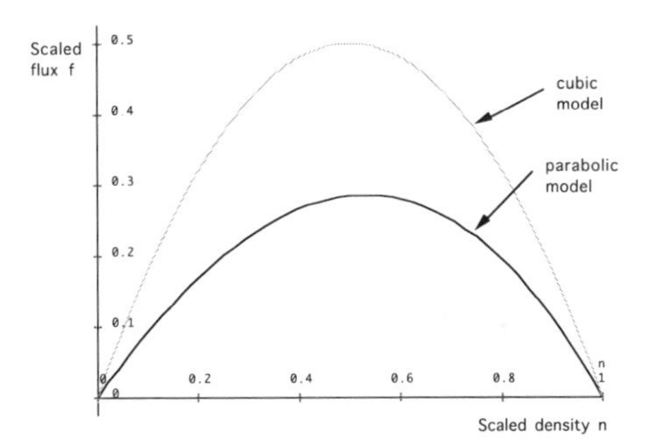

Figure 14.3: Flux vs. density models.

The results are displayed in Fig. 14.3 which includes the parabolic model for comparison purposes. For the purposes of later investigations you might like to make calculations based on your own observations.

Models of the above type can be used to determine appropriate speeds to impose to clear traffic through city centres, or for the appropriate speeds to enforce for travel through tunnels, see Exercise 14.2.

We're now adequately equipped to proceed to examine unsteady flow situations.

14.3 Modelling Unsteady Conditions

Let's now consider a unidirectional flow situation (preferably with more than one lane) in which the density varies with location along the road. How will the flow evolve? First we need to think about how the flow variables can be defined and measured under such varying conditions. A sensible scheme is as follows: select a number of points along the road and a number of consecutive time intervals and observe the number of cars passing each point and the number of cars between locations in each time interval. Using such observations one can define a flux and density level "at" each (x, t) location. If the time and space intervals are too short, however, (for example, fractions of seconds) the average will not be meaningful. If too large (1 hour say) then the average will certainly make sense but features of the flow that are of interest will be lost. Our interest here is not with the shuffling behaviour associated with individual cars but with the larger scale variations that affect traffic streams. It's not just convenient but *essential* to filter out the small scale individual car motions so as to arrive at a useful model; otherwise computationally intensive models will result and even if the results can be computed they'll probably be incorrect. Intuitively one expects that there will be a range of interval sizes (that involve several cars) that will lead to a sensible and interesting flow description. Thus:

Assumption 1: We'll assume that useful time and length intervals can be chosen for defining flow quantities.

Computer simulations of realistic dynamic traffic situations could be used to obtain useful practical information on the appropriate intervals to employ when collecting real data; of course a more sophisticated statistical/dynamical analysis of the situation would be preferable.

Aside: The situation is similar to the one that arises when describing gas motion. For situations in which the relevant length scale is large compared with the molecular separation distance (10^{-8} m) it pays to work with a (relatively slowly changing) velocity defined in terms of the net mass transport (metres per second typically), rather than the large and rapidly varying (18×10^2 metres/second) molecular velocities. If, on the other hand, one's concern is with questions such as how the transport properties change with temperature then a *Kinetic Theory* microscale approach is appropriate.

There is another assumption that we'll make that will greatly simplify matters:

Assumption 2: We'll assume that the steady flux/density relationship (14.5) also holds under the variable conditions of interest here.

Thus cars are assumed to *immediately* adjust to the local traffic situation. The advantage of this assumption is that it serves to filter out the transients associated with local (small scale) adjustments. Of course the assumption automatically excludes the possibility of describing the accelerations that occur when a light changes from red to green, or the inappropriate behaviour that leads to accidents, see Exercise 14.9.

Comment: Quasi-static models Such *quasi-static* models are often used in modelling to reduce the complexity. Underlying such models is the assumption that the dynamics avoided by introducing the assumption is of secondary importance as far as the process being described is concerned. Sometimes it can be shown that the quasi-static model equations and the more complete model equations are asymptotically equivalent in an appropriate sense for the circumstances of interest, but more often than not one simply proceeds with caution. In the present situation, intuitively one might expect that the small scale accelerations avoided by use of the quasi-static assumption are unlikely to significantly affect the global flow under *most* circumstances. If one is concerned with accident conditions, however, then it's clear the above *quasi-static* flow assumption needs to be abandoned in favour of an appropriate time delay model of driver behaviour, see Exercise 14.9.

A further assumption we'll use is not absolutely necessary but makes life much easier from a mathematician's point of view:

Assumption 3: We'll assume that a continuum description is viable.

As we well know continuous models are much easier to solve analytically than their discrete counterparts, so if possible (at least during the initial exploratory stages) one should examine a *continuum* model of the situation.

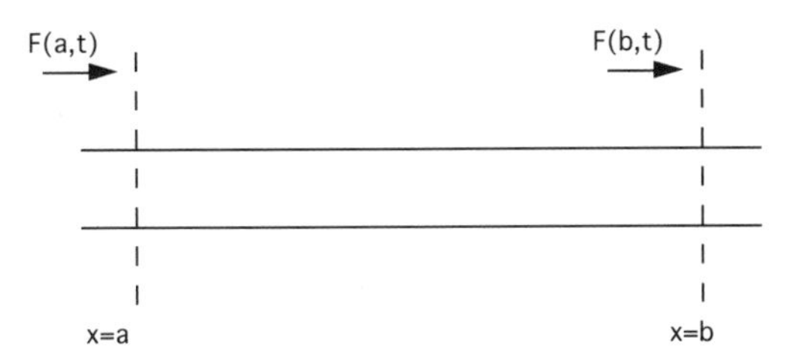

Figure 14.4: Car conservation.

The procedure for relating the discrete variables defined above to continuum equivalents is a familiar one to students of calculus. Thus, for example, we simply associate the flux measurement corresponding to a particular time and location interval with the midpoint of that time/space interval and interpolate a smooth function through the data. This procedure will work providing the intervals used for data collection are of an appropriate size, and the physics of the situation doesn't exaggerate small effects.

Comment: The assumptions introduced by pure mathematicians are often *purely technical* in the sense that they're introduced to avoid (mathematical) pathological situations that may destroy the validity of the theoretical results—no wonder mathematicians are misunderstood! The above assumptions are *not* of this variety, as we'll see.

14.3.1 Conservation principle for cars

If more cars per unit time enter a stretch of road than leave it, then the number of cars on that stretch of road will increase, and of course to understand changes in traffic conditions we need to quantify this car conservation result. If $(F(x,t), N(x,t))$ denotes the flux and density of the flow at a location x measured from a convenient origin at time t, then the total number of cars on the stretch of road between $x = a$ and $x = b$ at time t is given by, see Fig. 14.4

$$\int_a^b N(x,t)\,dx,$$

so that the rate of change in this number is given by

$$\frac{d}{dt}\int_a^b N(x,t)\,dx = \int_a^b N_t(x,t)\,dx.$$

Assuming there are no entries or exits, such an observed change can only be brought about if the flux levels $F(a,t)$ and $F(b,t)$ at the ends of the stretch differ, and conservation of cars requires that

$$\int_a^b N_t(x,t)\,dx = F(a,t) - F(b,t) = -\int_a^b F_x(x,t)\,dx,$$

$$\text{so that} \int_a^b [N_t(x,t) + F_x(x,t)]\,dx = 0.$$

If this is to hold for *any* arbitrary interval (a,b) the integrand must vanish identically so that

$$N_t(x,t) + F_x(x,t) = 0, \tag{14.11}$$

providing the functions are sufficiently smooth functions. Later we will see that the smoothness requirements are by no means simply mathematical technicalities. The above equation provides us with a relationship connecting N and F that ensures car conservation.

We now have two equations (14.5,14.11) in the two unknowns N and F, so we expect on mathematical grounds that (apart from appropriate boundary conditions) no further information is required to "close" the problem. Further simplification is achieved by invoking the assumption described earlier; namely that

$$F(x,t) = F(N(x,t)).$$

With this assumption the conservation equation becomes

the traffic equation

$$N_t + F'(N)N_x = 0, \tag{14.12}$$

where the prime denotes differentiation of F with respect to N. This first order nonlinear partial differential equation for the traffic density contains both the driver behaviour information and the conservation information required to determine how density changes occur in traffic. We'll refer to it as *the traffic equation.* Once the density is known, the speed and the flux are uniquely determined from (14.5). Of course we need to specify appropriate subsidiary conditions. On physical grounds one would expect that, given the traffic density at an initial time, the subsequent traffic behaviour (density, flow, speed) would be determined. Given the equation is first order in space and time a single initial condition makes sense, but one might also expect to need to specify the traffic state at one location, presumably the point of entry of the cars. The usual process of working with both the physics and the mathematics to arrive at a sensible formulation is evident here and is *required* for modelling.

14.3.2 Solution construction

The general theory of how to deal with partial differential equations of first order was developed by Cauchy and an account of this theory may be found in Carrier and Pearson (1976). It is not necessary, however, to invoke the general theory in order to find solutions in cases such as this. The following nice observation enables one to generate solutions. Any function, G, will remain constant along paths $x = X(t)$ in the (x,t) plane providing that its total derivative with respect to t vanishes i.e.

$$\frac{d}{dt}[G(X(t),t)] = 0.$$

Thus, recognizing explicitly the dependence of G on its arguments x and t and using the chain rule, this gives

$$G_t + X'(t)G_x = 0.$$

Writing this in the form

$$G_t + f(G,x,t)G_x, \text{ with} \tag{14.13}$$

$$X'(t) = f(G,X,t), \tag{14.14}$$

we can see that the implication is that level curves in the (x,t) plane for the solution $G(x,t)$ of the first order partial differential equation (14.13) are given by the equation (14.14). These curves are referred to as **Cauchy characteristics**. Normally G will be prescribed on some curve in the (x,t) plane (usually $t = 0$), so a sensible plan is simply to map out the level curves that originate from this initial state and thus determine the solution. Note additionally that since $G(x,t)$ is constant (and thus time and space independent) along these curves, the equation (14.14) simply amounts to a straightforward ordinary differential equation for the $X(t)$ curves. The special case in which $G = const$ in *regions* (rather than just along curves) in the (x,t) of the plane turns out to be particularly important.

In our traffic case, see equation (14.12), the Cauchy characteristics are given by the straight lines

$$\frac{dX(t)}{dt} = F'(N) \tag{14.15}$$

in the (x,t) plane. Using this information it's conceptually easy to determine $N(x,t)$, given the traffic conditions at an initial time. One can even graphically construct the answer in the following way, see Fig. 14.5. At each point $x = \xi$ along the $t = 0$ axis draw a straight line L with the required Cauchy characteristic slope $F'(N(\xi,0))$. Along this line L, N retains the value $N(\xi,0)$. A corresponding analytic description is given by

$$N(x,t) = N(\xi,0) \text{ along } x - \xi = F'(N(\xi,0))t. \tag{14.16}$$

This characteristic solution is ideal for graphical presentation, as we'll see with later examples, but is not at all in a convenient form for analytic work. For such work we'd prefer to have an explicit solution representation $N = N(x,t)$ rather than the above (deceptively simple) implicit representation. It's not even clear from the analytic form (14.16) whether the solution

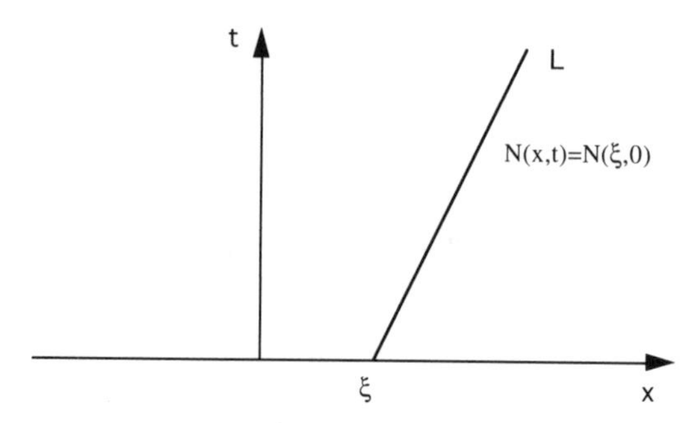

Figure 14.5: Cauchy characteristics.

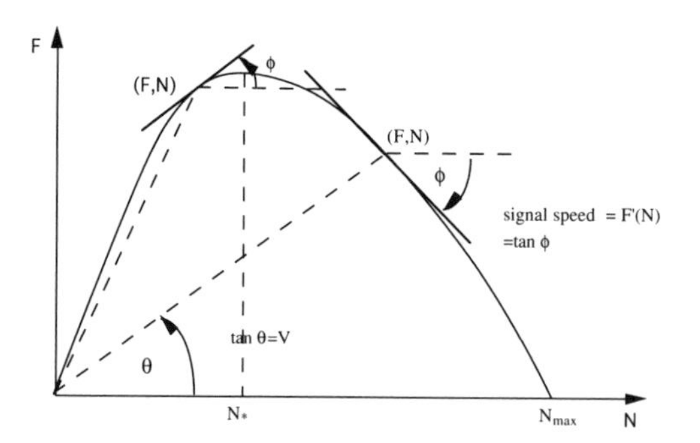

Figure 14.6: Signal speed.

is unique In fact it isn't for many circumstances of importance as we will see. It's a simple matter to show explicitly that the solution given by (14.16) satisfies the traffic equation (14.12) and that under certain conditions the solution is unique. The conditions under which one can display uniqueness are, however, unacceptably restrictive. The reason for these difficulties is not superficial, as will become evident as we proceed.

Aside: It's often useful to interpret significant mathematical simplifications so that corresponding situations can be identified and made use of in other contexts. Thus in the present context it's worth contemplating further the physical significance of Cauchy characteristics. We'll find the lessons to be learned in this particular situation are most valuable.

Signal speed

It's clear from equation (14.15) and also on dimensional grounds that $F'(N)$ represents a speed. The role played by this speed in the solution process indicates its central importance. We'll refer to this speed as the **signal speed** for reasons that will become evident. Note that the signal speed is given by the slope of the flux curve $F(N)$ at N and so does *not* coincide

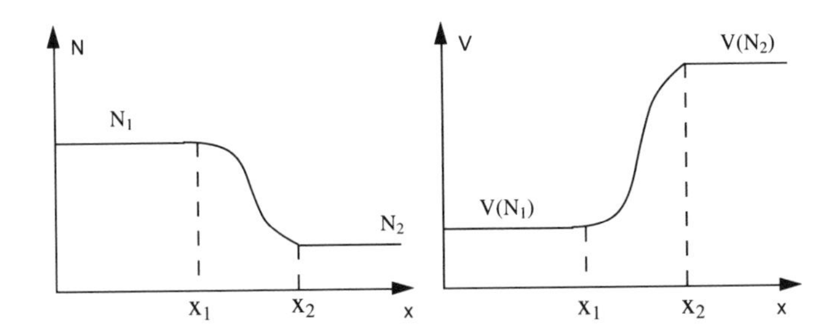

Figure 14.7: Example 1: The initial density and velocity distributions.

with the car speed which you'll recall is given by the slope of the chord from $(0, 0)$ to (F, N) on the flux curve, see Fig.14.6.

Example 14.3.1 *A First Example*

To see the relevance of the signal speed we'll graphically construct the solution corresponding to the initial density distribution displayed in Fig. 14.7. The situation envisaged is one in which initially the cars at locations $x < X_1$ in the traffic stream are moving at the speed $V(N_1)$ associated with the constant density N_1. These cars are heading into a traffic stream moving at a somewhat greater speed $V(N_2)$ corresponding to the smaller constant density N_2 in the zone $x > X_2$. There's a transition zone between the two streams in which cars are speeding up to V_2. How will the pattern evolve? The case in which $N_2 < N_1 < N_\star$ will be illustrated in the figures.

By referring to the flux/density graph for the road we can read off the relevant traffic speeds and fluxes, and signal speeds corresponding to N_1, N_2, and all the densities in between. The characteristics emanating from $x < X_1, t = 0$ all have the same slope $F'(N_1)$ and mark out a zone in the (x, t) plane in which $N = N_1$, see Fig. 14.8. Similarly the characteristics to the right of the $x = X_2$ characteristic mark out an $N = N_2$ zone. In the transition zone intermediate slopes corresponding to the intermediate density levels result. The signal speed decreases monotonically with increasing traffic density, see Fig. 14.6, so the characteristics will fan out as shown in the figure; the density profile spreads out. Using these characteristics the traffic density at some later time $t = t_0$ can be constructed and is as shown. *The thing to notice is that the traffic conditions at any initial location in the traffic stream are simply transported forwards with speed $F'(N)$ corresponding to the local N value.* Note that this *doesn't* mean that the traffic pattern at t_0 is simply a replica of that at $t = 0$ shifted forwards along the road; basically because the *local* signal speed varies with the *local* density, causing the shape of the pattern to change as shown. Often, however, one is interested in the changes in a uniform stream of traffic (density N_0 say)

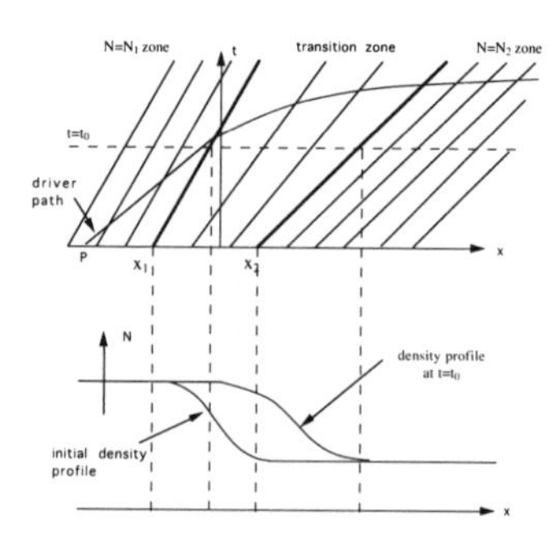

Figure 14.8: Example 1: The characteristics.

brought about by a perturbation. In such circumstances (to a high degree of accuracy) the changes *will* be propagated with constant speed $F'(N_0)$ and the pattern will be unchanging; so the simple pure translation description is accurate in this important special case. In other circumstances it can be useful to think of the changes as being due to a pure translation with speed $F'(N_{av})$, where N_{av} is the average density, with shape changes superimposed due to local density variations. Explicit solutions can be extracted corresponding to particular initial states; however we'll leave this until after we've explored the signal speed ideas further. We'll find, in fact, that the signal speed ideas will assist greatly in this solution process.

Aside: The signal speed is often referred to as the wave speed of the system by analogy with sound wave propagation in air, or gravity waves on water. Sound waves are small amplitude pressure variations in air propagating with unchanging shape through the atmosphere at the sound speed. Larger amplitude pressure variations travel with a speed that's dependent on the local pressure, so the pressure profile changes; this situation is similar to the traffic situation.

Now let's examine the situation from the driver's seat. Firstly note that it's evident from Fig. 14.6 that, because of the convexity of the flux/density curve, the signal speed is always less than the car speed. The implication is that drivers will experience traffic conditions created earlier by cars ahead of them in the traffic stream. This matches our experience. Thus in our example a driver initially located at position P in $x < X_1$ will be moving with speed $V(N_1) > F'(N_1)$ until he passes through the $N = N_1$ zone. He will then pass through the transition zone and eventually will be traveling at $V(N_2)$. His path in the (x, t) plane is shown in Fig. 14.8.

Example 14.3.2 *A Second Example*

In the first example we examined a situation in which cars were headed into less congested traffic conditions. What happens if the road ahead is more congested as in Fig. 14.9? The signal speed remarks are still pertinent,

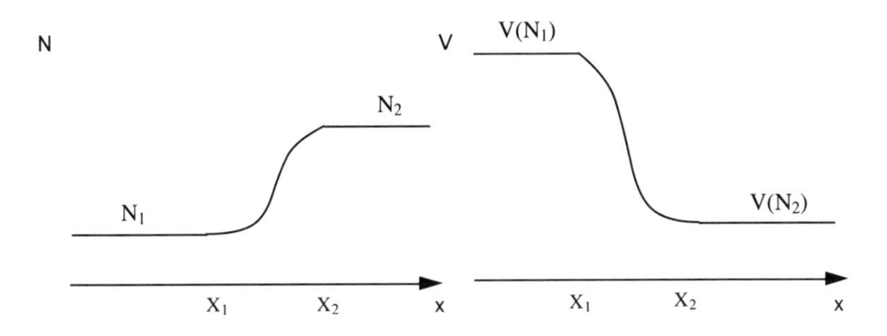

Figure 14.9: Example 2: initial profiles.

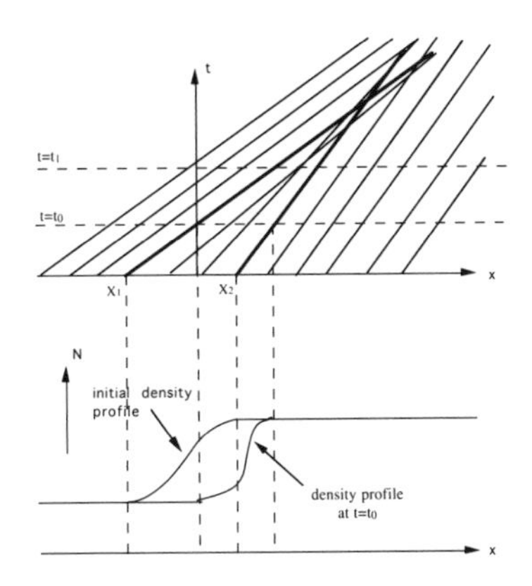

Figure 14.10: Example 2: solution using characteristics.

however, there are major differences in the implications in this case. Again we'll assume that initial conditions far upstream $x < X_1$, and downstream $x > X_2$, are uniform, and that there's a transition region. In this case, however, we'll assume $N_1 < N_2$ so that $V(N_1) > V(N_2)$; so cars in the transition zone are slowing down. In this case the characteristics in the zone $x < X_1$ will have a smaller slope (dt/dx) than those in the $x > X_2$ zone, again with intermediate slopes in the transition zone, as shown in Fig. 14.10. In this case, therefore, the characteristics will converge rather than diverge, so the density profile will steepen as shown.

Using the results of the two examples discussed above we can build up quite a good picture of the changing density pattern on a road system. Thus for example with the density distribution shown in the third situation shown in Fig. 14.11 the following changes will occur:

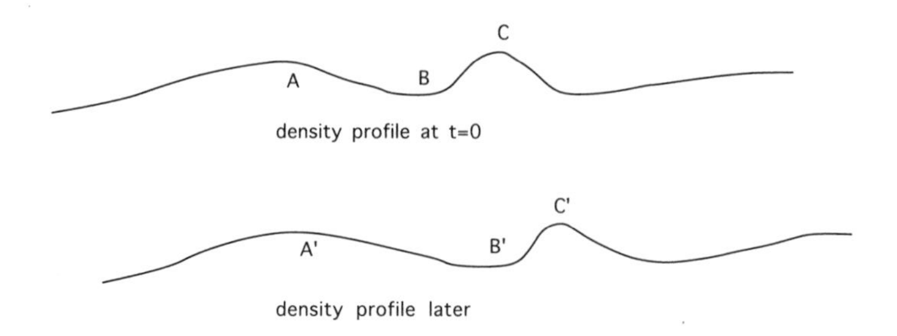

Figure 14.11: A third example.

- The whole density pattern will move with something like the average signal speed.
- Portions of the profile in which cars are entering a less dense traffic zone (for example AB in the figure), will spread out.
- In regions in which the density increases in the direction of travel (for example BC) the profile will compress.

It's a complicated picture and any word description is inadequate, but it's one that can be easily quantified by constructing characteristics etc.

There's one feature of the the second example that we've glossed over and that's disturbing. The characteristics in this example are convergent straight lines, so inevitably for some value of time $t = t_1$ they'll intersect and then for later time ($t > t_1$) they'll overlap, see Fig. 14.10. Each of the intersecting characteristics will have an associated N value and these N values will be different; so the theory predicts *three* values of density at locations in the overlap region. The predicted $N(x,t)$ for $t > t_1$ will be as shown in Fig. 14.12. In physical terms this doesn't make sense. The type of difficulty exposed here is one of major importance that was first understood in the area of gas dynamics (shock waves), but similar difficulties arise in water waves (wave crashing), hydraulics (hydraulic bores), combustion (detonation and conflagration waves), and in many other areas. Characteristically the basic physics of all these problems is described by nonlinear *hyperbolic* (wave propagation) equations, the simplest of which is the first order partial differential equation relevant here. Also, characteristically the nonlinearity causes a steepening of the solution under particular circumstances. Under such circumstances physical effects normally ignored in setting up the basic model (because of their relative smallness) assume a major role—yes, it's another *singular perturbation* situation! In spite of the flaws in the basic models it turns out that simple physical arguments, requiring no major model changes, lead to acceptable results in all well

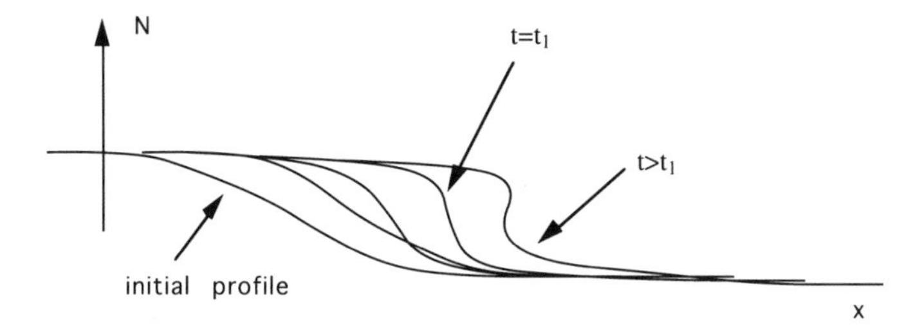

Figure 14.12: Profile steepening.

known cases. It's in fact easier to understand the nature of the difficulty in a specific context, so we'll defer discussion of this issue until later when examining a traffic jam.

Before producing explicit solutions to particular problems there are additional observations concerning the variation of signal speed with density that are evident from the flux/density graph, see Fig. 14.6, and need to be noted:

- Under open road conditions $N \to 0$ and the signal speed $\to V_{max}$, so that the car speed and signal speed coincide. Thus any disturbance will be carried out of the network with the cars causing or initially experiencing the disturbance, and cars further back in the stream won't be affected.

- As the traffic density increases the slope of the $F(N)$ curve reduces, see Fig.14.6, so that disturbances will travel more slowly and thus remain in the network longer; more and more cars will feel the effect of the disturbance.

- When the density reaches the value N_* associated with the maximum flux level F_{max} for the road, the signal speed reaches zero. The implication is that any disturbance will remain at the location of its initiation. Obviously this is a disastrous situation! Clearly it would be handy to have a traffic cop handy to change the circumstances! It's interesting to note that this awkward situation occurs at precisely the maximum flux circumstances that are desirable from an engineers point of view. Such is life!

- Things get even worse at densities greater than $N_\star$. The signal speed becomes negative so that information propagates *backwards* into the traffic stream! The speed of propagation is not small either. For example the parabolic model predicts speeds of V_{max} as $N \to N_{max}$. *This explains why (under congested conditions) minor disturbances can cause major disruptions throughout a traffic system.* We of course call such situations a *traffic jam.*

Comment: Characteristics play an important role in all wave propagation systems. Essentially the characteristics map out in space/time the region of influence of a disturbance. In higher order systems the situation is more complex because there can be more than one set of characteristics ("waves" of various speeds being possible) but the simple wave speed ideas described above can be used to great advantage both for conceptualization and for numeric and analytic work. It is a general feature of human intellectual abilities that when the situation becomes too complex, developing ideas or concepts in terms of simpler objects helps one to deal with the situation. The advice to *keep it simple* is one which is always worth trying to follow, even if that has to be achieved at the expense of hiding a complex process under a simple word.

The **only way** to test a model is to apply it to specific situations to see if it works. We'll now extract explicit solutions which you can check out yourself; fortunately observations are often easy to make in traffic studies.

14.3.3 Starting from traffic lights

Suppose we have a stream of cars lined up behind red lights and the lights turn green. Let's see if the model explains the features of the traffic flow that we're all well aware of. We'll assume the road ahead of the lights is free of cars. We well know that immediately the lights change the first driver in the stream moves off with as much horse power as his machine can muster (well, almost). Drivers further back in the queue may see that the lights have changed and that the first car is moving off, but of course can't move off for some time; the car in front blocks their path. In terms of the signal speed idea this is easy to explain. Cars at rest at the maximum traffic density will remain so until the system signal (as distinct from the visual signal that the lights have changed) reaches them, telling them that they may start to move. What is the signal speed for such cars? The cars behind the red lights are in the state of maximum density so that the signal will travel at speed, $F'(N_{max})$, which is negative. Thus the signal travels backwards along the queue from the traffic light with the speed $|F'(N_{max})|$. The further from the lights a driver is the longer he/she will have to wait before he/she can move off. Explicitly the driver at distance d back from the lights will commence moving $d/|F'(N_{max})|$ time units after the lights turn green.

Let's construct the solution. The lights are located at $x = 0$, and at $t = 0$ they change, so that in the (x, t) plane we have $N = 0$ for $x > 0$ and

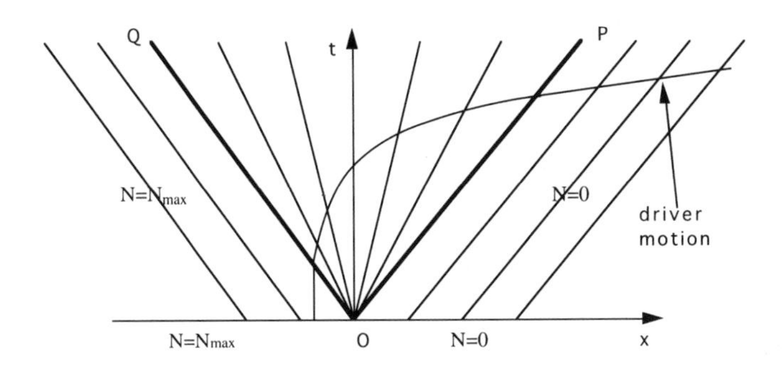

Figure 14.13: Traffic lights.

$N = N_{max}$ for $x < 0$ along the $t = 0$ axis. The characteristics emanating from these states are as shown, see Fig. 14.13. In the region of the (x, t) plane to the right of the characteristic OP given by $x/t = F'(0) = V_{max}$ (which represents the path of the first driver) there are no cars—the first car in the queue (as well as the others) can't travel at speeds faster than the maximum allowable speed represented by this characteristic. In the region to the left of the characteristic OQ defined by $x/t = F'(N_{max})$ the cars remain stationary. There's a wedge shaped region defined by QOP not covered by the above characteristics. What happens in this region? One might reasonably expect that there will be cars on the road with all speeds in the range between zero and the maximum value V_{max}, corresponding to the whole range of possible densities, so this region must be a variable density/velocity region. Now each density has an associated (unique) characteristic slope so that it makes sense that we cover the wedge region with characteristic lines emanating from the origin, each of which identifies its associated density. The resulting pattern is shown in Fig. 14.13. Explicitly along the ray $x/t = \alpha$, the equation $\alpha = F'(N)$ can be solved to determine the N value associated with α. For the parabolic flux/density model (14.6) this gives

$$\text{along } x/t = \alpha, \quad N/N_{max} = \frac{1}{2}[1 - \alpha/V_{max}]. \tag{14.17}$$

All possible states will thus be found in this region and conditions at the edges of this wedge will match those adjacent to the wedge. The only disquiet one might have with this solution is its behaviour at $(x, t) = (0, 0)$; all states exist at and near this location! Note, however, that immediately after $t = 0$ the solution is well behaved and satisfies all the required traffic flow conditions, so the flaw in the model is a minor one related to its continuum basis.

There is one practically useful model prediction that's worth noting.

Along the characteristic $x = 0$ (with $dx/dt = 0$) the characteristic condition (14.15) requires $F'(N) = 0$, which occurs when $N = N_\star$. The location $x = 0$ of course corresponds to the traffic light position, so the implication is that the traffic density at the traffic light remains fixed *for all* $t > 0$ and at a value equal to $N_\star$. Not only does this provide a test for the model, but once verified it provides a simple way for the engineer to determine the optimal speed to set for drivers in order to achieve maximum flux through the system. The authors in fact obtained the estimate for the speed associated with maximum flux used in the last section (to determine the steady state flux/density relation) by observing their own average speed as they crossed the lights while maintaining their place in the traffic. The figure of between 35 and 40 km/h was obtained in this way. You might also like to check your speedometer next time you pass by the traffic light under similar circumstances.

There's one obvious deficiency of the model, anticipated earlier. The first car in the queue has an empty road ahead and according to the model will immediately commence travelling with the speed appropriate for zero traffic density, namely V_{max}. Now of course even the most aggressive driver behind the wheel of the most powerful machine can't realize the acceleration required to do this. This is a minor matter as far as we're concerned, but such limitations of the model should always be borne in mind. Cars further back in the queue of course don't have the option of rapidly accelerating, so the model is likely to accurately describe their motion.

Do the predictions obtained from the model seem to be at least qualitatively correct? Whether they are quantitatively correct for some flux vs. density relation might be checked by timing how long it takes after the lights change before a car well back from the lights moves. In any case it is for the reader to judge whether they have a better understanding of why queues of cars take so long to clear.

14.3.4 A traffic jam

Consider the peak flow of traffic travelling along a two-lane carriageway with the traffic flux equal to the maximum capacity of the road. A ^{0} superscript will be used to distinguish initial conditions. Suppose an accident occurs which blocks one lane completely but does not interfere with the other lane. Further, suppose that the drivers are so co-operative that they filter smoothly into the open lane without affecting its traffic carrying capacity. The objective is to predict the traffic conditions while the lane is closed. For simplicity the parabolic flux vs. density relation (14.6) will be used, so that before the traffic jam the flux level is $F^0 = N_{max}V_{max}/4$. After the jam occurs only half the road is usable at the jam location so the flux level

Flux	$F^0 = V_{max}N_{max}/4$	$V_{max}N_{max}/8$
Density	$N^0 = N_{max}/2$	$N^- = N_{max}(1 + 1/\sqrt{2})/2$ $N^+ = N_{max}(1 - 1/\sqrt{2})/2$
Speed	$V^0 = V_{max}/2$	$V^- = V_{max}(1-1/\sqrt{2})/2$ $V^+ = V_{max}(1+1/\sqrt{2})/2$
Signal Speed	0	$-V_{max}/\sqrt{2}$ $V_{max}/\sqrt{2}$

Table 14.1: Traffic jam parameters.

reduces to $V_{max}N_{max}/8$ locally. The conditions on the problem are thus

$$F = F^0 = V_{max}N_{max}/4 \text{ at } t = 0, \quad \text{and}$$

$$F = V_{max}N_{max}/8 \text{ at } x = 0 \quad \text{for } t > 0.$$

The densities, velocities and signal speeds associated with the above flux levels are readily obtained from the flux/density and velocity/density relationships (14.6,14.4) and are displayed in Table 14.1. Corresponding to any flux level less than F_{max} there are two possible solutions for density and velocity. The superscripts $^-$ and $^+$ are used to distinguish the two possible solutions corresponding to the flux level at the jam location. Any attempt to set up a characteristic net runs into difficulties near $x = 0$. For a start the prescribed condition along the $x = 0$ line requires $N = N^+$ or N^-, which is clearly incompatible with the characteristic relationship (14.15) that requires $N = N_{max}/2$ along $x = 0$. Also, characteristics overlap in the neighbourhood of $x = 0$. The present situation is of course related to the situation examined in Example 2; cars behind the jam are heading into a region of congestion so that characteristics overlap and steep density profiles are to be expected.

Faced with the present difficulties one can either examine the basis of the assumptions leading to the inconsistent results, or use specific knowledge of the practical circumstances to see the nature of the difficulty. We'll adopt the second approach, but not before listing the assumptions that may have led us astray. Firstly note that the difficulty doesn't arise because our solution scheme for solving the traffic equation (characteristics etc.) is flawed. The difficulty lies at the more basic level of setting up the traffic equation (14.12). The assumptions made were:

- the continuum assumption,
- the quasi-static model assumption, and
- the smoothness assumptions on N.

You might like to speculate about the adequacy of these assumptions; we'll return later to examine these when we understand more about the physical nature of the difficulty.

Everyday experience suggests the following: The immediate effect of the obstruction will be to cause a bank-up of cars behind it, with the bank-up increasing as time goes on. Of course cars initially in front of the obstruction will be unaffected by it, and so will continue to move with speed $V_{max}/2$. After the drivers initially behind the obstruction pass by it they'll see a relatively clear road and will speed up until eventually they overtake the cars unaffected by the obstruction. They'll then be forced to slow down. Is it possible to fit together an acceptable solution that behaves like this?

Based on the above we'd expect a velocity and density discontinuity across $x = 0$, with the larger density (small velocity) occurring for $x < 0$. The solution (N^-, V^-) for $x = 0-$, and (N^+, V^+) for $x = 0+$ (both of which correspond to the prescribed flux level) thus represents a satisfactory solution locally. Note that in a straightforward trivial sense, namely $F^-(0) = F^+(0)$, car conservation is assured across $x = 0$, and the only other model requirement that the steady state flux/density relation is satisfied is also of course satisfied. So what went wrong with the traffic equation work? The traffic equation was derived under the assumption that N was differentiable; clearly this isn't the case at $x = 0$. The condition $F^-(0) = F^+(0)$ represents the appropriate replacement expression for car conservation under the discontinuous circumstances that occur at $x = 0$.

Comment: In spite of the fact that the traffic equation fails near $x = 0$, it's not likely to fail everywhere; in fact we might expect the equation to produce correct results providing the differentiability conditions are satisfied, and everyday experience suggests that smooth traffic conditions will occur almost everywhere.

Now the information that there's a jam will travel back into the advancing traffic with the signal speed $-V_{max}/\sqrt{2}$ corresponding to $N = N^-$ and, since the corresponding characteristics emanate from $x = 0-$ where $N = N^-$, we'd expect cars in this region to be also travelling with the jam speed V^-. The cars well back from the jam will on the other hand be completely unaware of its presence and will be travelling at the speed $V_{max}/2$ associated with their characteristics, until the signal that the jam has occurred reaches them. One might therefore expect there to be a sharp interface in the (x, t) plane; on one side of which cars are completely unaware of the jam's presence and are proceeding as normal, and on the other side of which they're reduced to travelling at the jam's speed V^-. Experience confirms this abrupt change in circumstance. The question thus becomes: at what speed does this interface travel? Recognizing that the traffic equation expressing car conservation is again inapplicable, we need to determine an alternative car conservation equation that is appropriate for the present (discontinuous) circumstance. It's as easy to develop an

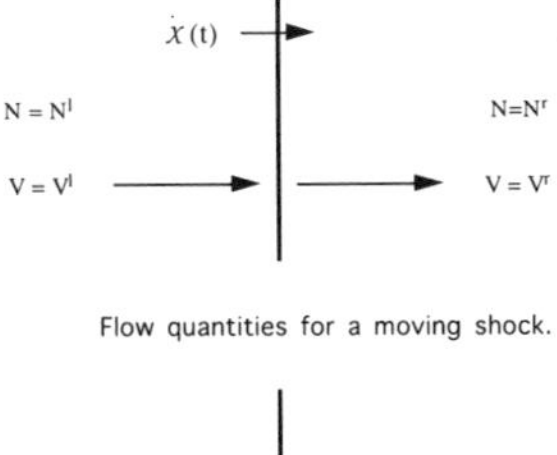

Flow quantities for a moving shock.

Flow quantities relative to the shock.

Figure 14.14: Shock conditions.

appropriate equation in the general case as in the case of specific interest here, so we'll do just this.

Shock conditions

Let (N^l, V^l, F^l) and (N^r, V^r, F^r) denote the (known) traffic states to the left and right of the assumed discontinuity moving with velocity $\dot{\mathcal{X}}(t)$ (unknown). All velocities will be measured in the same direction as the flow of traffic (to the right), see Fig. 14.14. As we've seen earlier in Example 2, steepening occurs only in the case in which cars are heading into a region of greater congestion, so we'd expect that discontinuities of the type being considered will only occur if $N^l < N^r$; we'll assume this is the case. Relative to an observer located on the interface, cars on the left hand side will be entering at speed $(V^l - \dot{\mathcal{X}})$, so that an observer moving with the interface will see $N^l(V^l - \dot{\mathcal{X}})$ cars entering the interface per unit time from the left. The same observer will see $N^r(V^r - \dot{\mathcal{X}})$ cars leaving the interface per unit time to the right. Car conservation thus requires

$$N^l(V^l - \dot{\mathcal{X}}) = N^r(V^r - \dot{\mathcal{X}}), \tag{14.18}$$

which yields the *shock speed*

$$\dot{\mathcal{X}} = \frac{N^r V^r - N^l V^l}{N^r - N^l} = \frac{F^r - F^l}{N^r - N^l}, \tag{14.19}$$

i.e. the shock speed is given by the ratio of the flux jump to the density jump across the shock. By analogy with gas dynamics the interface is referred to as a *shock wave* and the above condition (14.18) determining the shock speed as the *shock condition*. A simple geometric representation of the situation on the $F(N)$ state diagram is again available, see Fig. 14.16.

The shock speed is the slope of the chord joining the states on the two sides of the discontinuity. From this geometric result it's evident that the shock moves with a speed somewhat greater than the signal speed in the dense traffic zone and somewhat less than that in the less dense zone. Thus the path of the shock in the (x, t) plane intersects both sets of characteristics; a result implicitly assumed in the above calculations. It's also evident (either from the geometry or the expression we've obtained) that weak shock waves $(N^l \approx N^r)$ travel at the signal speed. It all fits together and makes sense!

Explicitly, substituting $(F^l, N^l) = (F^0, N^0)$ and $(F^r, N^r) = (F^-, N^-)$, see Table 14.1, we get $\dot{\mathcal{X}} = -V_{max}\sqrt{2}/4$.

Consider now the cars that have passed by the obstruction. The signal speed ideas and the above shock ideas suggest that these cars, receiving as they do information from the jam, will proceed with speed V^+ until they reach the stream of traffic moving with the normal speed V^0. They'll then pass through a shock separating the zones and once again move with the normal traffic speed. Thus, using (14.19) with $(F^l, N^l) = (F^+, N^+)$, and $(F^r, N^r) = (F^0, N^0)$, we find that the shock separating the two zones travels with speed $V_{max}\sqrt{2}/4$.

The characteristics pattern is shown in Fig. 14.15, where the car path of an individual driver is also shown.

Comment: shocks in other contexts An essential prerequisite for shock-like behaviour is that there exists more than one steady state solution satisfying the required conservation conditions. Under normal circumstances continuity determines the appropriate solution; however, if the dynamic behaviour of the system is such that profiles can steepen, then under the appropriate circumstances an abrupt transition will occur from one solution possibility to the other. This type of situation occurs in many circumstances and the modeller should be aware of when to expect such behaviour.

Comment: In many respects it's a very interesting story that's emerged from the above work. Firstly it should be recognized that the discontinuity represents a mathematical artifice introduced to cope with inadequacies of the model. In the real situation the change in traffic state occurs over a small (but finite) zone. Remarkably, global conservation principles together with signal speed ideas have led to an acceptable resolution of a major difficulty in the basic mathematical model. Elaborating further: the point is that the equation describing the traffic behaviour under smooth conditions cannot describe what happens when the abrupt shock conditions are set up; and given that shock conditions will almost assuredly occur in zones in which cars are entering regions of greater congestion, the problem is major. Now from a mathematical viewpoint the error arises because the smoothness conditions required to derive the basic traffic equation are invalid under such circumstances. Although informative, this provides little insight into the nature of the difficulty and how to deal with it. We've seen that invoking car conservation conditions *across* such shocks leads to a unique determination of the situation. When you think about it this is quite a striking result. *Without understanding what's actually happening to cars in the shock zone, we've been able to determine uniquely what happens across the shock!* This means that (within reason) the detailed behaviour in the shock zone is *irrelevant* as far as the behaviour outside the zone is concerned; so there's no real need to search for the correct defining equations in the shock region.

The above solution procedure manages to neatly sidestep major issues which need addressing, even if only to satisfy one's curiosity. What is the correct defining equation? Why does the basic traffic equation when appended with the shock condition produce correct results? Such issues are of paramount theoretical interest but also are clearly of practical importance. For example, accidents are likely to occur within shock zones so a better understanding of the shock zone structure is likely to be necessary to investigate such situations.

So what is likely to happen in practice when drivers encounter shock conditions? Under such conditions drivers are unlikely to simply *instantaneously* respond to the *local density* situation. For a start they are likely to anticipate the shock to some limited extent and then, after a time delay, they are likely to hit the brakes. Additionally, variability of driver behaviour could have a significant effect on the behaviour in the shock zone. Obviously the situation is complex and, although the introduction of equations describing such effects is in principle simple, the resulting equation set quickly becomes unmanageable. It's a matter of judgment to what extent one should even try to model the various aspects of this situation, especially given that we suspect that it's going to be difficult to distinguish between various models experimentally anyhow. *Bigger is generally worse, not better!* The appropriate model to use will also depend on the circumstance of interest. For example, any accident model should in some way model the response delay of the driver, whereas this may not be of significance for other circumstances (providing an accident doesn't result). Leaving aside the accident prediction problem the simplest model that takes into account the fact that drivers respond to traffic conditions some way ahead on the road is given by $F = F(N, N_x)$. In Exercise 14.12 we'll show that such a model leads to results that produce the shock condition, and produces solutions that asymptotically match onto the simple shock wave solutions.

Other situations

Although it's by no means a trivial matter to patch together solutions in various regions of a road system, the way to go about it is now clear. Working from an initial state one simply needs to (numerically) trace the characteristic network from an initial state. When characteristics intersect, as they will in regions of increasing congestion, shocks need to be intro-

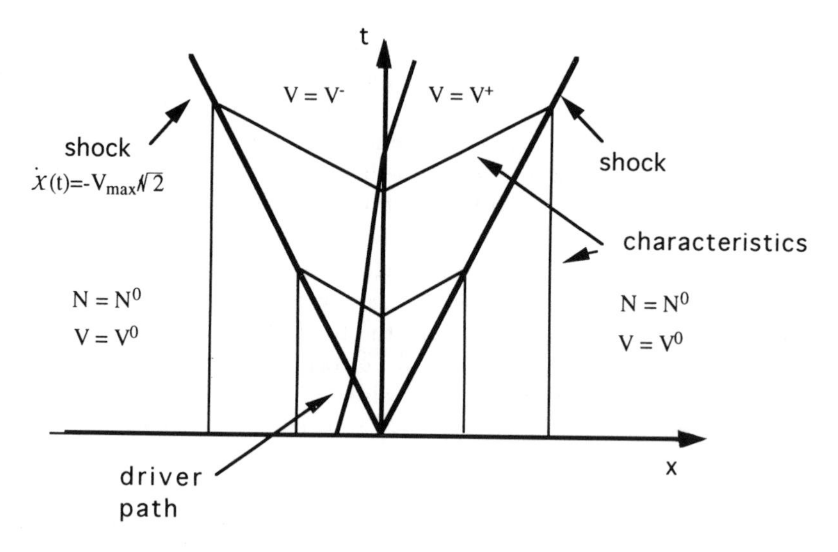

Figure 14.15: The traffic jam solution.

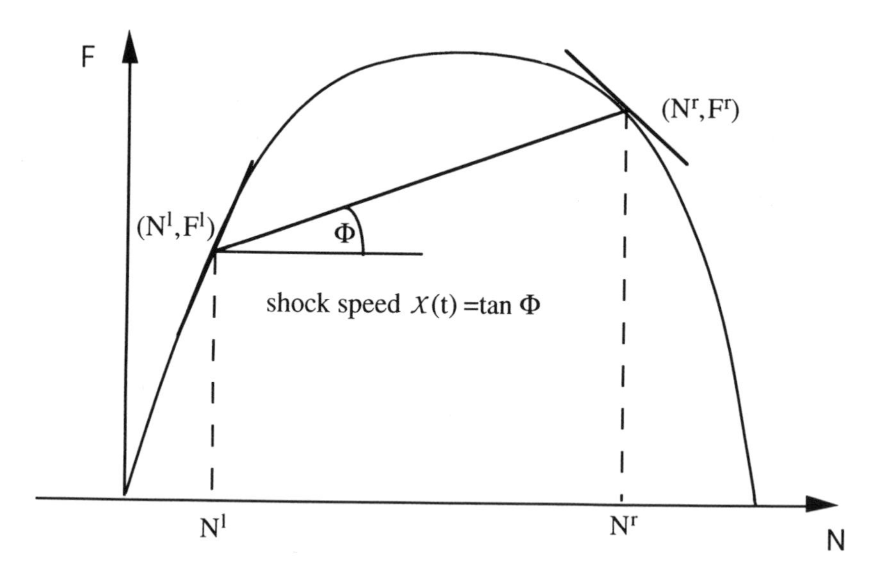

Figure 14.16: Shock geometry.

duced and their paths need also to be traced using the shock speed result (14.19). Exits and entries introduce no further conceptual difficulties. For a clear and elementary account of these and other aspects of traffic flow the reader is referred to Haberman (1977), and for an excellent account of wave phenomena generally see Whitham (1974).

14.4 Summary

A simple steady state flux verses density relationship based on driver behaviour led to two possible traffic states corresponding to any flux level less than F_{max}. The particular state realized in practice was seen to depend on earlier history, necessitating the study of unsteady traffic behaviour. Under dynamic circumstances the car conservation equation, when appended to the flux/density relation, led to *the traffic equation*; a first order partial differential equation which (after attaching appropriate subsidiary conditions) determined the traffic state under most circumstances. Simply interpreted this equation says that traffic flow information will propagate in the traffic stream with the local *signal speed*, and exact solutions were constructed by simply tracing the associated *characteristics* in the (x, t) plane. The traffic equation was, however, found to be inadequate (producing multiple solutions) in circumstances in which cars were moving into zones of greater congestion. Under such circumstances discontinuous *shock* solutions result, which were obtained by invoking the correct car conservation equation across the shock. Such shock solutions represent an acceptable representation of the traffic behaviour providing one's interest isn't cent red on issues that are closely associated with the shock structure (eg. accident situations).

The ideas presented in the above work are of broad applicability.

14.5 Exercises

Exercise 14.1 *Loose Ends*

Statistical variations are a feature of any real traffic situation so, even to just interpret data for use in a deterministic steady model, it's necessary to make decisions about how to determine averages. As indicated in the main text, for good physical reasons averages for (F, N, V) should be defined so that the relationship $F_{av} = N_{av}V_{av}$ holds. Spatial averages for velocity and density measurements, and time averages for flux do just this. To see how/why this is so, consider the situation in which there are two independent lanes of traffic (cars are prohibited from crossing from one lane to the other) with cars in the two lanes all travelling in the same direction but with different speeds and associated densities. Show that under these circumstances, with averages defined as above, $F_{av} = N_{av}V_{av}$.

Exercise 14.2 *A Braking Distance Flux Model*

Based on the assumption that the car will travel a distance given by $t_r V$ ($t_r \approx 0.3$ secs, the reaction time) before a driver reacts to a traffic change and applies the brake, and then will travel an additional distance $V^2/2b$ before the car will come to rest, a spacing of $Vt_r + V^2/2b$ between cars seems safe. (The assumption here is somewhat conservative—the driver in front of you is assumed to "stop dead").

(a) Using the above model and the rule of thumb "leave 1 car spacing for each 10 mph " (that is meant to apply for average city speeds only), determine an appropriate value for b and thus determine a velocity vs. density relationship. Plot it and check to see if it fits in terms of your own experience on the road.

(b) Determine from this relationship the speed at which maximum flux levels are attained. This is the speed engineers would like to see prevailing in situations such as a tunnel so that the flux through the tunnel is maximized. For many drivers this speed is uncomfortably slow. There are other difficulties. The signal speed work suggests that if drivers did travel at maximum flux levels instabilities would develop in the traffic flow; a result that is borne out in numerical studies and in field measurements. Higher speeds with associated lower flux levels thus seem more desirable.

Exercise 14.3 *Loose Ends: Traffic Lights*

(a) Using the result (14.17), which determines the traffic density along rays from the origin, see Fig. 14.13, determine the density and velocity distribution as a function of location at a given time t_0 after the lights turn green for a range of values of t_0, and plot out the results. (Choose appropriate scales.) Use the parabolic flux/density model, see (14.6).

(b) $\star$ Using Maple repeat the above calculations with the improved flux/densi relationship (14.10) developed in Section 14.2.4, and make comparisons.

Hint: You'll need to write a procedure to determine N and V as a function of the ray slope.

Exercise 14.4 *Similarity*

There is an alternative method of solution for the traffic lights problem based on similarity arguments. The natural density and flux scales are the maximum density N_{max}, and the maximum flux level F_{max}; so the scaling $F = F_{max}f(n)$, where $N = N_{max}n$ is sensible.

(a) Determine the equations governing the scaled density n and flux $f(n)$.

(b) Using dimensional arguments show that $n = n(\xi)$, where $\xi = x/[(F_\star/N_{m\iota}$
and determine the equation for $n(\xi)$.

(c) Write down the solution and make appropriate comparisons.

Exercise 14.5 *Peel Inlet Pollution Problems I*

Excessive algal growth in the Peel Inlet in Western Australia has caused damage to estuarine marine life, and an unpleasant stench (of rotting algae) for residents. As you can well imagine the loss to the tourist industry is also considerable. The environmental imbalance arises basically because, in an attempt to increase production levels, farmers in the area have used excessive fertilizer levels. Rain and irrigation water carry the excess superphosphate into the groundwater below, which subsequently flows into the estuary. The build up in phosphate leads to algal blooms. The problem is basically a political/social/economic one (how to persuade farmers to use less fertilizer) but the modeller can contribute by determining concentration levels, time spans, and by commenting on the effectiveness of possible remedies.

Background: Moist soil consists of a mixture of soil particles, air and water. Under saturated conditions water fills the voids not occupied by solid particles. In regions in which the soil overlays impermeable rock (typical of catchment areas) rainfall and irrigation water (together with soluble contaminants) percolate through the soil to the saturated groundwater below which is gravitationally driven to the ocean. The velocities of flow (defined to be the volume of water crossing unit area in unit time) are small; of the order of metres per month.

Here we model the groundwater transport aspect of the problem. We'll assume

- the groundwater flow is one-dimensional and uniform,
- the phosphate is carried along with the groundwater,
- no phosphate is absorbed by soil particles and it moves without diffusion.

The phosphate concentration $c(x,t)$ is measured in kg/m^3 of groundwater. A known quantity of phosphate per unit area per unit time Q kg/m^2/s is released into the groundwater at $x = 0$ for $t > 0$, see Fig. 14.17.

(a) Show that phosphate conservation requires

$$c_t(x,t) + uc_x(x,t) = 0, \qquad x > 0,$$

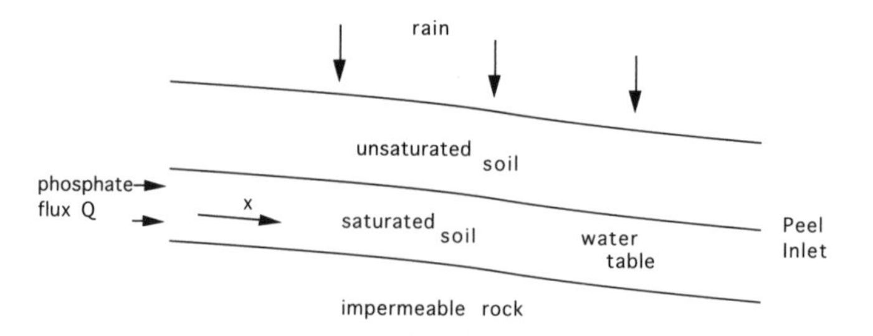

Figure 14.17: The Peel Inlet pollution problem.

where u is the groundwater flow velocity, and that the conditions envisaged above require

$$c(0,t) = Q/u, \qquad c(x,0) = 0.$$

(b) Find the solution using characteristics.

(c) Using the typical value $u = 10^{-4}$m/s (for sand) determine how long it would take for the phosphate to reach the Inlet (distance 2 km from the source) and the concentration levels that are to be expected there.

Exercise 14.6 *Peel Inlet Pollution Problems II*

One might expect diffusive effects of various types to simply smooth out the step function concentration profile obtained in the previous exercise. Thus one might expect a profile of the form

$$c(x,t) = C(\xi,t), \text{ where } \xi = x - ut$$

where C is "close" to the step function solution obtained above. To check this out we'll examine the model equation

$$c_t + uc_x = \kappa c_{xx},$$

in which a diffusion type term has been added.

(a) Show that if $-\kappa c_x$ is the diffusive flux of phosphate, then the above equation does in fact represent an accurate description of the phosphate conservation equation. In practice κ would be experimentally determined, and would be u dependent.

Hint: Examine a volume element moving with the groundwater.

(b) Substitute the above form into the model equation and thus show that a solution of the expected form exists if $\mathcal{C}$ satisfies

$$\mathcal{C}_t = \kappa \mathcal{C}_{\xi\xi}.$$

(c) What initial conditions on $\mathcal{C}$ are appropriate? Determine $\mathcal{C}$.

Exercise 14.7 *Pollutant Dispersal*

In addition to the convective and diffusive effects discussed in the last two exercises, chemical or biological decay and/or deposition usually play a role in determining the concentration levels of pollutants in the environment. Often the assumption is made that the loss rate is proportional to the concentration. We'll examine the case of pollutant dispersal in a river flowing with velocity u. The pollutant will be assumed to be uniformly distributed with depth and across the river. Diffusive effects will be ignored in this one-dimensional model.

(a) Show that pollutant conservation is assured if

$$c_t(x,t) + uc_x(x,t) = -\alpha c(x,t),$$

where αc is the loss rate per unit volume of water if the concentration of pollutant is c.

(b) We'll assume the river is initially pollutant free and a prescribed volume of pollutant is then instantaneously released at $t = 0$ over the interval $-L < x < L$, so that

$$c(x,0) = f(x) = \begin{cases} c_0 & -L < x < L \\ 0 & x < -L, \text{ and } x > L. \end{cases}$$

Using characteristics determine the solution in the $\alpha = 0$ case.

(c) In the $\alpha \neq 0$ case, given the equation is linear, one might expect an exponential decay in pollutant levels so it makes sense to look for a solution of the form

$$e^{-\lambda t} C(x,t)$$

By choosing λ appropriately remove the "sink" term from the equation for C and thus show that the solution for c is given by

$$c(x,t) = f(x-ut)e^{-\alpha t}.$$

You'll recall that this is a trick often used to simplify ordinary differential equations. Interpret this result.

(d) ⋆ Suppose now that we have a situation in which the pollutant release *rate* at $x = 0$ is specified by

$$R = r(t) = \begin{cases} r_0 & 0 < t < T_0 \\ 0 & t > T_0. \end{cases}$$

Determine concentration levels downstream from the release point. How far downstream does one need to go for concentration levels to drop to 1% of those at the source?

Hint: Modify the trick used in (c) to obtain an appropriate solution form for this situation.

Exercise 14.8 *Traffic Lights Timing*

In order to decide on the correct apportionment of green light time to red light time for inner city conditions it's necessary to determine the length L of a queue of cars that will be cleared over any prescribed time interval. Let's assume all cars (an infinite number) are initially at rest behind the red light and the lights turn green at $t = 0$, and stay that way for a period of time T. For simplicity we'll use the parabolic flux/density model. Consider a car at arbitrary distance D back from the lights. Follow the indicated path of calculations:

(a) Calculate how long the car remains at rest.

(b) Using the solution obtained in the main text, see equation 14.17 and Fig. 14.13, determine the density in the "fan" region corresponding to a particular (x, t) value, and thus show that the car speed in this region is governed by

$$\frac{dX(t)}{dt} = \frac{V_{max}}{2}[1 + \frac{X(t)}{V_{max}t}],$$

where $X(t)$ is the location of the car at time t

(c) What is the appropriate initial condition for this ordinary differential equation?

(d) Using Maple solve for $X(t, D)$, and plot the position and velocity for drivers (identified by their D values) for a few values of D.

(e) Derive a condition for a car to reach the lights at time T, and hence find the length of queue which can be cleared. Why is this likely to be an overestimate of the length cleared? How might one improve the estimate?

(f) There is a much easier way to obtain L: What is the flux at the lights? How many cars cross the lights in the time interval T? What length of roadway would this number of cars have occupied while waiting for the lights to change?

Exercise 14.9 *Accidents*

Any accident model must take into account the delayed response of drivers. Let's assume that there is a desired speed U^D that drivers would like to travel at, and that drivers will try to eliminate any difference between their actual speed and their desired speed. Two simple models of driver behaviour suggest themselves:

$$\text{Model A}: \quad \dot{x}_n(t+\tau_a) = U^D(t),$$

$$\text{Model B}: \quad \ddot{x}_n = \frac{U^D(t) - \dot{x}_n(t)}{\tau_b},$$

where x_n denotes the location of the front of the nth car measured from a datum, see Fig. 14.18. In the first model the driver makes adjustments τ_a seconds after observing the velocity discrepancy, whereas in the second model the driver responds immediately by accelerating at a rate proportional to the velocity discrepancy. In both models the driver's response may be too late to avoid an accident. Of course the desired speed will be spacing dependent; let's use the linear relationship

$$U^D = \lambda[x_{n-1} - x_n - L],$$

where L is the car length. This corresponds to the continuum linear model (14.4) with parabolic flux/density (14.6) used in the main text.

(a) Show that, after appropriate scaling, the model equations become

$$\text{Model A}: \quad x_n'(s+\epsilon_a) + x_n(s) = x_{n-1}(s) - 1,$$

$$\text{Model B}: \quad \epsilon_b x_n''(s) + x_n'(s) + x_n(s) = x_{n-1}(s) - 1.$$

Our interest is in determining the size of the response parameters ϵ_a, ϵ_b corresponding to collisions.

(b) What mathematical statements indicate a collision?

(c) Maple can handle Model B. Consider the case of just two cars, and examine the situation in which initially both cars are moving at their desired speed for their spacing and the leading car (at location x_0) "stops dead". Determine the path $x_1(s)$ of the trailing car for a variety of values of ϵ_b, and plot the results. For what values of ϵ_b is a collision expected?

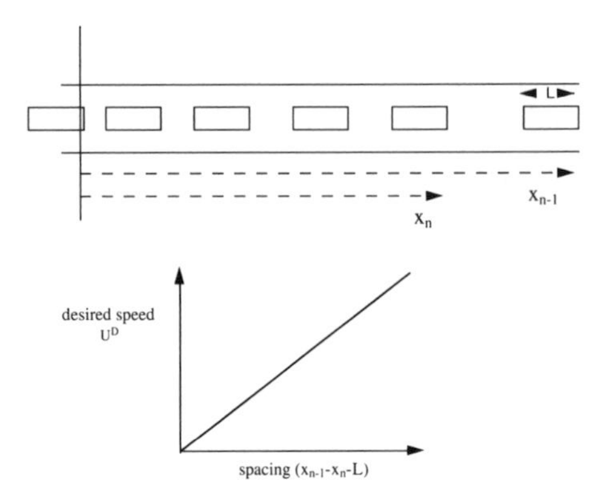

Figure 14.18: Accident models.

(d) ⋆⋆ To process Model A one needs to either use Laplace transforms or use numerical methods. If you have experience in these areas process this model.

Exercise 14.10 *Another Jam*

The situation envisaged is as in the main text except that the cars are initially moving at speed $\frac{3}{4}V_{max}$ and an obstruction forces drivers to reduce their speed to $\frac{1}{8}V_{max}$.

(a) Determine the solution.

(b) Determine the path $X(t)$ of an individual driver.

(c) Will a velocity reduction at an obstruction always obstruct traffic (i.e. reduce flux levels)?

Exercise 14.11 *Hydraulic Jumps*

A rather remarkable phenomenon is often observed in rivers under flood conditions. The onrush of flood waters results in a sudden build up of water which does not settle down as one might expect, but rushes forward with little change in form towards the sea. A similar situation occurs in estuaries when large tidal effects cause hydraulic bores to propagate. The situation can be quite spectacular under the right set of circumstances. (Many observers each year at the appropriate time line the banks of the Severn estuary in the British Isles to see the annual bore.) We'll examine the flood waters situation under the steady conditions displayed in Fig. 14.19.

If water rushing down a slope (defined by $H(x)$) is partially blocked by an obstruction on the bottom, the water piles up as shown and a steady

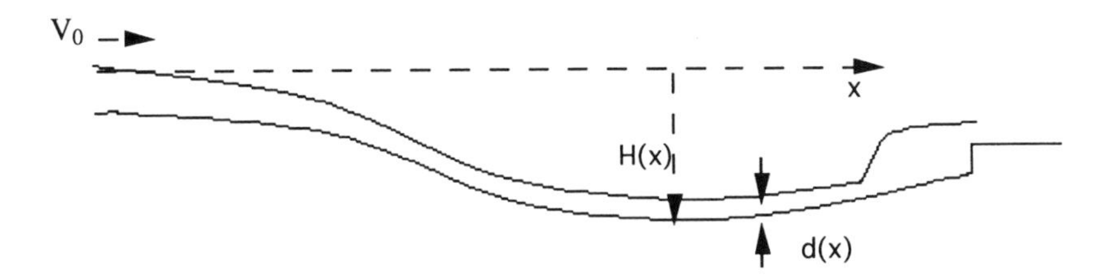

Figure 14.19: The hydraulic jump.

state jump is observed; a similar situation is observed when water from a tap flows into a sink. We'll assume the stream width w remains fixed. The velocity $V(x)$ and depth $d(x)$ of the flow will alter with location in accordance with the relevant conservation laws:

$$Mass\ conservation: \quad V(x)wd(x) = Q(const)$$

$$Momentum\ conservation: \quad \frac{1}{2}[V^2(x) - V_0^2] = \gamma g[H(x) - d(x)],$$

here γ is a dimensionless parameter introduced to account for turbulent momentum losses, Q is the volume flux carried by the stream, and V_0 is the stream velocity at a selected datum for H, see Fig. 14.19.

(a) Eliminating $V(x)$ between the above equations, show that the flux vs. depth relationship can be written in the form

$$Q = \sqrt{2\gamma g}[\sqrt{H_0 - d(x)}]d(x),$$

with H_0 appropriately defined. The algebra is tedious but the implications are interesting.

(b) Thus show (plot graphs etc.) that there are 2 values of depth $d(x)$ with corresponding velocity $V(x)$ associated with a fixed flux level Q allowed by the above conservation laws at each location x; a fast (" shooting ") flow of shallow depth, and a deep and slow ("tranquil") flow. This situation is analogous to the traffic situation, and to the gas dynamics shock situation. A transition from one allowed state to another can occur abruptly when conditions are right. In the present case the obstruction can cause a transition from shooting flow to tranquil flow—one might expect this given our experience with the traffic jam situation.

Exercise 14.12 *Shock Structure*

The flux/density model $F = F(N)$ leads to discontinuous solutions under increasing congestion circumstances. We'll look for a better model under such circumstances. Explicitly we'll seek a better description of the constant strength travelling shock, as obtained in the traffic jam situation examined in Section 14.3.4, see (14.19).

A basic flaw in the flux/density model is that it assumes that drivers purely react to *local* density conditions. Clearly drivers respond to a certain extent to the conditions *further ahead* in the traffic stream, and the density gradient provides a measure for such changes in traffic state. We'll thus examine the simple modified model

$$F = F(N, N_x) = F(N) - \gamma N_x,$$

with γ constant.

(a) The coefficient γ should be positive. Why?

(b) Show that the car conservation requirement leads to the modified traffic equation

$$N_t + F'(N)N_x = \gamma N_{xx}. \tag{14.20}$$

Note that the steepening and breaking associated with the nonlinear term $F'(N)N_x$ examined in earlier work will be moderated by the (smoothing) diffusive γN_{xx} term.

Comment: The opposing tendencies of nonlinear steepening and diffusive smoothing occur in many areas of science; herein lies the importance of the above equation referred to as **Burger's equation.**

(c) Given the behaviour of the jam solution obtained in Section 14.3.4, it makes sense to look for a steady profile solution of the form

$$N = N(\xi), \text{ where } \xi = x - Ut$$

travelling with the constant speed U, to be determined. Substitute this form into equation (14.20), and show that the resulting equation integrates to give

$$F(N) - UN + c = \gamma N_\xi, \tag{14.21}$$

where c is a constant of integration.

(d) The solution of interest should smoothly match onto the constant state conditions $N \to N^l$ as $\xi \to -\infty$ and $N \to N^r$ as $\xi \to +\infty$. This is a classic two-point boundary value problem with solutions only existing under special circumstances. By examining (14.21) (there's no need to actually solve the equation) show that the boundary conditions are satisfied only if

$$U = \frac{F(N^r) - F(N^l)}{N^r - N^l}.$$

Note that this is precisely the shock speed result we arrived at with the simple flux/density model, see (14.19)!

(e) Exact solutions for the density profile can be obtained for the parabolic flux/density model (14.6). Show that for this case, after appropriate scaling, equation 14.21 reduces to

$$(n - n^l)(n^r - n) = \nu n_\xi,$$

and use Maple to plot out the solutions for fixed n^l, n^r and a variety of small ν values. Observe and comment on the effect of ν on the shock structure.

Comment: The implication of this work is that the traffic equation with appended shock condition gives "correct" results except within a thin shock zone.

Exercise 14.13 *Jam Aftermath*

⋆⋆⋆ Suppose the accident considered in the text in Section 14.3.4 reduces the traffic flow for a time interval t_1, and then both lanes are clear to resume natural traffic flow. How will the traffic situation evolve, and how long will it take for normal flow to resume? To examine this situation first specify the traffic state immediately after the road is open.

Based on our experience of such situations and the ideas developed in the text our expectations for the situation may be:

- Immediately after the obstruction is removed there will be a "partial vacuum" to the right of $x = 0$, and locally the situation will be very similar to the one encountered in the traffic light problem. Thus we might expect an "expansion fan" to spread into the vacuum region causing an increase in density in this region.

- The expansion fan will also spread into the choked-up region $x < 0$, thereby relieving congestion in this region.

- Eventually the edge of the expansion fan spreading into the vacuum region will reach the traffic travelling at normal speed and a shock will be created, whose strength and location will be determined so that the car conservation (or shock) condition (14.18) is satisfied.

- Similarly the edge of the fan spreading into the congested zone will reach the zone corresponding to normal flow, and again a shock might be expected.

- After a period of time we might expect all cars to be moving at normal speed so that all characteristics will asymptote to $\frac{dx}{dt} = 0$.

Using the above as a guide:

(a) Roughly sketch the expected characteristics network.

(b) Determine the locations in the (x, t) plane where the edges of the expansion fan intersect the regions of normal traffic flow.

(c) Using the shock condition and the (known) traffic states in the regions bordering the shocks, determine differential equations for the locations of the shocks, and specify appropriate "initial" conditions for these equations.

(d) Solve the resulting equations and check to see that the asymptotic solution behaviour is correct.

Bibliography

Carrier, George, F. and Pearson, Carl E. (1976). *Partial Differential Equations*, Academic Press, New York.

Cole, J. D.(1968). *Perturbation Methods in Applied mathematics*, Ginn and Company, Boston

Haberman, Richard.(1977).*Mathematical Models*, Prentice-Hall, New Jersey.

Lighthill, M. J. and Whitham, G. B.(1955). *On Kinematic waves: I. Flood movement in long rivers; II. Theory of traffic flow on long crowded roads.* Proc. Roy. Soc. A 229 p 281-345.

Whitham, G. B. (1974).*Linear and Nonlinear Waves*, John Wiley and Sons, New York.

Part IV

Solutions

Hints and Answers

Part I

Chapter 2

Exercise 2.3

(a) *Hint:* Substitute $\theta(t) = \alpha\Theta(t')$, $t = Tt'$, and choose T so that the new equation has coefficients unity.
$\tau \approx \pi/2\sqrt{l/g}$.
The period is independent of m.

Exercise 2.5

(a) *Hint:* Replace $\sin\Theta$ by the first two terms of it's Taylor expansion, and use the approximation procedure described in the text.
(b) *Hint:* Solve the equation analytically (use `dsolve`) and use `fsolve` to determine a numerical solution for the period. You can use a `proc` to define, $\tau(\alpha)$, and `plot` the results.
(c) *Hint:* Use `dsolve,numeric` and then extract the solution from the procedure (`F`, say) thus defined, using `op(2,[F(1)])`.

Exercise 2.6

(c) By altering S the diver can effect a $1/\sqrt{S}$ change in his terminal velocity.

Exercise 2.7

(a) A has units m^{-1}, T has units kg m/s^2, μ has units kg/m/s. Time scale $= \sqrt{m/Tl}$.
(c) We require the μ term to be relatively small compared with the retained terms. This requires $\mu l/\sqrt{mT} \ll 1$.

Exercise 2.8

(a) $t_0 = R^2/\kappa$. $t_c = (R^2/\kappa)t_c^{'}(\gamma)$, where $t_c^{'}(\gamma)$ needs to be determined from the scaled equations.
Note that the cooling time increases with the square of the sphere radius. The dependence on γ can't be determined without solving the equations, but clearly it's the combination γ that characterizes surface cooling.
(d) About 3 years.

Exercise 2.9

(a) $\eta_0 = pl^2/T$. $\mu = \rho\omega^2 l^2/T$.
(b) The results *suggest* that the amplitude of motion increases in proportion to the amplitude of the applied force p and varies inversely with string tension T.
maxamp $= \eta_0 fn(\mu)$.
Hint: Note that with μ equated to zero the defining equation is effectively an ordinary differential equation for η and so can be easily integrated. The integration 'constants' arising will, however, be functions of $t^{'}$.

Exercise 2.11

(a) Guitar players tell us that the temperature and humidity matter. Possibly these effect a change because of the resulting T change. The amplitude of vibration could also matter. Additionally the guitar's body could be important. Clearly careful experiments are necessary to isolate out particular effects.
(b) The combinations in the brackets are dimensionless.
(c) The experiments imply that the function gn is uninfluenced by changes in the combination El^2/T, but must be affected by the value of σ or some other physical feature of the situation not anticipated in our discussion.

Chapter 3

Exercise 3.1

(a) *Hint:* Note that, apart from the moment equation associated with legs #2, #4, and the translational equilibrium equation, the defining equations are unaltered.
(b) Rocking will only occur if leg #2 clears the floor and if an external force greater than $4P$ causes the table to rotate about diagonal 1-3.

Exercise 3.3

Hint: If E_s is the Young's modulus of the sleeve material and $(t_i, t_i^{'})$ are the stressed (with N_i acting) and unstressed thicknesses of the sleeves, then

$[(t_i - t_i')/t_i]E_s A = N_i$, where A is the effective sectional area of the sleeves.

Exercise 3.4

(b) $N'_i = N_i + w$.
(d) The total length change in the leg is given by

$$-\int_0^l F(z)/(kl)\,dz.$$

(e) The change in leg length is given by

$$-[N'_i - w/2]/k.$$

Note that change in leg lengths brought about by the leg weight terms is the same for all legs, so that the value of Δ is unaffected by the value of the leg weight w.

Exercise 3.5

(a) The symmetric solution is accurate *only* if $k\Delta/W \ll 1$. Perfect symmetry is a mathematical ideal rarely exhibited in the real world. The arguments will often work, however, providing the situation is *almost* symmetric in an appropriate (context defined) sense. The relevant length scale in our problem is W/k (*not* l); so almost symmetric tables, almost never exist.
(b) If the table is perfectly rigid it will rock unless perfect. If the table is geometrically perfect it won't rock. These limiting cases are, however, not physically speaking relevant. It's the combination $k\Delta$ that determines the table's behaviour.

Chapter 4

Exercise 4.1

Hint: For the optimal solution $T(y)/A(y) = \alpha$.

$$A = \frac{Mg}{\alpha}\exp(\frac{\alpha}{\rho g}(l_0 - y)).$$

Thus the sectional area at $y = l_0$ has to be sufficient to sustain the load Mg. The required area increases exponentially with distance from the end of the chain with the length scale $\rho g/\alpha$ dependent on the strength of the material.

Exercise 4.2

Hint: Note that the paths initially drop vertically, and then flatten out to reach the terminus. Paths asymptote to c and so we'd expect solutions for large b to be given to first order by $y' = \sqrt{(d-y)/y}$.

Exercise 4.3

(b) *Hint:* Break the first order variation up into an integral contribution (of the type arising in the archetype problem), and a boundary contribution. Why should both contributions vanish independently?
(d) *Hint:* Note that the Euler-Lagrange equation is singular at $y = 0$, so the slope here will exceed the allowable limit α.

Exercise 4.6

(b) *Hint:* $T = EA(l - l_0)/l_0$. Use a Taylor series approximation based on the result that $\int_0^l \frac{1}{2}u_x^2 dx \ll (l - l_0)$. Remember to determine the change in elastic energy from the stretched string equilibrium configuration.
(d) The Euler-Lagrange equation is

$$\frac{\partial}{\partial x}[F_{u_x}] + \frac{\partial}{\partial t}[F_{u_t}] = 0.$$

Exercise 4.7

Hint: Proceed as for the archetype problem, introducing a possible candidate $U(x,t)$ with the variations $\eta(x,t)$ satisfying the required zero boundary and initial conditions. Integrate by parts the required number of times to obtain an expression similar to the δJ of the archetype problem.

Exercise 4.10

Hint: To test your results examine solutions corresponding to a sphere or a plane.

Exercise 4.11

Hint: In a constant v region we expect a straight line path solution. In circumstances in which v is constant in patches we'd expect the required solution to spend as much time as possible in large v patches.
It's possible to solve the equations exactly if $v = ax$. In some cases a phase plane analysis may be useful.
(e) It's harder to run uphill, so the speed of travel will be direction dependent. Thus the scalar function $v(x,y)$ needs to be replaced by a vector function $\mathbf{v}(x,y)$.

Part II

Chapter 6

Exercise 6.1

(a) *Hint:* If $\phi(x,t)$ is the solution of $L(\phi(x,t)) = 0$, with $\phi(0,t) = \phi(x,0) = 0$, and $\phi(L,t) = 1$, then $\phi_0 = T_0\phi(x,t)$ also satisfies the same linear operator equation. What boundary conditions does $\phi_0 = T_0\phi(x,t)$ satisfy? Relate this to possible rod experiments.
If a material has "grain" in it then there's likely to be a preferred direction for heat flow (along the grain usually) and so one might expect non-isotropic behaviour eg. wood. Nonlinear behaviour occurs if the chemical status of the material changes with temperature. Material that dries out as the temperature rises exhibits such behaviour.
(b) $T = T_0 + (T_1 - T_0)/4$.
(c) No.

Exercise 6.2

(a) One would expect the rod's temperature to increase in time, with the maximum temperature being reached at the rod's centre. Because of the rise in the temperature of the rod, one would also expect the heat loss through the boundaries to increase with time. Eventually one might expect the heat loss rate to reach the heat input rate. If this happens one might expect a steady state situation to result.
(b) $T(L/2,t) = T_0 + [q_0L^2/(kA)]T(\kappa t/L^2)$.

Exercise 6.5

1. $T_a - T_r = \frac{P}{S}\frac{d}{k}$. The temperature drop increases in proportion to the efficiency of the conditioner and the thickness of the walls.
2. $T_a - T_r = \frac{P}{S}[\frac{d_m}{k_m} + \frac{d_i}{k_i}]$.
3. The dimensionless group $\frac{S_g}{S_i}\frac{k_g}{k_i}\frac{d_i}{d_g}$ determines the effect of windows on insulation.

Exercise 6.7

(c) $U = q/[\rho c(T_{ign} - T_{inf})]$.

Exercise 6.10

(a)(i) $\omega = n\pi$. Thus to double the sound frequency one needs to halve the string length, with the string tension remaining fixed. To generate a particular mode one can arrange the initial displacement profile to be

as prescribed. In practice one maintains zero displacement at a location corresponding to a node of the required vibration mode; but other modes also have zero displacement at this location, so a pure tone may not be generated.
(ii) The pattern repeats over a time span determined by the fundamental $n = 1$ mode.
(b)(i)(ii)$\eta(x,t) = \sin n\pi x[e^{-\gamma t/2)} \cos(\sqrt{1-(\gamma/2n\pi)^2 t})]$.
Note that the mode is damped, its spatial shape remains unchanged, and different modes (with different n values) vibrate with slightly different frequencies.
(c)*Hint:* Substitute the expression directly into the equation and cancel the common exponential term.

Chapter 7

Exercise 7.2

(b) Note that $e^{\xi^2/4} >> e^{a^2/4}$ as $\xi \to \infty$ for any finite a.
(d) *Hint:* For large ξ write

$$\mathcal{I} = \int_0^a e^{y^2/4} dy + [2/\xi - 2/a] + [(2/\xi)^2 - (2/a)^2] + O(2/a)^3$$

and choose a to make the error small.

Exercise 7.3

(e) $f = 1 - \text{erf}(x/2\sqrt{\kappa t})$.
Heat flux$=-k/\sqrt{\pi\kappa t}$. Note that the flux is singular as $t \to 0$, but that the heat transferred per unit area into the body over a finite time interval is finite.

Exercise 7.4

(a-c) *Hint:* Write $\mathcal{H} = \mathcal{H}_0 + [\mathcal{H}_1 - \mathcal{H}_0]H$ and show that for large ζ, H to first order satisfies the same equation as obtained in (7.3), so

$$H \approx A \int_\infty^\zeta e^{-\zeta'^2/[4\kappa(\mathcal{H}_0)]} d\zeta',$$

with A arbitrary.
(d) An exponential $\kappa(\mathcal{H})$ would be appropriate. You could then easily explore the affect of varying the κ dependence.

Exercise 7.5

(b) $k_1 T_x(0-,t) = k_2 T_2(0+,t)$.
(d) *Hint:* Note that $T_c = T_c(t, k_1, K_2, \kappa_1, \kappa_2)$ and check out the dimensionless groups.

Exercise 7.6

(a) $T_{max} = 4\mu H/\sqrt{\Delta}$.
(b) The solutions are all of the form $f(n)\mu H/\sqrt{\Delta}$, as one might expect on dimensional grounds, where $f(n)$ varies by a factor of about 2. The temperature reached is strongly dependent on Δ (note the instantaneous heat input result), and weakly dependent on the shape of the heating rate function.
(c) Low conductivity, low ignition temperature and rapid heat input are required for combustion. Note that at the high temperatures required for ignition, heat losses will be large if the time span of heating is not short. It's thus doubly important to apply the heat in as short a burst as possible for combustion.

Exercise 7.9

(c) Steady state will be reached when the total radiative heat input to the body per unit time E_0 balances the heat being expelled by conduction from the body's surface per unit time i.e. $kT_x(0)$. After integrating the equation one obtains the result

$$T(x) - T(0) = \frac{E_0}{k\mu}[1 - e^{-\mu x}].$$

There will be a net conductive transfer of heat into the depth of the body until the temperature at depth is greater than it's surface temperature. Eventually the temperature within the body will be sufficiently higher than the surface temperature that a balance between radiative heat supply and conductive loss is realized.
(d) Materials with low conductivity will (given sufficient time) actually reach higher temperatures than low conductivity materials, but the time required for low conductivity bodies to reach equilibrium is large (of order L^2/κ) so that we don't normally encounter steady state conditions in such bodies. Note also the results of Exercise 7.5.

Exercise 7.10

(b) $c(z,t) = \frac{M}{2\sqrt{\pi\kappa t}} e^{-z^2/(4\kappa t)}$.
We'll denote this solution by $C(z,t)$.

(c) Significant errors arise when the $c(h,t)$ is significant i.e. about a 2% error when $h^2/(4\kappa t) \approx 4$, i.e. $t \approx h^2/(16\kappa)$.
(d) $c(z,t) = \mathcal{C}(z,t) + \mathcal{C}(z-2h,t)$. The solution is useful up till $t \approx 4(2h)^2/(16\kappa)$.
(f) $c = M/h$.

Exercise 7.11

(b) *Hint:* Because of linearity, solutions can be added, so the required solution can be obtained by adding together solutions we've already obtained.

Exercise 7.14

(a) The similarity equation is

$$\mathcal{C}_{\xi\xi}\xi + \mathcal{C}_{\xi}(1+\xi) + \mathcal{C} = 0.$$

(b) The solution is $Ae^{-\xi}$.
(d) The concentration increases initially, reaches a maximum level after a time of order $r_0^2/4\kappa$, and then decreases slowly.

Exercise 7.16

(b) The solution is linear in the two regions.
(f) *Hint:* Note that the boundary condition required by the physics is

$$k\frac{dT(0)}{dx} = q.$$

Chapter 8

Exercise 8.1

(c) *Hint:* Note that

$$\frac{\partial \mathcal{E}}{\partial a_j} = \langle 2(f - \sum_{n=0}^{N} a_n u_n) \cdot (u_j) \rangle,$$

and use the orthogonality result.

Exercise 8.2

(a) The error is of the order of the first neglected term, so roughly 200 terms are required.
(b) The approximation is particularly poor at $x = 1$. Note that the Fourier terms all vanish at 1 so this is not surprising.

(c) Over the range $-1 < x < 1$ the function $f_{ext}(x)$ is linear. Elsewhere $f_{ext}(x+2) = f_{ext}(x)$ determines f_{ext}, except at $-3, -1, 1, 3$ etc. where $f_{ext} = 0$.
(d) (i) The poor convergence arises because the periodic extension has a discontinuity of size $2(e-1)$ at $x = 1$.

(ii) Note that $h(0) = h(1) = 0$ so the periodic extension is continuous. The slope of the periodic extension has jumps at $-1, 1$ etc. so the convergence is like $1/n^2$. To plot the approximation either use the result obtained in (i) or determine the expansion for $h(x)$ directly.

Exercise 8.4

(a) Note especially that as $a\alpha$ gets large $\lambda_1 \to \pi/2$. The appropriate dimensionless groups are $t_{ign}\kappa/a^2$ and $a\alpha$.
(b)*Hint:* Replace sin and cos by Taylor approximations centred on $N\pi$.

Exercise 8.5

(a)(ii)

$$T(x,y) = \frac{4T_0}{\pi}\sum_{n=0}^{\infty}[\frac{1}{2n+1}][\frac{\sinh(2n+1)\pi(y/l)}{\sinh(2n+1)\pi}]\sin\frac{(2n+1)\pi x}{l}.$$

(b) *Hint:* Break the problem up into 4 problems and use the results of (a).

Exercise 8.6

(a)(i) A steady state temperature of order $q_0L^2/(\rho ck)$ will be reached in a time of order L^2/κ.
(ii) In scaled variables a solution form $\mathcal{T}(t)\cos \pi x/2$ is required. The resulting differential equation for $\mathcal{T}$ with the prescribed boundary conditions and initial conditions has the solution

$$\mathcal{T} = \frac{4\gamma_0 \cos\frac{\pi x}{2}}{\pi^2}[1 - e^{-(\frac{\pi}{2})^2 t}].$$

(iii) Use the orthogonality condition to determine γ_n. The procedure works because the heat equation is linear so superposition is possible. Also Sturm-Liouville theory tells us that the eigenfunctions used form a complete orthogonal set, and thus can be used to represent any (reasonable) function f. Notice also that each of the individual terms satisfy the required homogeneous boundary conditions so their (convergent) sum also satisfies these conditions.
(b) Add to the above solution, the solution to the homogeneous heat equation and boundary conditions with the prescribed initial condition.

Exercise 8.7

(a) The boundary condition at $x = 0$ is given by

$$-kT_x(0,t) = r + \beta[T_0 - T(0,t)].$$

The steady state solution is

$$T(x,t) = -\frac{r + \beta T_0}{k + a\beta}(x - a).$$

(b) The eigenvalues are given by

$$\tan(\lambda a) = -\frac{\lambda a}{\alpha a} \quad \text{where } \alpha = \beta/k.$$

(c) Only the first eigenfunction is significant.

Chapter 9

Exercise 9.1

(a) One possible $T(t)$ to work with is given by

$$T(t') - T_0 = \Delta + (T_h - T_l)[-1 + e^{-10t'} + e^{10(t'-1)}],$$

where $\Delta = T_h - T_0$ and $t' = t/t_0$.
(b) We've seen that the nature of the heat equation is such that fluctuations are smoothed out, so one would not expect small changes to persist.

Exercise 9.3

(a) The analysis parallels that of the text.
(b) The test functions $[x(x-1)] \cdot [A + B(x-1/2)^2]$ can be used (we'd expect symmetry about $x = 1/2$).

Exercise 9.4

Note that, as expected, the cube has the maximum cooling time. The cooling time goes down asymptotically like $1/S^2$ for large surface areas S.

Exercise 9.5

(a) *Hint:* Note that the volume element of the shell defined by $(r, r + dr)$ is $4\pi r^2 dr$.
$\lambda^2_{approx} = 9.875$ compared with the exact value $\lambda^2 = 9.869$.
(b) The test functions $f = A + B - (Ar^2 + Br^4)$, are sensible.

Exercise 9.6

(a) $\gamma = 2\pi k/\log(\frac{r_2}{r_1})$
(b) The length scale is $Q\rho c/\gamma$.

Chapter 10

Exercise 10.1

(a) With 1 year as the time scale, $\omega_1 = 2\pi$ and $\omega_2 = 365 \cdot 2\pi$.
(c) *Hint:* The surface temperature has components of frequency $(\omega_1, \omega_2, \omega_2 - \omega_1, \omega_2 + \omega_1)$.

Exercise 10.2

(b) The atmosphere heats more rapidly than the land, which in turn heats more rapidly than the ocean, according to the model.

Part III

Chapter 12

Exercise 12.1

(a),(b) Note that over a long initial time interval (of order $1/(1-\varpi)$) the near resonance and resonance solutions are close, and the external force performs positive work on the oscillator. For larger times the solutions separate.

Exercise 12.3

(a) *Hint:* The equations

$$\ddot{X} + \omega_0^2 X = e^{i\omega t},$$

$$\ddot{x} + \omega^2 x = -\epsilon X,$$

provide a useful approximation for the initial energy build up stage.
(b) The response amplitude is of size $\epsilon/(\gamma\omega\nu)$ where $\gamma = \omega_0^2 - \omega^2$, and so can be large if this combination is large eg. if the damping is small compared with coupling in the sense defined by this expression.

Exercise 12.4

(b) *Hint:* Use `int(eqnexp*sin(sigma*t),t=0..2*Pi/sigma);`.
(c) *Hint:* Examine the coefficients of the powers of ϵ. Note that the A of the example and the A used in the text are *not* the same.

Exercise 12.6

If $\alpha = a/\omega^2$ (the response scale at non-resonant frequencies) then the amplitude at resonance is $\alpha^{1/3}$. Note that although this is large compared with α, it's small compared with 1 for small α.
Hint: The ratio of the neglected term in the representation of $\sin\theta$ to the retained cubic is of order $\theta^2/20$.

Exercise 12.7

(b) The size of the restoring force depends on the sign of the displacement x.
(c) The relevant cubic is

$$A[\delta - \epsilon^2 \frac{10}{3} A^2] = \frac{1}{2},$$

and a balance between the all three terms is realized if scales are as indicated.

Chapter 14

Exercise 14.4

(a) The equations are

$$n_t + \frac{F_{max}}{N_{max}} n_x = 0,$$

with

$$n(x,0) = \begin{cases} 1 & \text{if } x < 0 \\ 0 & \text{if } x > 0 \end{cases}$$

(b) *Hint:* Note that $n = n(x, t, F_\star/N_{max})$ from the above equations. The equation for $n(\xi)$ is $n_\xi[\xi + f'(n)] = 0$.

Exercise 14.5

(c) 230 days.

Exercise 14.7

(d) *Hint:* Look for a solution of the form $e^{-\mu x} C(x - ut)$.

Exercise 14.8

(c) The car moves off at time $t = t_0$ given by $t_0 = D/V_{max}$, so $X(t_0) = -D$ is the required initial condition.

Exercise 14.9

(b) An accident will occur if $x_{n-1}(s) - x_n(s) - 1 \leq 0$.
(c) For $\epsilon_b > 1/4$ the solutions are oscillatory in character so $x_1 - 1$ might be expected to become negative, and a collision is likely. Solution plots also suggest this. It's not so easy to actually prove this is the case (one needs to determine the location of the roots of a transcendental equation).

Exercise 14.10

The shock speeds are given by $(-1/8, 5/8)V_{max}$, with car speeds

$$(3/4, 1/8, 3/8, 3/4)V_{max}$$

in the various zones from left to right.

Exercise 14.12

(a) Drivers will reduce their speed to account for increasing densities ahead. Also a negative γ corresponds to a negative diffusion coefficient, leading to unstable behaviour.
(d) *Hint:* $N_x \to 0$ as $X \to \pm\infty$.
(e) The shock thickness is of order ν. Since $\nu \propto \gamma$, the model predicts a smoother shock transition if drivers are more sensitive to conditions ahead in the traffic stream.

Index